FIRE IN THE PINEY WOODS

THE TEXAS EXPERIENCE
*Books made possible by
Sarah '84 and Mark '77 Philpy*

FIRE IN THE PINEY WOODS

Texas Lookout Towers,
the Forest Service, and
the Civilian Conservation Corps

EDWARD CAVALLERANO

Texas A&M University Press
College Station

COPYRIGHT ©2025 BY EDWARD CAVALLERANO
All rights reserved
First edition

∞ This paper meets the requirements of ANSI/NISO Z39.48-1992 (Permanence of Paper).
Binding materials have been chosen for durability.

LIBRARY OF CONGRESS CATALOGING-IN-PUBLICATION DATA
Names: Cavallerano, Edward, 1980– author.
Title: Fire in the Piney Woods : Texas lookout towers, the forest service, and the Civilian Conservation Corps / Edward Cavallerano.
Other titles: Texas experience (Texas A & M University. Press)
Description: First edition. | College Station : Texas A&M University Press, [2025] | Series: The Texas experience | Includes bibliographical references and index.
Identifiers: LCCN 2024056583 | ISBN 9781648432811 (hardcover) | ISBN 9781648432828 (ebook)
Subjects: LCSH: Civilian Conservation Corps (U.S.)—History. | Forest fires—Texas, East—Prevention and control. | Fire lookout stations—Texas, East—History. | Fire lookout stations—Texas, East—Design and construction. | Fire prevention—Texas—History. | Forests and forestry—Texas, East—Safety measures. | Forest management—Texas, East.
Classification: LCC SD421.32.T4 C39 2025 | DDC 363.37/9097642—dc23/eng/20250210
LC record available at https://lccn.loc.gov/2024056583

Contents

Acknowledgments vii
A Note on Nomenclature and Field Ethics ix

Introduction *1*

1 · The Breaking Wave: Forestry in Texas *5*
The Timber Problem: The Future Is Forbidding *10*
Awakening Conservation Consciousness *14*
The Careless Habit *18*

2 · Neutral Ground: Geography and Climate *21*
Dissecting the Piney Woods *21*
Fire Season *26*

3 · Forest Protection and Fire Control *31*
Tools of the Trade *34*
Fixed Point Detection: The "Eyes of the Forest" *36*
O. C. (Oley Cecil) Braly *46*
Reconstructing the Network *50*
Nonproductive Time *59*
Seen Area *60*

4 · Cussed and Discussed: The 1920s *61*

5 · Come Up Sometime: The 1930s *68*
Texas National Forests *83*
Camp Segregation *89*
Pioneering Units *92*
University of the Woods *105*
Building the Core Network: 1934 *107*
Extending the Core Network: 1936 *114*
Firefighting *118*
Planting *121*
Mapping: Visibility Surveys, Geography, and Forest Types *122*
The Lookout Yard: Forest Infrastructure and Forest
   Improvements *125*

6 · Chestley Dickens and Betty Huffman: The 1940s *137*
Aircraft Warning Service *144*
"Emergency" Roles for Women *145*
Administrative Changes *149*
Best and Worst Fire Years *152*
Albert D. Folweiler *158*

7 · Guardians of the Piney Woods: The 1950s *161*
Lookout! Deaths and Near Misses *166*
Employment Opportunities *167*
Reliance on Air Patrols *170*
Uncertainty and Renovations on Texas National Forests *175*

8 · Unit 308: The 1960s *182*
Maintenance and Network Expansion *184*
"A Crewleader's Wife" *190*
Looking Ahead in Fire Control *196*

9 · A Lot of Smoke but No Fire: The 1970s *197*
Mr. Dependability and Other Remaining Old Hands *209*

10 · Potshots or Pearls: The 1980s *211*

11 · Epilogue *218*

Appendices
1. Beaumont Quadrangle *223*
2. Palestine Quadrangle *255*
3. Tyler Quadrangle *287*
4. Texarkana Quadrangle *303*
5. Lost Pines *309*
6. Pier Dimensions *311*

Notes *313*
Index *349*

Acknowledgments

WILLIAM OATES, the associate director emeritus of the Texas A&M Forest Service, and his staff entertained my visits to College Station, making accessible volumes of material for study. More important, the associate director patiently shared his knowledge of southern forestry and the history of the Texas A&M Forest Service. His enthusiasm and dedication to resource management, history, and the citizens of Texas are exceptional.

Collections preserved at the offices of the Texas A&M Forest Service contributed extensively to this effort. Included are memoranda, personnel files, blueprints, and photographs taken by Sherman ("Jack") L. Frost throughout the mid-twentieth century. The compilation of known lookout observers has been assembled by reviewing issues of the *Texas Forest News*, personnel files, and *Have You Heard*, a newsletter containing items of interest for Texas Forest Service employees. Unquestionably, there are anonymous, dedicated staffers who have been unintentionally omitted. For this, an apology is offered.

Patrick Ebarb shared valuable memories from his early career and tenure as the head of Fire Control, significantly adding to the narrative during the final decades of the fixed point system. Thanks are also extended to Betsy Deiterman and the Polk County Memorial Museum for graciously sharing information on the Livingston Civilian Conservation Corps (CCC) camp. Larry Shelton, a dedicated naturalist with an eye for the subtleties of the East Texas landscape, kindly provided copies of enigmatic maps and reports and shared his extensive knowledge with me in the field. His curiosity and commitment to preserving Texas' open spaces are unsurpassed.

Dan Utley offered enthusiastic support and advice while editing the manuscript, and Simon Winston unselfishly invited strangers to his ranch to study the Elysian Fields Lookout and his exceptionally managed timber farm. Jonathan Gerland at the History Center, Diboll, saw potential in the project and provided guidance and encouragement. Collections there, at the Montgomery County Memorial Library System, the Kurth Memorial Library in Lufkin, and the Crockett Public Library also contributed to the completeness of the story. The National Archives in Fort Worth, Texas; Atlanta, Georgia; and College Park, Maryland, located primary documents. US Forest Service personnel, including Joshua O'Banion and David Foxe, aided in researching lookouts on the National Forests and Grasslands in Texas. Finally, digital archives from the "Portal To Texas History" and the "Walker County Treasures and County Historical

Commission" augmented this effort. Reviews by William Oates, Pamela Cavallerano, and Jerry Cavallerano Jr. have clarified the content, while editorial contributions by Cynthia Lindlof have resulted in a more concise narrative.

This manuscript is dedicated to my wife and children, in appreciation of their interest and decade-long tolerance for family activities that centered around visits to fire towers. This is a human story, demonstrating both the best and worst of our past. It is filled with imperfect, but nevertheless, remarkable characters. My hope is that my children, and yours, learn from this history and mature in a society filled with responsibility and civility. Publicly accessible lookout sites still encourage curiosity, allow adventure, and build self-confidence. They also offer an opportunity for families to explore together.

A Note on Nomenclature and Field Ethics

TERMS FOR POPULATIONS, places, groups, and organizations have changed through time. Where possible, this book maintains period-specific nomenclature to convey aspects of culture and geography despite trends in modern usage. Importantly, readers will also observe changes in the official names for organizations like the Texas Forest Service and Texas National Forests, which have rebranded through time. Note that the US Forest Service (USFS) tendency was to refer to agency lookouts by the standardized drawing numbers without regard to the manufacturer or the manufacturer's designation. This can result in multiple names being applied to the same structure. When a tower's manufacturer is unknown, the author resorts to inventorying the lookout using the recorded drawing numbers.

Finally, fieldwork is central to an understanding of the geography, utility, and history of the lookout network. Paradoxically, however, these activities have the potential to accelerate the demolition of standing lookout towers and create access restrictions. Readers are reminded that publicly accessible lookouts (and former lookout locations) are unevaluated archaeological sites and that landowner permission should be obtained before entering private property. Preservation depends on ethical exploration.

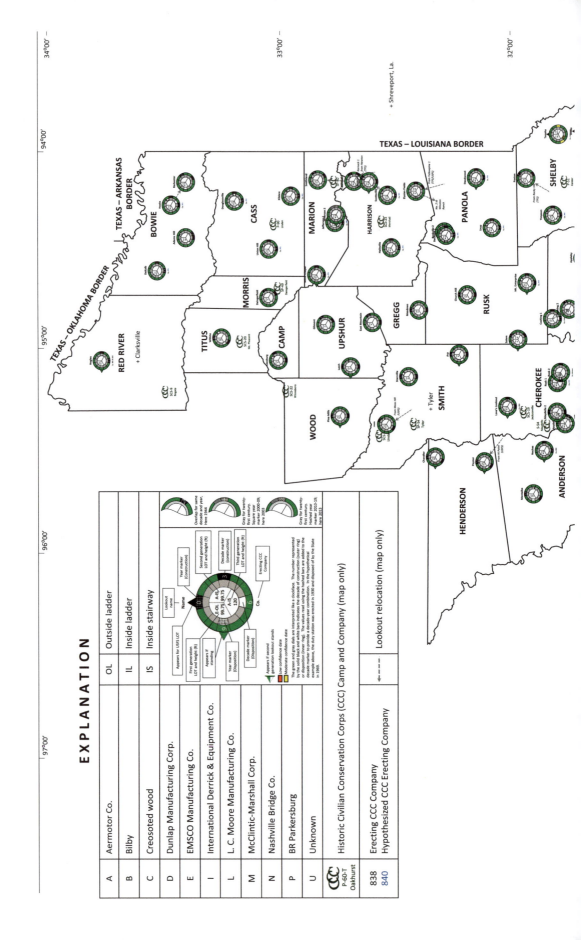

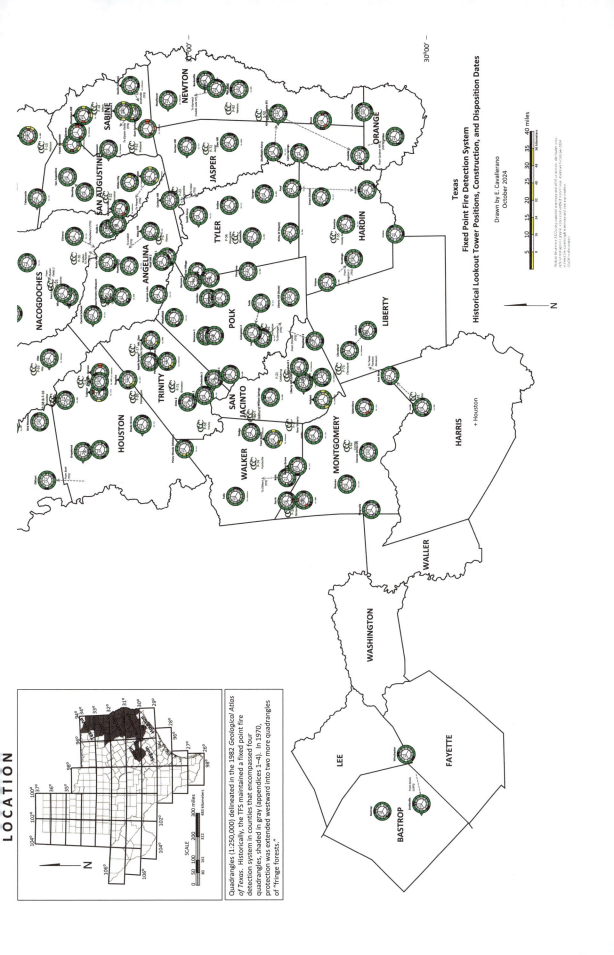

FIRE IN THE PINEY WOODS

Introduction

IT'S AN EASY WALK from an established trailhead on the west side of Texas Farm-to-Market (FM) Route 149. Instead of following the main footpath—the Lone Star Trail—into the Little Lake Creek Wilderness Area, a turn toward the northwest along an easement leads up the flank of a broad hillside. At its crest, the easement intersects a largely forgotten, grass-covered woods road. Four hundred feet to the right is the rushing traffic of FM 149. A few paces away to the left, the road bifurcates around the spot where a fire tower once stood. The brushy, westerly trail leads past two sets of tower piers and a stainless-steel latrine before terminating at a fence and gate post. Beyond the latrine are a thicket and a second, angled line enclosing a once-cultivated garden. Back at the junction, the second track drops slightly as it rounds the piers and artfully contours the southeastern flank of the hillside as it progresses deeper into the woods, passing the remains of the ranger's dwelling and a garage.

Painstakingly measured and oriented on the hand-drawn map in my pocket, I am here to verify several bearings with my compass and remeasure the distances between a few important landmarks. With an appreciation that has developed alongside my map, it is clear that the lookout compound was built with purpose, economy, and aesthetic appeal. Unquestionably, the Civilian Conservation Corps (CCC) builders from Company 2892, stationed at nearby Forestry Camp F-19-T, were preceded by surveying crews and an architect following the US Forest Service's most modern design policies. The improvement plan accepted for the site could only have been selected because of the fashion in which it neatly conformed to the sandy hillside.

Two black-and-white photographs in Carolyn Frances Hyman's thesis regarding the nascent history of the Texas National Forests (TNF) complement the map in my pocket and bring life to the hillside.[1] The first photograph, labeled only as "lookout tower," reinforces the linear, surveyed geometry of the lookout compound. In the foreground, a straight, rutted, sandy road bathed in morning sunlight and fringed by young pines leads directly to a battered steel lookout tower in the mid-ground. Likely built by the EMSCO Corporation, a ladderway rises up the frame of the northern face and is punctuated by resting platforms. The silhouette of an informational sign is visible below the tower braces, partially blocking the bright clapboard walls and roof of a cabin discernible along the same surveyed line in the background.

The second image, identified as "a ranger's home," appears to have been taken at the same location. Concrete steps lead upward to a screened porch on which sits a single wooden chair. Right of center is an additional room with a single, shaded window and, at the margin of the frame, a cistern connected by gutters to the split-shingle roof. Downslope, across the yard to the left and angled with respect to the dwelling, lies a simple garage. Farther downhill, out of frame, lay a cow barn and chicken house.

Access road, lookout tower, dwelling, garage—all located on a southwestward surveyed line along the apex of the hillside and connected by a contoured track road. These design attributes are in agreement with recommendations W. Ellis Groben outlined in *Acceptable Plans, Forest Service Administrative Buildings* to avoid excessive daytime heat by reducing the number and size of windows and positioning living and working rooms to maximize comfort and utility.[2] All dimensions and orientations are consistent with my hand-drawn map.

History jumps out of the hillside in this context and makes the curious wonder about the spirits that may linger in the shadows. Philip Conners expressed his affinity for Apache Peak after examining the names of former observers penciled in the cab of his tower,[3] considering that each hinted at stories and created myriad unanswerable questions. Despite having reconstructed so much from observation and primary documents, my list of questions remains expansive. Today I wonder about the fire spotters who spent solitary days at the station and daydream about sitting behind the wheel of a surplus Jeep as it passes around the tower and continues along to the sparsely furnished cabin set back from what must have been a far less traveled farm road. Alone with my compass alongside the stem-wall foundation of the dwelling, it's easy to get lost in Robert Frost's "Ghost House"[4]—an abandoned New England farm—that was situated down an overgrown, neglected road. There we find the speaker, amid the ruined cellar walls and dancing shadows of the farmhouse, reflecting on the "tireless," but anonymous, inhabitants who rest in nearby moss-covered graves.

Together, these thoughts are linked to a burgeoning but distinctively American cultural thread with roots in the nation's conservation philosophy. For many, lookout culture is associated with self-reliance and living symbiotically with nature. Best articulated in *Poets on the Peaks*, John Suiter suggests that as "solitude, self-reflection, and plain silence have become ever more fleeting and rare in our lives, the lookouts have emerged as symbols of a nearly lost and longed-for American simplicity and integrity."[5]

The aim of this book is to develop several parallel stories. The first is to describe the Piney Woods and the antecedent conditions that necessitated fire prevention, forest fire detection, and the development of Texas' fire lookout network. Second, through collected stories, poetry, and primary accounts, it aims to pay tribute to a generation of CCC enrollees, foresters, surveyors, and self-sufficient lookouts who redeveloped or protected the forestlands of Texas and proudly educated rural communities about the value of their natural resources. Finally, it endeavors to describe the location and design attributes of the lookout towers and associated structures in Texas by consolidating a decade of observations. In this effort, methodologies are introduced to assess tower characteristics that may be useful to workers in other regions.

The storyline progresses linearly to achieve these objectives. Chapters 1 and 2 investigate aspects of Texas geography, forest communities, climate, and fire weather, while also discussing the unsustainable logging practices at the beginning of the twentieth century. With this foundation, the narrative transitions to Texas' response to forest protection and fire control in chapter 3, pages that explore the inauguration of the fixed point detection system and the technologies utilized to achieve these objectives. Lookout suppliers, methods of lookout construction, and variations in lookout architecture are central themes. Chapters 4 through 8 outline the development and continued expansion of the lookout network between 1920 and 1960, focusing on the machinery and personalities that helped defined the state's management response.

The decade of the 1930s (chapter 5) receives comprehensive treatment because of the revolution in conservation and fire protection that accompanied federally funded Depression-era relief programs. Simultaneously, this period also witnessed the maturation of the Texas Forest Service (TFS) and the establishment of the TNFs. Critically, cultural biases that limited opportunities for many East Texans are examined, raising awareness for the experiences and limitations of minority Americans during the "make-work" era.

Though lookout tower construction projects continued into the 1970s, chapters 9 and 10 emphasize evolving management philosophies, advancing technologies, and changing public perceptions that led to the abandonment of the fixed point fire detection system. Finally, appendices 1 through 5 detail the attributes and known locations for Texas' lookout towers, while appendix 6 provides engineering details useful for workers in the field. The inventories in appendices 1 through 5 have taken over a decade to compile and are meant to complement the map compilation and assist in visualizing the historic fire control program.

It is hoped that by sharing an appreciation of the past and celebrating the architecture and regional heritage these sites *still* possess, we can preserve the history of forest fire protection in Texas, enrich our recreational experiences, and build a compelling case among Texans to preserve the lookout towers that remain.

1 The Breaking Wave
Forestry in Texas

THE PATCHWORK OF SUNLIGHT beaming through the canopy strikes the young, ovate beautyberry leaves and makes them translucent. Nearby, myrtle and yaupon—with thicker, leathery leaves—compete with tender sassafras saplings to create a shrubby understory. Overhead, the reverberating whine of cicadas overwhelms the sound created by the breeze stirring the loblolly canopy. It's difficult to remember this is Texas, the frontier land known for independent personalities, desert vistas, and oil. But forests, and forest products, have been a Texas-sized industry in the eastern, aptly named Piney Woods portion of the Lone Star State for nearly two centuries.

The development of lumbering in Texas is similar to the exploitation of other commercial timberlands in the United States. Following destructive harvests in the northern United States, operators left the area by the late 1800s and turned to the Reconstruction-era South for new sources of wood.[1] The combination of evolving technologies and large corporate holdings nearly led to the collapse of complete forest ecosystems. Like conditions in the wake of northern lumbering, bare landscapes succeeded the initial southern harvests and devastated Texas' forestlands. Eventually, there would be calls for conservation, but it ultimately required a profound economic collapse and aggressive educational effort to produce lasting conservation measures. When these policies finally arrived and the country reoriented, fire lookout towers became not only utilitarian structures but also "an important point of public contact"[2]—recognizable symbols of forest protection and the nation's budding conservation consciousness.

Today's sunlit forests are evolving within a complicated framework of natural and cultural controls. Leveled roadbeds and overlooked tramlines through mature pine stands, winged ditches at the edge of an overgrown field that, even today, marvelously create conduits for water to escape, and leaning fences that snag the unsuspecting and compartmentalize the woodlands all hint at this history. Species diversity and forest composition have changed too, altered by the removal and girdling of "weed trees" in favor of more economically valuable varieties.[3] The large forested tracts in East Texas today are a mixture of both federal and private holdings that collectively span all or parts of forty-eight counties.[4] They are managed to support multiple beneficiaries, sometimes with conflicting visions of preservation and use.

Settlers in the Piney Woods experienced very different forest conditions, and Robert Maxwell and James Martin suggest that the "pinelands remained virtually untapped" as late as 1880,[5] even though small water- and steam-powered mills had appeared as early as the 1820s.[6] William L. Bray was one of the first to offer a comprehensive description of the early twentieth-century East Texas timber belt, characterizing it as the westernmost edge of a once uninterrupted forest of "the same general character" that expanded across the Atlantic and Gulf margins of the United States: "halting," in his words, "like a vast wave that rolled on upon a level beach" by the drier, more arid climate west of the Brazos River.[7] He recognized several climax forest associations within the territory, fully one-fifth of the state,[8] which extended west from the Louisiana border to a loosely interpreted transition zone running "somewhat short of the 96th meridian."[9] This meridian, more conveniently, runs northward from Houston near the communities of Tyler and Clarksville and then northward to the Texas-Oklahoma border. West of this boundary lies a mosaic of scrubby post oak forests and grasslands. Eastward, the district is home to swamp and bayou forests, hardwood forests, mixed loblolly pine and hardwood forests, and longleaf and shortleaf pine stands. Because several species of southern yellow pines dominate the forest, homesteading settlers commonly referred to the area simply as the Piney Woods.

If there were many forest communities within the Piney Woods, some were more valuable than others, and there was strong market demand for longleaf pine heartwood and naval stores like turpentine or pitch. The species was big business in the southern United States, and within an area of about 5,000 square miles occupying all, or parts, of nearly a dozen Texas counties, where it "thrust itself like a broad wedge southwestward from the Louisiana border to the Trinity River."[10] The first longleaf harvests were taken from a centuries-old fire climax succession composed of nearly pure stands that were capable of reaching skyward 150 feet.[11]

The species once dominated between 60 million and 91 million acres of the country from North Carolina south to Florida and westward into the Gulf states.[12] The tree's persistence was partly due to its preference for sandy, well-drained soils that lacked the nutritional requirements for less hardy species.[13] This, coupled with an evolutionary reliance on fire, developed because of the frequency of lightning and anthropogenic fires throughout prehistory. Some estimates even suggest that a fire occurred every three to four years, and the longleaf's deep taproot ensured that it could flourish without being outcompeted by other *Pinus* species that might otherwise shade it or crowd it for space. The ecosystem that resulted from these conditions extended across the Southeast and promoted open forest stands with a patchwork of legume grass savannas that were often described as "regal" or "parklike."

Pure longleaf stands were ubiquitous but tended to offer unpredictable harvests that often disappointed operators. Driven by forest economics and a desire to understand lumber yields within the virgin stock, Herman H. Chapman, professor of forest management at Yale University, traveled to Tyler County, Texas, with a group of forestry students around the turn of the twentieth century to understand the capacity and productivity of the land.[14] Chapman was immediately frustrated by failing to find suitable research plots of even-aged, merchantable timber to study, and he was forced to rethink his methodology upon arriving. Settling on nine large, 40-acre study areas, his students spread out with calipers to measure the diameter of each tree and record its merchantable height. "Veteran," "mature," "young merchantable," and "immature" age classes emerged from this effort, and the observations were grouped with growth data obtained from resinous stumps to tally heights and lumber volumes. The area occupied by each tree class was plotted, and the map that emerged was a complex profile of the age distribution of the virgin forest in which fully one-fourth of all the pines were immature trees.

The results were surprising but became the foundation for understanding fire climax systems and longleaf ecology, whereby fires, tornadoes, or hurricanes expose spots for seedlings to develop, germinating from older survivors that serve as seed trees. This view contrasted with more-established uniformitarianism concepts that predicted any events acting to change the stability or equilibrium of a stand would destroy it. Rightly, Chapman recognized the impact logging had on removing prolific seed trees while also destroying or leaving the immature "badly injured, with no thought of its potential value" in restocking future forests.[15] Based on these conclusions, he was also one of the first to advocate the use of fire for restoring longleaf tracts as early as 1926.

Chapman also understood the impact feral hogs were having on longleaf pine reproduction.[16] Traditionally, East Texas ranchers allowed rooter hogs and woods cattle to fend for themselves on the open range, wandering between river bottoms during the colder winter months and the pine uplands during the spring and summer.[17] Omnivorous hogs survived on acorns, berries, carrion, mussels, and insects. But Chapman's field observations revealed an almost complete absence of young longleaf pines in their seedling and grass stages, a phase when the tree focuses energy on growing strong roots at the expense of upward growth. With so few young trees around, Chapman hypothesized that for over a century rooting razorbacks had been devouring the young, succulent pines to the extent that they were limiting future growth of the species. One hog alone, Chapman observed, could completely denude 40 acres of longleaf seedlings in a season, eating at a rate of six trees per minute for "as long as his appetite lasts."[18]

Loggers *and* hogs, then, were responsible for altering longleaf habitat and modifying the East Texas landscape.

As the virgin stands were cut and seedlings failed to take hold, the composition of the forest began to change. Pure longleaf forests, "the monarch of the wood,"[19] were being replaced by longleaf-loblolly, longleaf-shortleaf, or longleaf-hardwood types when partial or total harvests left space for an influx of other species.[20] Without ground fires to promote regenerative longleaf growth or kill competing species, other varieties—and the communities *they* supported—began to outcompete the longleaf.

Even during Bray and Chapman's time, East Texas sawmills were expanding from small, temporary facilities to large commercial operations using band saws capable of processing astounding volumes of pine that "appeared overwhelming and of little value" to early residents.[21] Initially, the larger mills were confined to the major river corridors, where water and steam power were easily obtained and logs could be floated downstream for processing. Later, however, more intensive lumbering expanded away from the primary drainages, supported by a dendritic system of short-line, narrow-gauge railroads that linked industrial-scale mills directly with the raw materials in the Piney Woods. These "railroad loggers" laid hastily constructed tracks deep into the backcountry, on average, at quarter-mile intervals.[22] From these, steam-powered skidders mounted on special railroad cars could then stretch 1,000-foot-long cables out to an awaiting felled tree. Crews would then "choke" the trunk of the fallen tree by clamping the line around the pine's circumference before winding the cable and dragging the tree back across the forest floor to the railroad-mounted loader.

When the longleaf tracts in the central Piney Woods were depleted, lumbermen shifted their harvests to other available yellow pines like the loblolly and the remaining shortleaf. Shortleaf ecosystems have the widest range of any southern pines and occur in more than twenty states from southern New York and Illinois southward to Florida, Texas, and Oklahoma.[23] In Texas, the species dominates about 30,000 square miles of hilly terrain north of the longleaf belt in forests that are often interspersed with various hardwood species such as post oaks, hickory, elm, and sweetgum.[24] Requiring more fertile soils than their longleaf cousins, the trees competed for space with ideal agricultural lands and, as a result, were quite literally uprooted from regions with early settlement.[25] Especially in Texas' northeastern counties, smaller landholdings were logged quickly to supply local sawmills with the materials necessary for community consumption. Though the timber yields were less attractive to large-scale operators, observers noted that the shortleaf was "pretty thoroughly cut out" by the beginning of the twentieth century.[26]

While the longleaf and shortleaf were named for the characteristic appearance of their needles, the loblolly's (*Pinus taeda*) tolerance for wet feet is what lends the tree both its common and scientific names. At one time the largest continuous tracts of loblollies were found south and west of the longleaf belt in forests interspersed with hardwoods—either on dry ridges or the moist, lower slopes of drainages and low, boggy "loblollies."[27] Today's Piney Woods are dominated by this species, which now extends well beyond the estimated 7,000-square-mile area of San Jacinto, Walker, Montgomery, Harris, Jefferson, Liberty, Orange, Hardin, Grimes, Newton, Jasper, and Chambers Counties, where it once thrived.[28] In part, this is due to the efforts of early foresters, who were focused on establishing productive forests on cutover lands. Characteristics like survivability, adaptability, disease resistance, and fire tolerance made the loblolly, a native species, a practical choice for tree improvement projects. In addition, the tree's early rapid growth, yield characteristics, shorter rotations, and high market value offered a faster return on investment.[29] So pronounced was the conversion that James Cruikshank and I. F. Eldredge predicted that the loblolly, in the southeastern part of the timber belt, was "destined to have an increasing acreage."[30] Even though markets and interests are changing today, many East Texas plantations still go through cycles of land preparation, loblolly planting, thinning, and harvesting,[31] in similar fashion to other crops like cotton or corn.[32] As one forester notes today, it is "*the* species that recovered Southern forests and made us the 'wood basket' of the world."[33]

Texas logging was booming as the twentieth century began, and all was "activity, rush & good cheer in the piney woods."[34] Large regional operators like the Angelina, Southern Pine, and Kirby Lumber Companies—along with others like the Carter Brothers and Central Coal and Coke Company—had established themselves as the state's barons. Alone, between 1900 and 1918, the 4-Cs, as the Central Coal and Coke Company was known, cut over 120,000 acres of land without much oversight or pause to reforest.[35] It turns out that Chapman's 1907 field season in Tyler County coincided with the year of peak timber production. That year, Texas was the nation's third-largest timber producer and provided 2.25 billion board feet of lumber.[36]

These days, one can walk through the grounds of two former mill sites to appreciate the scale of industrial milling that led to the perception of "activity" and "good cheer." Just west of the Ratcliff Lookout tower in the Davy Crockett National Forest lies the Ratcliff Lake Recreation Area. Both the tower and the lake were constructed by Company 1803 near the site of the 4-Cs sawmill. Though only brick-and-mortar foundations remain in the woods today, the town had a population of nearly ten thousand in 1910, and the 486-foot-long sawmill and 450-foot-long planer mill were once the largest in the South. One thousand employees labored in the woods and mill, which was capable of producing 300,000 board feet of lumber during each eleven-hour workday.[37] Surrounded by competitor landholdings, however, the 4-Cs struggled to purchase enough lands or stumpage to keep the mill operational, and by 1918 the facility was sold.[38] After purchasing the assets, the Houston County Timber Company downsized and dismantled the large mill. Soon, only the mill pond and sturdy foundations were left, but David A. "Andy" Anderson recalled that they were still visible during the 1930s,[39] as they are today.

More remains at the Aldridge Mill site in the Angelina National Forest. In the shadows of towering pines are the ghostly ruins of the iron-reinforced and hand-poured concrete mill and engine buildings that were built in 1912 to replace structures destroyed by fire.[40] Both the 4-Cs and Aldridge mills relied on tramlines to funnel sawlogs to company millponds, like the much-smaller modern-day impoundment at Ratcliff Lake, where they were stored until continuous chain conveyors hauled the logs into the mill for processing.[41] Though Aldridge was contemporaneous with operations at Ratcliff, daily production by the Aldridge Lumber Company was far less, a mere 125,000 board feet per day.[42]

The Timber Problem: The Future Is Forbidding

Unsustainable forest harvesting "hacked, scorched, & wasted" the timber belt and foreshadowed "the passing of the pine."[43] Incredibly, Texas' most passionate, early advocate for forest rehabilitation and sustainability was a worldly and broadly educated banker named William Goodrich Jones, who was born in New York to John Maxwell Jones and Henrietta Offenbach in November 1860.[44] Though William Goodrich was born in the Empire State, his father had established himself as a prominent Galveston, Texas, jeweler, and it was during a business trip that W. G. Jones was born. Alone, Jones's father returned to Galveston shortly afterward, apprehensive about his business in the period before the Civil War.

Within six months, Henrietta and the couple's three children traveled south to reunite with John Maxwell. The passage sent them by train to Ohio, New Orleans by riverboat, and Galveston aboard a schooner that evaded the Union blockade.[45] Worried by threats of bombardment in the coastal city, the family moved frequently during the conflict, first to Houston, then Austin. William Goodrich and Henrietta were in poor health by the end of the war, compelling John Maxwell to sell his businesses and move closer to family on the East Coast.[46]

The family traveled to Germany in 1873,[47] where Henrietta had been born. While the trip was meant for Jones's mother to recuperate, it also allowed William Goodrich to experience Dresden, Germany's Black Forest, and Paris.[48] Maxwell points out that it was during walking tours through the Black Forest with his father that the young man "developed a deep appreciation for both the beauty and the commercial value of well-managed forests."[49] The "perpetual forest and a perpetual forest harvest" Jones observed were the results of scientific forestry practices and restraint.[50] This progressive conservation philosophy sustained livelihoods and supported communities. In fact, European field excursions by Jones, and other contributors to the American forestry movement such as Carl Schenk, Bernard Fernow, and Gifford Pinchot, ultimately helped shape important aspects of US forest policy.[51]

After graduating from Princeton in 1883, Jones returned to Texas and spent his first five years as a young professional working in Galveston and San Marcos banks.[52] Perhaps just as important as his time in the Black Forest a decade before, Jones embarked on a survey of East Texas by horseback during this period to locate some of his father's landholdings.[53] Unquestionably, his ride through the patchwork of forests and clearcuts influenced his conservation philosophy. After settling down as the president of a Temple, Texas, bank in 1888, Jones found that the treeless landscape left him with an "unfavorable" impression of the city. The silviculturally minded Jones became so "obsessed" with the desire to plant trees for shade and aesthetics that he began cultivating pecan trees on his hotel windowsill to add greenery to his surroundings.[54] Later, he hired laborers to transplant native, river-bottom hackberry trees along city avenues and promoted one of the first urban forestry projects in Texas.[55]

Jones encouraged others to join his efforts and was so impassioned that he became known as the "tree crank" of Temple. These local attempts evolved from community projects to lobbying campaigns that challenged the state's approach to forestry. His influence prevailed in creating a statewide Arbor Day in 1889 and several forest associations.[56] The activities attracted the attention of Bernard Fernow, the first director of the US Division of Forestry, who traveled to Temple to mentor "Hackberry Jones" on a road map for securing conservation legislation at the State Capitol in Austin.[57] At Fernow's suggestion, Jones returned to East Texas to report on the extent and condition of the existing forest, but his Piney Woods tours were in contrast to the perpetual forests he admired in Germany. While Jones was not critical of consuming the "God-given wealth" of the region,[58] he was appalled by "what is wasted. The tree is cut from two & a half to three feet above the ground. Above the stump, say four logs, each 16 feet long are sawed. From the lowest branches to the top, say thirty to forty feet, is rejected & left on the ground to rot & be destroyed by fires & worms."[59] Ultimately, Jones concluded, private corporations should secure and manage large forest tracts and harvest them judiciously, under the supervision of either state or national government agencies.[60]

A different metaphorical wave was crashing. As Jones lobbied and resources dwindled, the toll on the Texas forest economy was becoming increasingly more obvious. Of the 673 sawmills operating during the peak year of timber output in 1907, only 466 were running by 1910.[61] Mirroring conditions in Texas, the nation's "inexhaustible" forest resources were also proving to be nonrenewable as apathy and reckless production practices prevailed. Forests were deteriorating quickly in the wake of lumbering and subsequent, uncontrolled fires as timber operators migrated across the country harvesting the materials necessary for industrialization. In response, President Theodore Roosevelt, who was heavily influenced by Gifford Pinchot's utilitarian or "wise use" forest philosophy,[62] assembled a White House Conference of Governors on Conservation in the spring of 1908 to address forest and water conservation policy. Jones, of course, had been invited as an adviser representing Texas, along with two others in the absence of Governor Thomas Campbell.[63]

Speaking eight years after the Great Storm of 1900, Jones acknowledged the aid rendered to Texans as "Galveston lay wrecked and bleeding" in the aftermath of the hurricane and pledged to cooperate "in any movement" to conserve the productiveness of the nation's natural resources. Sympathizing with territories in the forested Appalachian and White Mountains as well as the headwater regions of Lake Michigan and the Everglades, Jones wanted "to change the present method[s] of wholesale cutting, waste and fire-swept sterility" that devastated the country.[64]

Conference attendants issued a declaration at the end of the sessions, calling for the establishment of state conservation commissions, the extension of forest policies, and the conservation of soil and water.[65]

The delegates concluded with an elegant, but pragmatic, opinion that the nation should strive to "conserve the [resources that are the] foundations of our prosperity."[66]

Despite the conference's good intentions, progress was painstakingly slow both nationally and in Texas. Even though Massachusetts senator John Weeks introduced a bill in Congress fourteen months after the convention enabling states to cooperate with the federal government in watershed protection,[67] it would take a disastrous 1910 fire season before enough support allowed for its passage. Finally, on March 11, 1911, the legislation, which famously became known as the Weeks Act, allowed the secretary of agriculture to purchase lands along the headwaters of navigable streams for the purpose of conserving perennial flow and navigation. Importantly, other provisions of the law encouraged cooperation in the area of fire prevention, with the federal government making available matching funds to any state that organized a conservation commission overseeing fire prevention and resource management. When this condition was met, up to $10,000 a year in federal aid was available to finance the salaries of patrol officers who canvassed the territory within qualifying watersheds to smother fires and educate the public.

Jones returned from Washington energized and quickly organized a coalition of USFS officials, Gulf Coast lumbermen, and interested citizens into the Conservation Association of Texas.[68] The initiative advocated natural resource sustainability and called for a state agency to supervise the work.[69] Though many states created legislation in favor of conservation following the Conference of Governors on Conservation, the Texas response was, in Maxwell's words, "abortive," and the association faltered within a few years. Jones and his allies were resilient, however, and inaugurated the Texas Forestry Association (TFA) in November 1914.[70] This new organization had the expressed goal of creating a state department of forestry and a comprehensive forest conservation program. After decades of concentrated effort and a refreshed charter, Texas became the thirty-fifth state in the country to organize a forestry department in 1915 when Governor James E. Ferguson signed narrowly supported forest legislation into law and appropriated an annual $10,000 budget to the agency.[71] Later, Ferguson reflected that there was "no incident greater than my signing as Governor the Forestry bill during the first year of my administration."[72] Jones wrote the job description for the state's forester, seeking a "Chesterfield!, an orator, a lecturer, a mixer, a highly trained specialist in the theory of forestry."[73]

There could be no better commemoration of Jones's slow influence and "ceaseless enthusiasm for conservation" than the forest itself.[74] Quickly becoming an urban oasis, the W. Goodrich Jones State Forest in Montgomery County pays tribute to the "Father of Texas Forestry."[75] In addition to protection of red cockaded woodpecker habitat and offering people an opportunity to be outdoors, the experimental forest cultivates curiosity and promotes education, especially among children. Managed by the Texas A&M Forest Service, the property serves as a primer to laypeople and is carved into numerous plots of loblolly and shortleaf pines, where different growing methods are demonstrated.[76]

Awakening Conservation Consciousness

Once the Department of Forestry was authorized, John Foster, a Yale graduate and former Weeks Law inspector, answered Jones's advertisement and became Texas' first state forester during the summer of 1915.[77] With insider knowledge of the USFS and the provisions of the Weeks Law, Foster understood that fire control was at the heart of Progressive Era policy and a key mechanism for securing supplemental financing. To subsidize the department's budget with additional federal funding, Foster embarked on a reconnaissance trip through the Piney Woods with J. Girvin Peters to inspect forest conditions in December 1915.[78] Peters, who worked for the USFS as the chief of state cooperation, had also helped draft the forestry bill with Jones in 1914.[79] As the seasoned professionals went afield, they were proficient in the provisions of the law and understood the significance of defining the boundaries of navigable streams and developing a detailed fire protection plan.[80]

Foster was a prolific writer and worked hurriedly to develop his observations into a series of bulletins and annual reports. In support of his efforts to attract federal attention, the Department of Forestry's first bulletin, published in the spring of 1916, was titled *Grass and Woodland Fires in Texas* and was devoted entirely to that topic.[81] Other bulletins and field excursions followed quickly. Taken together, Foster's efforts appear as an early one-two punch, highlighting the state's timber problems and justifying the necessity of the department. Within months of Foster's becoming the state forester, the department succeeded in establishing a cooperative agreement under the Weeks Law on February 1, 1916, and secured a $2,500 supplement to the state's $10,000 budget.[82]

On the same day that the cooperative agreement became effective, Foster appointed Tenaha resident George W. Johnson as an agent of the state forester to supervise fire protection work.[83] The inaugural fire control program began on September 1, 1916, and provided rudimentary protection to thirty eastern counties encompassing the navigable waterways of the Red, Sabine, Neches, and Trinity River watersheds,[84] where yellow pine prevailed and was of economic importance.[85] State and federal funds were immediately used to pay the salaries of six forest patrolmen, who worked from headquarters in Lufkin, Livingston, Jasper, Longview, Tenaha, and Linden. Each received $90 per month plus expenses.[86]

The state forester followed USFS guidelines when hiring Texas' patrolmen. Each was expected to exhibit "intelligence, dependability, and sobriety" while interacting with people and performing protection work within a patrol area that extended outward in a 25-mile radius from their homestead headquarters.[87] These large units included over 1.2 million acres, and the men were required to ride on horseback between 15 and 25 miles per day to extinguish fires, post fire prevention notices, and act as community advocates. Bulletin 9, *Forest Fire Prevention in East Texas*, outlined patrolmen responsibilities. Each was expected to work

> each day except Sundays and National Holidays. . . . Not less than eight hours will be acceptable as a day's work. You are employed on a monthly basis and every working day must be properly accounted for. . . . Cover your district in a studied and systematic manner. . . . You will find your chief usefulness as a patrolman to be in pointing out to people the importance of fire prevention and in convincing them of the soundness of your arguments. . . . You will be given a notebook in which to take notes during the day for your daily reports and to record other information of value concerning forest and fire conditions. . . . Bear in mind that this is a new work in Texas, although well established in many other states. When you discover a small fire burning it is your duty to extinguish it; try to get neighboring residents to help extinguish larger fires. Your judgement will determine whether or not it is practicable to extinguish the fires found burning. . . . Once there is a sentiment against fires, fewer fires will occur and persons will be more readily extinguish those which do occur.[88]

By all accounts, the Department of Forestry achieved success within its first two years. Fire protection work had been initiated and forest losses had decreased, surveys had commenced to inventory resources and industry, and farm forestry experiments had begun. Still, there was work to be done, and conditions on the ground were not encouraging. In Polk County, virgin stock stood scattered, and all of the "timber easily accessible to the railroads" had been cut out.[89] More damaging, Foster also noted that the woodlands were "practically barren of pine reproduction as a result of clear cutting and fires."[90]

To attract attention and plea for increasing allocations in advance of the new legislative cycle, Foster used the second page of his *First Annual Report* to showcase states with "well advanced" forestry programs, pointing out that each on average invested $64,884 to support patrolmen, fire control efforts, and nurseries and provide management advice.[91] Even "treeless Kansas," he pleaded, appropriated $3,700 for maintaining a forest nursery. In addition to this aggressive writing campaign, Foster taught five forestry classes in the Department of Horticulture and lectured externally.[92]

Despite the department's efforts to describe the state's forest conditions and promote fire protection and reforestation, the 1917 Senate Finance Committee almost eliminated the new agency's budget. The trouble arose as a consequence of Governor Ferguson's impeachment and the political desire to investigate all state offices and educational institutions associated with his administration.[93] When a report was filed with the legislature regarding the Department of Forestry, it suggested that the work was inconsequential and maintained that "the public takes little interest" in forest conservation.

Swift petitioning by Jones, Foster, and his supervisor, William Bizzell, the president of the Agricultural and Mechanical College of Texas, successfully saved the Department of Forestry's legislative appropriation and preserved the federal allowance available under the Weeks Act. But the move frustrated Foster, who vocalized his disdain in a "redhot and much needed roast" published by the *Southern Industrial and Lumber Review* in March 1918.[94] In it, Foster pointed out that the federal government had assisted in drafting the Texas Forestry Law and proudly listed the department's accomplishments. He also charged that the "real lack of interest is on the part of the legislative committees which tie our hands through their failure to give financial support." Afterward he resigned and returned home to New England to become the state forester of New Hampshire. As Chapman noted, Foster was immensely successful and "left behind him the foundation of an operational forest conservation program where none had existed before."[95]

Eric O. Siecke succeeded Foster as state forester and served twenty-five years through the Depression until the brink of war in 1942.[96] A Nebraskan, Siecke came to Texas as an experienced hire, having worked under Pinchot in the USFS and as Oregon's deputy state forester.[97] Siecke had two primary objectives when he took over the agency in May

1918: he endeavored to develop a cooperative relationship with lumbermen, and he desired to increase the department's "pitiful" budget.[98]

Like Foster, Siecke was initially burdened by the teaching responsibilities that accompanied his role as chief forester. Nonetheless, he kept pace with Foster's writing program, and the department published additional bulletins, including *Forest Fire Prevention in East Texas and Forestry Questions and Answers*.[99] The latter was described as a "brief and rather elementary bulletin" designed to increase awareness and educate Texans about their "forestry problems." Both reports emphasized that fire prevention, reforestation, demonstration, and education were priorities for Siecke after 1918.[100] To advance these causes and increase visibility, the Department of Forestry, together with the TFA, began publication of the *Texas Forest News* (hereafter the *News*) in June 1919. Initially a short newsletter, the publication aimed to disseminate forestry information to members, the Texas legislature, and the general public when the press picked up and republished key forestry stories.[101]

Unquestionably, the *News* contributed to a conservation awareness among Texans, and legislative appropriations increased incrementally from $12,000 in 1918 to $64,000 in 1932.[102] The additional funding made it possible to establish the Division of Forest Protection in 1922 to sharpen focus on preventing and combating wildfires. The organization of the new division came two years before liberal revisions to the Weeks Act. The passage of the Clarke-McNary Law expanded federal land purchases away from the headwaters of navigable streams and increased federal matching dollars and incentive programs for private landowners to participate in fire prevention, thus expanding fire protection.[103] As Ronald Billings noted in *A Century of Forestry, 1914–2014*, "Preventing and suppressing forest fires would prevail as the most publically recognized responsibility" of the recently renamed Texas Forest Service.[104]

Annual reports and bulletins published during the mid-1920s demonstrate the commitment and excitement of the maturing agency. In 1926, for example, progress highlighted by the Division of Forest Protection included the "awakening of the public to a fire consciousness and thus the reduction . . . of man-caused forest fires."[105] This awareness was more tangibly demonstrated by a 28 percent reduction in fire occurrence and a 90 percent reduction in fire loss since the previous year.[106] Enforcement of fire laws, used in combination with education, also initiated change. The first forest fire convictions were secured in 1924, following the passage of fire laws in 1923,[107] and by 1936 fines were routinely being assessed against those found violating the law.[108] To Siecke, emphasizing law enforcement was a valuable deterrent against arsonists and timber loss.[109] Finally, Siecke also successfully established relationships with important timber interests and was able to demonstrate the corporate value of fire protection. These efforts secured cooperative agreements between the TFS and several lumber companies for the extension of the fire patrol district boundaries.

Before Siecke's retirement, a TFS publication, *Pineywoods Pickups*, summarized his career: "The Director has worked hard and judiciously to mold the Service into the fine organization it is today."[110] David Anderson, a Texas A&M Forest Service notable himself, remembered Siecke's disciplined but compassionate personality during a 1980 interview. Above all, Anderson recalled, Siecke "knew politics and how to get things done."[111] Maxwell and Martin agreed, noting that the director earned the respect of the state legislature because of "his ability, his devotion to the cause of conservation, and his staunch character."[112] Siecke's bold, flowing signature appears on much fire protection and CCC correspondence, and it is noteworthy that the core fire lookout tower network was built during his tenure. Through Siecke's leadership, the TFS "expanded in size, responsibility, and importance to become one of the best organized state forestry services in the South."[113]

Presently, there are five state forests in Texas. One is named for Buna-area tree farmer and land donor Leonora Masterson. Others honor the "Father of Texas Forestry," conservation-minded state senator I. D. Fairchild, and visionary lumberman John Kirby. Another, formerly State Forest Number 1, commemorates the efforts of E. O. Siecke. If the Jones Forest provides a tangible space to reflect on the benefits of Texas forests, then the Siecke Forest celebrates the director's commitment to Texans and the maturation of the agency. More broadly, the forest is also wrapped in the legacy of the New Deal. The space offers a convenient place to consider the impacts of twentieth-century conservation philosophy and digest the successes and institutional shortcomings of the CCC. That Siecke was part of all of it is clear. One needs to look no further than his name stenciled in the concrete landing platform below the first lookout tower ever constructed in Texas or at the forested panorama from the top of the replacement, CCC-era Aermotor tower,[114] which still stands by the district office today.

The Careless Habit

Foster recognized that fire should not be completely excluded from Texas forests, as it was "an indispensable agent when rightly controlled."[115] But, he reasoned, the "promiscuous, unrestrained burning" of timberlands must be stopped to protect the value and productivity of the land.[116] Keeping unregulated fires out of the woods was partly a cultural problem, though, as southern traditions viewed fire as a progressive tool for prosperous farming and ranching.[117] Throughout much of the agrarian South, fire was considered an agent in eradicating pesky arachnids, encouraging the growth of fresh grass, killing dangerous snakes, and controlling the proliferation of understory vegetation with the understanding that it would prevent malaria. Remarkably, these traditions are not dissimilar to those of prehistoric cultures or, arguably, even those in use today. Regardless, the 1919 publication of *Forest Fire Prevention in East Texas* went as far as to recognize a "class of citizens which frankly believes that the losses incurred through burning are more than offset by advantages gained."[118]

Contemporary harvesting practices exacerbated the fire problem, producing a wasteland in East Texas at a time when T. S. Eliot was conjuring up his own in verse. Young growth was destroyed by skidders during the harvesting process, and few seed trees were left standing for regeneration. During the early "bonanza era,"[119] Jones estimated that nearly 40 percent of each tree's mass was discarded as slash, increasing fuel loads and making the fires that started hotter and more devastating. If the impact of fire was not enough, the practice sterilized the land, altered the nutrient cycle, reduced soil moisture, and accelerated soil erosion. In fact, the majority of ephemeral streams observed today draining the uplands are a product of this unchecked erosion.[120] As early as 1919, Lenthall Wyman and L. Goodrich Jones acknowledge in Bulletin 9 that the situation could only be "redeemed now by keeping out fires," and only if "artificial propagation . . . be resorted to."[121]

Interviews during the 1970s with Roy Gay, a Liberty Hill crewleader, and TFS foresters like Maurice V. Dunmire preserve stories from a time when people set the woods on fire as "a recreation on Sunday afternoon,"[122] before hunkering into a swampy place to watch the firefighters work. Or other times when pyromaniacs tied "lighted kerosene rags to a buzzard's tail and let it fly around starting fires until the buzzard—and the woods burn[ed] up."[123] In another incident, TFS forester Bill Hartman even remembered that "young girls were setting fires so they could meet the CCC boys."[124] Sometimes, the CCC enrollees were setting the fires themselves, as was the case when small fires were spotted in the same vicinity every Sunday afternoon. The fires continued for nearly a month until the crews noticed they were coming back to camp with one additional rider. The enrollee, on weekend leave, thought he had discovered a quicker way to return to the barracks.[125]

Each of these fires was incendiary in nature, classified as having been intentionally started by a person without authority. "It is immaterial," the *Manual of Practical Forestry for East Texas* explained, "whether the object be malice, gain, or amusement."[126] By 1938, the *Manual* had been completed to provide a printed guide of relevant practices and definitions, including one that defined a forest fire as any "fire, large or small, of unauthorized origin occurring on forest type lands," which could include any "young or old growth timberland, cutover, brush, or grassland." These fires were only exacerbated during the Depression. "The principle cause" of forest fires, the *Seventeenth Annual Report* unabashedly proclaimed at the height of the economic downturn in 1932, was "attributed to the many unemployed people who roamed the woods to fish, hunt, and gather fuel."[127] Severe drought was also just as likely to blame.

For statistical purposes, the TFS categorized all fires during the 1930s into one of nine possible ignition sources, including lightning, railroads, lumbering, brush burning, campfires, smokers, incendiary, miscellaneous, and unknown.[128] The Texas Forest Service's investigation over a five-year period between 1933 and 1937 revealed that incendiary fires accounted for 33.7 percent of the total. Smokers' matches and cigarette butts contributed to another 37.3 percent during the same period. Together, these two categories made up more than 70 percent of all fires in the protected area. Lightning, in contrast, was blamed for only 0.3 percent of fires. Likewise, on the Davy Crockett National Forest in 1936, over 50 percent of the reported fires were caused by smokers, while just over 20 percent were incendiary in nature.[129] Though methods for defining fire origins evolved along with the use of more descriptive fire terminology, "carelessness and indifference" continued to be recognized as the leading causes during the first decades of fire the prevention program.[130]

The scientific treatment of fire data led to other discoveries. In 1937, for example, TFS analysis of the number of fires, the acreage burned, and the size of each blaze was studied in comparison to the area protected by the Division of Forest Protection. A compilation of the year's fire report indicated that over half of Texas' fires occurred in just four Piney Woods counties.[131] That more fire observation towers had been erected in Polk and Tyler Counties than in any other counties by July 1935 then comes as little surprise.[132]

Neutral Ground
Geography and Climate

NATIONAL FOREST SERVICE landscape architect and recreational planner Hunters Randall wrote in 1936 that "the Texas National Forests boast no scenic grandeur. They have no snow-capped mountains, no waterfalls, no trout streams. . . . Whatever natural beauty exists here is due to the vegetation."[1] Forest supervisor Loren L. Bishop added to the pessimistic assessment the next year, writing that "recreational possibilities of the Texas National Forests are not outstanding."[2] At face value, the geomorphology of the Southern Coastal Plain and the spirit of the day contributed to these views. Randall's awareness that the area's significance was due to the vegetation cover, however, is an early recognition that East Texas and the Piney Woods, once the unoccupied Neutral Ground contested by several cultures, was also a biological crossroads.

The ecotone exists because of the climate, which Randall might have considered to be one of the region's more interesting attributes in addition to the vegetation. In this part of deep East Texas, subtropical maritime air masses influenced by the Gulf of Mexico square off against cooler, drier air descending from more continental sources. The collisions that occur when these air masses interact creates an invisible battleground in the skies. In fact, the frontal concept used by meteorologists today was developed with direct reference to the shifting battle lines in the aftermath of World War I.[3] In Texas, the clashes for atmospheric control are most intense in the spring when the winds pick up ahead of narrow, energetic boundaries that produce short but intensive rains, rattling thunder, and incredible temperature differentials. These storms, together with the passage of more moderate summertime fronts, convective showers, and periodic hurricanes, create a precipitation isopach where the rainfall amounts diminish from an annual average around 56 inches near the Texas-Louisiana border to about 45 inches near Tyler.[4] Flora and fauna from east and west mingle as the climate dries and Bray's wave crashes, most notably in the Big Thicket portions of Hardin County, but also in the Post Oak forest's transition to the Blackland Prairies.

Dissecting the Piney Woods

A topographic map is an engineering marvel and an example of human ingenuity. Much like language, with a bit of understanding and a little practice, it becomes easy to measure distances and visualize three dimensions—distance, elevation, and relief—on foldable, easily stored papers that fit into the corner of a shelf or an envelope between car seats. In the United States, topographic maps produced by the US Geological Survey (USGS) reduce the land surface into a series of quadrangles, which are rectangular areas bounded by lines of latitude (on the top and bottom) and longitude (on the left and right edges).

Outdoor enthusiasts rely on large-scale maps, like the popular 7.5-minute quadrangles, because they portray information about trails, campgrounds, and points of interest that are useful for navigation. Because there are different expressions of scale, these maps can be thought of as representing 7.5 minutes of latitude and 7.5 minutes of longitude, or having a 1:24,000 fractional scale, in which one unit of measure on the map corresponds to 24,000 identical units on the ground. Perhaps more simply though, the expression of scale on a 1:24,000 map can also be communicated by writing that 1 inch on the map represents 2,000 feet on the ground.

The USGS also publishes other maps expressing both small and large scales. Small-scale maps, like those at 1:250,000 scale, are often used as base maps in geologic mapping or for detailing transportation routes. At this scale, each map covers 1 degree of latitude and 2 degrees of longitude, and 1 inch on the map represents about 4 miles on the ground. Because each page covers over 4,500 square miles, a 1:250,000 map is restricted to depicting major features like roads, boundaries, and rivers.

It turns out, however, that of the thirty-eight 1:250,000 maps available in the 1982 *Geological Atlas of Texas*, only four cover the most heavily timbered regions of the state (fig. 1),[5] making them an ideal base for locating primary cultural centers, studying the region's natural history, and systematically evaluating and categorizing temporal changes in Texas' lookout tower network.

From south to north, these four quadrangles—Beaumont, between 30 and 31 degrees latitude; Palestine, between 31 and 32 degrees latitude; Tyler, between 32 and 33 degrees latitude; and Texarkana, between 33 and 34 degrees latitude—begin at the Texas-Louisiana line and

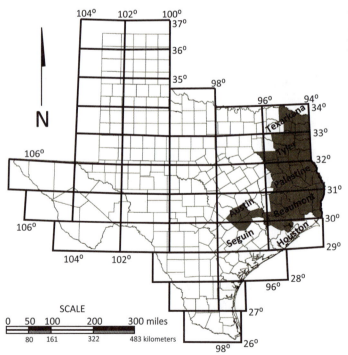

Figure 1. Quadrangles (1:250,000) delineated in the 1982 *Geological Atlas of Texas*. Historically, the TFS maintained a fixed point fire detection system in counties that encompassed four quadrangles, shaded in gray (appendices 1–4). In 1970, protection was extended westward into two more quadrangles (Austin and Seguin) of "fringe forests" (appendix 5).

extend to 96 degrees west longitude. The maps include portions of four major drainage systems that wind quietly across broad meandering belts, incising into a thick wedge of coastal plain sediments that began filling in an embayment on the young Texas coast.

These infilling sediments occur in linear belts that become progressively older away from the coast. The change in age reflects the way the belts grew, or prograded, farther and farther oceanward as clay and sand from the continent were deposited increasingly outward into the Gulf of Mexico. The orientation of the belts, which trend from the southwest to the northeast, are nearly perpendicular to large-scale changes in vegetative cover, which respond most directly to changes in precipitation from east to west. Yet the geology is a controlling factor in the geomorphology, and well-consolidated rocks dissected by drainages provide convenient promontories for lookout towers. Likewise, coast-parallel faults, like the one near Mt. Enterprise, create scarps with broad views on which towers sit. Locally, the geology also influences soils and creates microenvironments that favor certain species, ultimately predisposing some sites for farm forestry or acquisition of land for national forests.

Seventy-six percent of the state's lookout sites disproportionately lie in two of the four quadrangles. The Beaumont quadrangle alone hosted sixty lookout tower sites throughout the life of the protection network (appendix 1), and it includes communities such as suitably named Woodville, where in 1940, "the high-pitched whine of sawmills, the odor of raw pine, and talk of lumber, dominate[d] the town."[6] Woodville, the seat of Tyler County, is also near the geographic center of the forest region in which longleaf pine savanna prevailed historically. These stands were so important that the region was once visited by university professors to better understand the intricacies of longleaf forest succession. From here and Polk County, magnificent longleaf forests once extended eastward to the Louisiana border, covering about half of the Beaumont quadrangle.

The Sam Houston National Forest is also encompassed by the western portion of the quadrangle, and while the proclamation boundary of the forest acquisition area spans Montgomery, Waller, and San Jacinto Counties (fig. 2), the nucleus of government holdings lies within the Willis Formation outcrop belt. This formation consists primarily of clean, fine- to medium-grained sand, with lesser amounts of interbedded clayey sands, making it a well-drained substrate suitable for pine growth. Adjacent lands within the proclamation boundary, underlain by more clayey formations, were largely omitted in the acquisitions because of the difference in their productivity for farm forestry.

Directly north, the Palestine quadrangle encompasses the forest product hubs of Angelina and Nacogdoches Counties, as well as

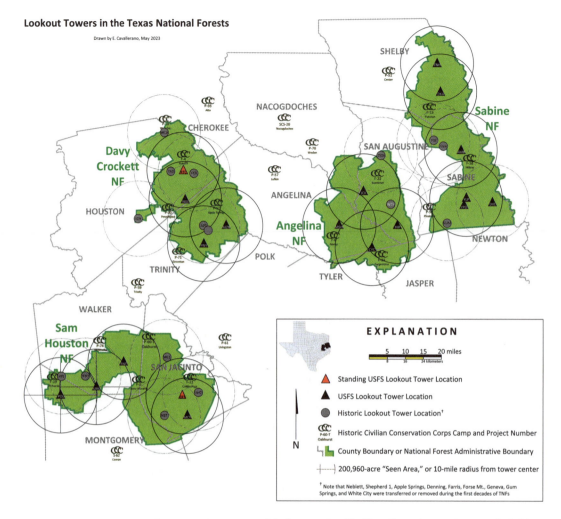

Figure 2. Map of the Texas National Forests, including the historical position of lookout towers.

important commercial forests in Shelby, San Augustine, and Sabine Counties. The southeastern corner of the quadrangle was once a continuation of the longleaf belt, but it also suffered from intensive early logging. Lufkin, in midcentury, with "lumber piled in vacant lots,"[7] is in the south-central portion of the sheet and served as the command center for state and national forestry activities during the height of the Great Depression along with the district headquarters for the US Army's Eighth Corps Area.

Sixty-three lookout sites have been identified in the Palestine quadrangle (appendix 2), no surprise given the forest cover and level of activity during peak CCC years. Importantly, the quadrangle also encompasses three national forests—Davy Crockett, Angelina, and Sabine—each of which once boasted its own fire control systems. Together, the Texas National Forests included eighteen permanent lookout sites.

While a handful of prefabricated lookout structures had been completed by the late 1920s (fig. 3), the majority of towers in the Beaumont and Palestine quadrangles date to 1934. That 76 percent of all the state's lookouts are confined to these two quadrangles is a consequence of protective efforts driven by forest cover and resource development. In some ways, it also reflects a learning knowledge that developed as the fire control program advanced with time across the state.

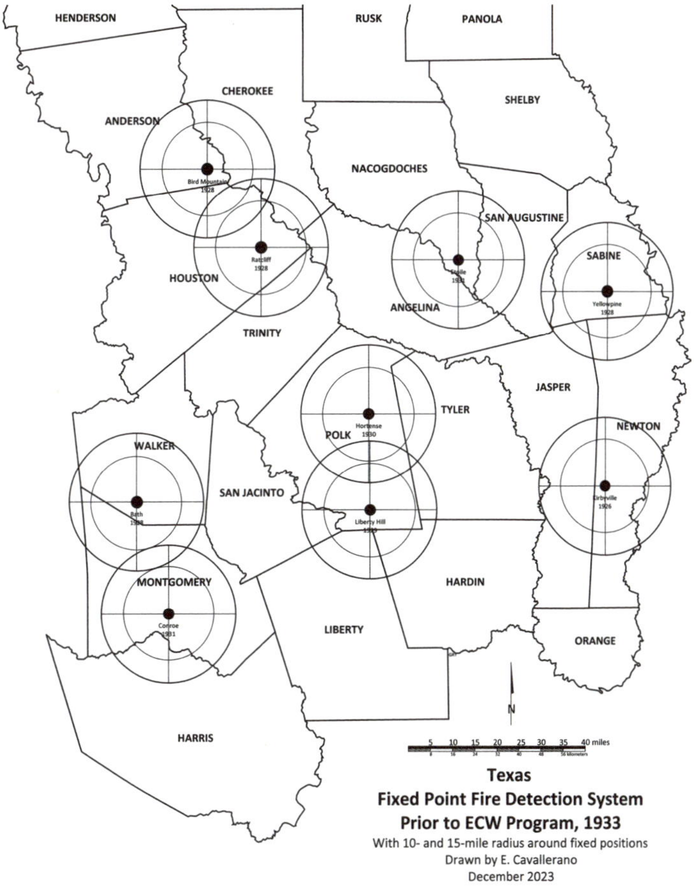

By 1936, towers were being constructed in the "northeast Texas forest," an area described by Cruikshank as second-growth "found in small tracts intermingled with cultivated land."[8] Conveniently, this entire area is covered by the remaining two 1:250,000 maps. In comparison to the network in the Beaumont and Palestine quadrangles, the lookout system encompassed by the area of the Tyler and Texarkana quadrangles is much more economically organized. The Tyler sheet, dominated by the Sabine River valley, includes twenty-six lookout sites (appendix 3)—half the number utilized in quadrangles to the south. Fewer still were necessary north of 33 degrees in the Texarkana quadrangle (appendix 4), where only ten sites protected the shortleaf and scrub-hardwood forests.

Peppered throughout this landscape are standard brass discs, or benchmarks, secured into concrete or bedrock and left by surveyors as they methodically traversed the country measuring distances and angles to build maps of the territory. Accompanying each of the markers are written descriptions on how to locate the station, offering a tantalizing look at the landscape, the people, and the past. Because surveyors historically relied on lines of sight for their measurements, the height of forest lookout towers made them attractive positions for working out the details of Texas geography. Thankfully, many long-gone lookouts still have fresh

Texas
Fixed Point Fire Detection System Prior to ECW Program, 1933
With 10- and 15-mile radius around fixed positions
Drawn by E. Cavallerano
December 2023

Figure 3. Prior to emergency conservation measures during the Depression, the TFS could secure funding for only nine lookout towers. These were distributed throughout the Piney Woods in state forests or protection units sponsored, in part, by industry.

Neutral Ground 25

descriptions written by professionals with an eye for describing the landscape. When studied alongside the topographic maps, benchmark descriptions become provocative breadcrumbs that provide details and fill voids in our understanding of the tower network.

For example, records for some of the earliest federal lookout locations are difficult to locate or hard to access, making maps and benchmark descriptions valuable firsthand accounts. A period of refinement followed the initial stampede of fire tower construction on the national forests, perhaps because of unrealized plans to secure expected acreage or the high density of understaffed primary and secondary lookouts at the close of the public works era. The benchmark description for the Neblett Lookout, an ambiguous federal tower appearing on some of the earliest maps of the Sam Houston National Forest, is one of the few written records available for the site. The descriptions confirm that the structure was standing in 1942, reached by turning west from "the main travelled road . . . at a sign 'Neblett Lookout Tower.'" Within a decade however, subsequent recovery notes indicate that the previous instructions to locate the benchmark were adequate but that "the last 1.85 miles is over bad road not used very much and hard to find in places, [as] the Neblett Lookout Tower (USFS) has been removed."[9]

While relative ages capture the early disposal of the Neblett station, a 1942 surveyor's note describes a lookout erected at Moores Grove (1939). The relatively late construction date for the steel tower the USFS had "on hand" teases at undocumented transfers that may have been necessary to secure better agency protection. After land acquisitions began on the Sam Houston National Forest's Raven District, for example, the only two active lookouts west of what became Interstate 45 were at USFS Pool and Bath, a condition that would have required dependable interagency cooperation between the state and federal Forest Services.

There are indications from a TFS memorandum written somewhat later in 1949 that there was difficulty finding seasonal personnel for the Bath Tower. Without the additional hands, the service reported, "we are using the Willis crew to handle suppression on lands for which the TFS is responsible in the area covered by Bath tower."[10] A tower positioned at Moores Grove would have allow USFS personnel there, and at Pool, to triangulate fires together, even when the TFS tower at Bath was unstaffed.

Fire Season

In March and early April, East Texans are watching for rain. Gardens have been started, and the plants are still tender and young. Inevitably, residents find themselves standing outside, supplementing any rainfall with a garden hose while being bitten by the first mosquitoes of the season. This is also the time when last year's grasses, fallen limbs, and autumn's leaves have had time to dry in winter's low humidity. It is the height of spring fire season, an interval that extends with some flexibility from February 1 until April 30. Usually less intense, another fire season occurs in the fall, beginning about August 1 and lasting through November 15, after a period of quiescence during the wetter, more humid summer months.[11]

These days, the spring fire season is also the time when state and national forest managers initiate prescribed burns to reduce the fuel load created by dead vegetation, controlling the landscape for people and wildlife alike. Ignited on blue, cloudless days behind a passing cold front, the winds are favorable and the air is dry. In these ideal conditions, the forest is burned, compartment by compartment, by fires that consume hundreds to thousands of acres per day. Viewed alongside such a burn, as I had recently while traveling southbound on Interstate 45, a column of hazy, bluish-white woodsmoke lofted steadily upward to low altitudes before being swept away by the northwestward winds in a triangular plume still discernible 12 miles away. The smokes observed from a prescribed fire are so large, in fact, that they can be seen on visible satellite images of East Texas and often diminish air quality in downwind cities such as Houston.

Historically, seasonal variations in fire prevalence have been superimposed by shorter-period daily schedules that changed little over forty years. The 1938 *Manual of Practical Forestry for East Texas* documented the fire occurrence pattern for the 1937 calendar year, revealing that Monday, Tuesday, and Sunday were the most likely days for fires to start in the area under protection. Weekday fires were often ignited by residential trash burning or boiling water for wash pots on laundry day, while weekend fires were attributed to arsonists and campers. Much later, an analysis of the fire history on the Sam Houston National Forest over a five-year period between 1972 and 1976 indicated that the majority of fires were ignited on Monday, Thursday, and Saturday between 2:00 and 7:00 p.m. In the latter study, further analysis demonstrated that local residents were, not surprisingly, responsible for weekday fires, while forest visitors were primarily responsible for weekend fires. Together 92 percent were anthropogenic.[12]

Nowadays, Texans generally have a fickle response to fire. On some occasions, we drive past roadside warnings alerting us "not to report" the controlled burn outside our window without much consideration, while at other times we pull to the shoulder of a highway to help quash the flames using backseat blankets or shovels, compelled to help until emergency crews arrive. Our inconsistent response hints at a somewhat confusing fire education. In some measure, this is due to the legacy of Smokey Bear and the "Ancient and Honorable Order of Squirrels" campaigns that advocated the complete and rapid suppression of all wildfires and required children to pledge that they would to be responsible with fire (fig. 4).[13] It also arises as we rediscover that some fires are beneficial to the ecosystem after years of fire suppression created more severe wildfires that were fueled by decades of accumulated debris.

Chapman and Foster also understood the ecological function of lightning-kindled and anthropogenic fires for maintaining the health of longleaf savannas. For his part, E. O. Siecke even approved the experimental use of prescribed fire on 35 acres of longleaf pine at State Forest Number 1 (Siecke State Forest) in 1935. At the same time, similar tests were also taking place on the Angelina National Forest, administered by District Ranger Alonzo Jared and the Southern Forest Experimental Station. Each experiment suggested fire was beneficial to second-growth longleaf forests, and today's maturing longleaf stands near Aldridge Mill and Boykin Springs are a testament to the success of these trials.[14]

Interestingly, influential geologist Charles Lyell also speculated as early as 1850 that the hillside longleaf nurseries near Tuscaloosa, Alabama, propagated because of indigenous burns that kept brush at bay.[15] Joe Truett and Daniel Lay treated these hunts vividly in "Before Cattle,"[16] an imaginative realistic fictional piece about an Ais people's fire-kindled bison hunt along Ayish Bayou.

Even so, midcentury messages called for complete wildland fire suppression, largely to protect young forests, prevent erosion, and

Figure 4. The Honorable Order of Squirrels program encouraged youth groups to climb lookout towers and obey outdoor burning requirements. The initiative dates to 1927, when Minnesota ranger Allan W. Stone sponsored, and paid for, the first supply of membership cards. In Texas, TFS personnel distributed membership cards like this one to anyone who reached the top of a fire tower as part of local educational campaigns.

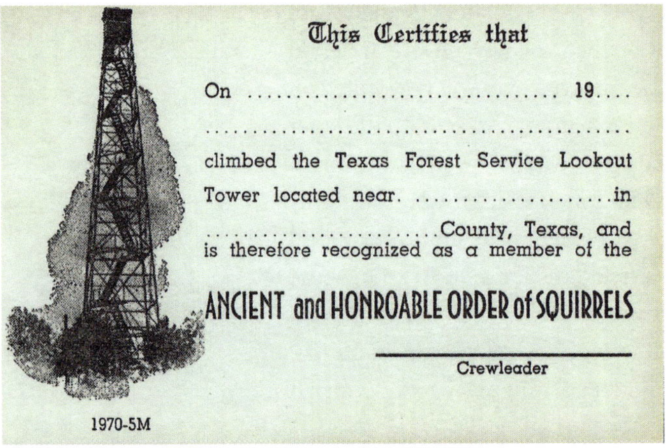
This Certifies that

On .... 19....

....

climbed the Texas Forest Service Lookout Tower located near. ....in

..........................County, Texas, and is therefore recognized as a member of the

ANCIENT and HONROABLE ORDER of SQUIRRELS

Crewleader

1970-5M

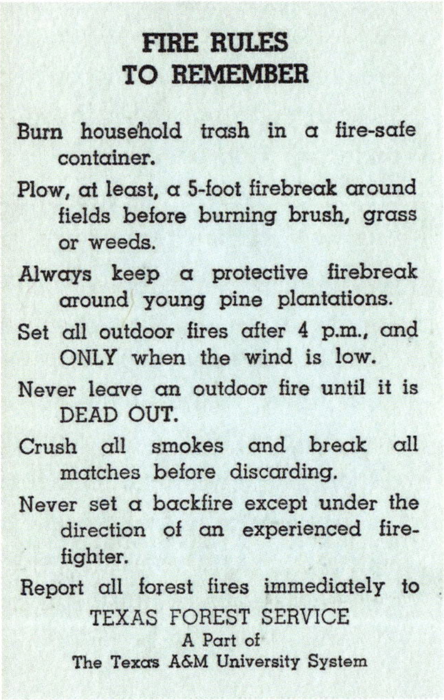
FIRE RULES TO REMEMBER

Burn household trash in a fire-safe container.

Plow, at least, a 5-foot firebreak around fields before burning brush, grass or weeds.

Always keep a protective firebreak around young pine plantations.

Set all outdoor fires after 4 p.m., and ONLY when the wind is low.

Never leave an outdoor fire until it is DEAD OUT.

Crush all smokes and break all matches before discarding.

Never set a backfire except under the direction of an experienced firefighter.

Report all forest fires immediately to

TEXAS FOREST SERVICE
A Part of
The Texas A&M University System

conserve water resources. Federal aid also encouraged state agencies to snuff out every fire, and a militarized attitude developed around fire response. This philosophy is emphasized clearly by the language of the trade: *strike teams* or *suppression forces* were *mobilized* to *surround* and fight fires, sometimes with supporting aerial operations, and tactical success was measurable by the *speed of attack* and *containment*. It is no coincidence that lookouts and patrol officers reported to the head of the Fire *Control* Department (FCD).

Within any given fire season, studying fire weather helped lookouts predict the likelihood of a blaze. It was the lookout's responsibility to observe the condition of the vegetation and record rainfall totals and temperature. Lookouts also watched for lightning and monitored visibility so that if a fire occurred, they would be able to better predict behavioral elements such as the rate of spread, the ignition risk, and the fuel energy.[17] In early years, the data were tabulated in the towers, and the observers were responsible for submitting monthly reports to the district. The rigor with which these records were kept is called into question in a 1942 roast that appeared in *Pineywoods Pickups*.[18] Bob Williams, a long-serving employee who began staffing the Willis Tower during 1934, was playfully called out by the editor, who wondered if Bob "sent in 29 reports for February like he did last year." Despite the oversight, Williams was a dedicated patrolman and was only five years away from fighting some of the most severe wildfires in Texas history—with hand tools.[19] Disappearing tower reports were in vogue during 1942, and the same issue also reported that Lookout John Huffman complained that the false teeth belonging to another patrolman chewed some of his Kirbyville Tower reports to bits.

More rigorous techniques evolved specifically for monitoring fire weather during the mid-1930s.[20] In addition to a rain gauge, psychrometer, anemometer, and wind vane, the toolkit expanded to include a duff hygrometer and fuel moisture sticks. These could be used to judge the inflammability of surface fuels and larger slash like dead trees or branches.[21] The instruments were housed in out-of-the-way shelters at lookout compounds or near state forest residences and were monitored several times a day to understand changes in relative humidity and calculate the fire risk (fig. 5). The TFS established its first danger rating stations during 1943 and, by December the following year, had built five, with a combined value of $250.[22]

The observer calculated the fire danger measurement and telephoned the results back to the division office, where the dispatcher could then alert other lookouts in the district to the local conditions. The consistent methodology allowed the TFS to better anticipate the likelihood of fire, and the quantitative approach became an important standard for

Figure 5. The Kirby State Forest danger rating station in January 1951. Theodore Busselle is measuring fuel moisture to calculate the risk of wildfire. In later years, monitoring fire weather was essential in determining when lookout towers were staffed or aerial patrols flown. Image courtesy of Texas A&M Forest Service.

coordinating with other firefighting organizations. In fact, the memorandum of understanding signed between the TFS and TNFs in 1949 included provisions based on these danger ratings, and both organizations agreed to staff their lookouts "when Class III and greater fire danger occurs."[23] Richard Malouf, citing Richard Baumhoff, described how the danger rating system was used in Missouri's national forests:

> In the periods of "1" and "2" hazard the lookout towers are not in use. When the danger rises to "3" the 27 primary towers, each a sturdy steel structure 100 feet high, are manned in the four eastern divisions.... As visibility decreases, along with increase of hazard, a series of 80-foot secondary towers are put in use, and on a "7" day, even the numerous 50-foot poles, surmounted by light crow's nests, are pressed into service.[24]

Correspondingly, the FCD coordinated with the Louisiana Forestry Commission during the early 1960s so that danger rating broadcasts made to the lookout tower in Greenwood, Louisiana, could be relayed to Texas personnel in Districts 3, 4, and 5.[25]

3 Forest Protection and Fire Control

SIECKE'S DEPARTMENT OF FORESTRY and its subsidiary, the Division of Forest Protection,[1] were both maturing organizations with expanding budgets at the time the economy collapsed during the autumn of 1929. The state had erected its first lookout tower, an 80-foot Aermotor LS-40, in 1926 at the recently acquired State Forest Number 1 in Kirbyville, and the network slowly expanded to include eight others by 1932 (see fig. 3).[2] More important, the public was beginning to view the towers as an "intelligent, profitable investment,"[3] as forestry publications, engagements, and propaganda campaigns reached more Piney Woods inhabitants.

A two-tiered scheme developed early during the history of the Department of Forestry to create a defensive ring around the Piney Woods. The "blanket patrol" system was initiated when Foster hired the agency's first patrolmen using Weeks Act funds to keep officers on the move during dangerous periods, covering ground and responding to fires with local help as best as possible. During wetter times, these same patrolmen hung informational posters and quietly worked to modify the public's behavior by visiting schoolhouses and corner stores to advocate fire prevention. These blanket patrols contrasted sharply with more rigorously monitored "protection units" that became a popular mechanism for forestry services and landowners to more fully safeguard private holdings.

The protection units were an offshoot of the Clarke-McNary Act (1924), which expanded conservation efforts away from the boundaries of watersheds and allowed the government to purchase lands that could be used for forestry.[4] The objective of the legislation was to increase wood production for future demands, and expanding fire prevention policy was eyed as a means of regenerating forests and sustaining timber harvests. The expansion not only provided matching funds to state conservation departments but also incentivized private landowners to enter into cost-sharing fire protection agreements with the states.[5] As part of the program, the Division of Forest Protection prepared statements for interested landowners on how they could more aggressively guard their individual or corporate holdings against fire.

Often, these recommendations included estimates for the equipment and personnel necessary to supplement the state's somewhat spotty blanket patrols. The additional resources meant that the regular district patrolman would be assisted by lookouts and smokechasers who worked exclusively to discover and chase the source of smoke within the unit before any fires caused extensive damage. To achieve "unit" status, however, a block of land greater than 100,000 contributing acres had to be available for protection. Crucially, the owners also had to agree to offset the financial burden of the effort. When these provisions were met, natural boundaries were selected to define the unit margins, and a special field force was hired to oversee the subsidized forestlands. By 1927, the Louisiana Division of Forestry had 3 million acres under this scheme, and landholders were contributing two cents per acre to underwrite the program. This support also meant Louisiana could afford constructing and maintaining fourteen lookout towers.[6]

Somewhat behind our eastern neighbor, the first protection unit in Texas was established in 1927 when the Southern Pine Lumber Company contributed to the cost of monitoring its corporate forests. The following year, the Houston County Timber Company, Gilmer Company, and Temple Lumber Company inaugurated similar programs on their holdings.[7] At the time, the Division of Forest Protection estimated that protection costs would range between two and four cents an acre in shortleaf and loblolly pine regions and between four and six cents in longleaf tracts after the initial infrastructure costs were satisfied.[8] With the promise of financial support and the obligation of intensively guarding the units, the Division of Forest Protection continued to expand the use of lookout structures, erecting four in 1928 to supplement the regular blanket patrol forces.[9]

The protection unit in Sabine County serves as an early example of progress. The Temple Lumber Company cooperated with several other timber interests early in 1928, allowing the state to install an 87-foot-tall galvanized lookout tower in the vicinity of the logging community of Yellowpine. Blueprints for the International Derrick and Equipment Company (IDECO) tower still exist in the Texas A&M Forest Service files at College Station and highlight the trade-off between technology and cost as the lookout program evolved. After scrambling up a series of short ladderways into the weatherproof room that awkwardly extended beyond the frame of the modified derrick, lookout personnel at Yellowpine could monitor Temple and Kirby landholdings within a 12-mile radius of the tower. If the lookout observed smoke, the officer could use the protection unit's private telephone network to communicate with other patrolmen, smokechasers, or emergency personnel located in less expensive "tree cabs" at the top of tall pine trees near Bronson and Milam. With the other observer on the line, the pair could quickly triangulate, or get a "cross-shot," on the position of the fire by using an alidade and the map board located in the tower before rushing to the plotted location of the blaze.[10]

Figure 6. (*Left to right*) William E. White, Walter Bond, Isaac C. Burroughs, and William O. Durham sit at the base of the Ratcliff Lookout during a Forest Guard meeting in July 1929 one year after the original tower was erected. During the ECW period, these men crossed the state and surveyed the majority of the lookout tower sites. They also drafted the initial fire protection maps, consolidating transportation routes and delineating duty station locations. Image courtesy of Texas A&M Forest Service.

The scheme met with industry approval. "It is expected," the *News* reported in January 1928, "that several lookout towers will be erected by private companies in cooperation with the State Forest Service within the next year or two."[11] They were, and protection units were established in 1930 to guard a 111,293-acre unit owned by three lumber companies in southern Walker and northern Montgomery Counties near land destined to become part of the Sam Houston National Forest. The following year, a smaller 70,000-acre tract known as Protection Unit 7, was approved in Angelina and Nacogdoches Counties. The smaller size of Unit 7 was likely allowed because of its proximity to already established Unit 2. Regardless, the addition necessitated construction of two extra tree cabs on Unit 2, two new tree cabs on Unit 7, and one new lookout tower. William E. White and Isaac C. Burroughs conducted the reconnaissance survey of Unit 7 in the heat of early July (fig. 6). Zachary T. Stripling and Ivan H. Jones returned three weeks later, along with Landman Jimmie Jacobs, to drive pine stakes into the Frost Lumber Industries property and define the parcel for what was originally described as the Camp Pershing Tower. It would later be known as the Etoile Lookout. A week later in short succession, the Huntington and Rocky Hill Tree Cabs were also delineated.

When White went afield with Burroughs in July, he was the relatively new, but skillful, chief of the Division of Forest Protection. He had been hired as a division patrolman in 1927, after already working for a decade at the USFS and as a forestry engineer overseas during World War I.[12] Quickly he was promoted to lead the division in Lufkin, a position he

Forest Protection and Fire Control   **33**

held through the Depression. His correspondence during this period is exacting and courteous, and it is extraordinary that he was able to oversee an immense portfolio of CCC work projects. As CCC work plans were submitted, White also helped coordinate the development of the lookout network and corresponded with lumber managers to build support for cost-sharing programs. That he was in the field on a steamy July day makes it easy to believe he was a quiet but persuasive advocate for intensive protection units, even convincing his industrial peers that there was value in participating during some of the worst years of the Depression.

Tools of the Trade

"Less determined boys would have given it up," Isaac Burroughs wrote of his characters in *Davids of Today Slay the Giant, Forest Fire*.[13] When not out surveying the Piney Woods, Burroughs had taken his pen to write a descriptive portrayal of an East Texas community in a style much different from the precise surveying descriptions he was responsible for submitting. Young Red and Tom, two firefighting protagonists, could fill the dreams of any schoolchildren. Together, they were magical propaganda. Relatable, strong, and willing to take on the "big tasks . . . to be done by the Davids of today," Burroughs began, with obvious reference to David and Goliath's biblical story.

As it happened, Red and Tom were walking home from church on a "fine" Sunday when, suddenly, they spotted smoke from a forest fire "mighty close" to Tom's home. The fire was burning briskly by the time the boys arrived, and they wasted no time cutting a couple of pine tops with their jackknives to swat the flames close to the head of the fire. The "red demon" greedily devoured the dry rough and young trees in its path, and the boys had to fight bravely, Burroughs explained. A car drove up and stopped just as the situation appeared "hopeless," and the boys immediately "recognized their forest patrolman who had visited their school several times and was a friend of all the children." He was here to help, he reassured them, handing them both fire swatters to replace the pine tops that were "about worn out."

These swatters, the pamphlet educated, consisted of "wide pieces of belting about two feet long, fastened on long wooden handles." They were used "in the same manner as pine tops" to suppress the flames. When Burroughs wrote his story in 1929, the swatter, along with other simple instruments like burlap bags, shovels, axes, hoes, council tools, and spike-toothed road or asphalt rakes, were the only resources available to patrolmen tasked with defeating a fire. The most sophisticated

Figure 7. Timpson Lookout Emory Covington displays the firefighting resources available to him in 1936. Image courtesy of Texas A&M Forest Service.

device, a five-gallon canister worn as a backpack with a hand pump, could be used later for mopping up fire lines or extinguishing snags and burning stumps.[14]

At first, blanket patrolmen carried whatever tools they could on horseback and borrowed what was available from residents near a fire. As automobiles slowly replaced horses, it became easier to transport equipment to a fire. A photograph of Emery Covington standing next to the collection of tools he carried in his personal vehicle demonstrates the primitive nature of early firefighting efforts (fig. 7). Once fire tower construction began in earnest during 1934, CCC laborers often constructed toolboxes to cache supplies below many of the lookout towers.

Along with tower piers, many of the original toolboxes deteriorate silently in the woods and are common associated structures. One or two show remarkable preservation. All share the same design, measuring 7.75 x 3.75 x 2.5 feet, with walls 4 inches thick. The depth of the box tapers several inches along its width to create a sloping surface onto which the lid was placed with simply fashioned hardware. At most sites, the covering was an early target of vandalism and rot, and the protective aluminum veneer that was once tacked to the dovetailed planks can be found at the base of the box or in the forest litter.

Garages and warehouses were also built to store tools and equipment, especially at the national forest lookout compounds. The 1949 *Texas Almanac and State Industrial Guide* pointed out, for instance, that "sufficient equipment such as rakes, council tools, flaps, shovels, backpack water pumps, saws, axes, and other hand tools are maintained in standard fire tool caches at strategic points to equip 1,000 men."[15] As time

Forest Protection and Fire Control   **35**

went on, these structures housed not only a collection of hand tools but also the mechanized plows and tractors that slowly replaced them. Representative examples include the USFS and FCD warehouses in Lufkin, as well as other administrative centers such as those at Four Notch, Jackson Hill, and Dreka.

Fixed Point Detection: The "Eyes of the Forest"

The Division of Forest Protection recognized that "a complete system of lookout towers" was required for its presuppression activities.[16] The elevated observation points would allow service personnel to quickly identify the location of a fire and travel to the source to extinguish the flames before they had an opportunity to spread. When more than one vantage point was available and a fire occurred, two observers could use their maps and alidades to independently establish compass bearings between their position and the smoke. When the two vectors were plotted on a map, a triangle formed, two points of which represented the position of the fixed observers, and the third represented the location of the fire (fig. 8). This procedure could be supervised by a dispatcher at district headquarters, who was communicating with the towers by telephone, or by the lookouts themselves using radios and map boards. In such fixed point systems, the observation point, just like a surveying benchmark, is established in relation to others. Essentially, each tower in the statewide network could then be connected by lines with known angles and distances to create a series of triangles, like a giant puzzle, across the Piney Woods.

Figure 8. Henry Cutler, the observer at Weches, establishes the location of a "smoke" in January 1951. If a fire was spotted from the lookout, observers like Cutler could determine the compass bearing of the smoke using an alidade in the tower. Then, with a similar bearing from another tower, the two vectors could be "crossed" using the azimuth rings and strings attached to a map board to locate the fire. Image courtesy of Texas A&M Forest Service.

Securing funds to complete the envisioned network was slow, even though protection units were gaining popularity. Not surprisingly, corporate interest fluctuated with financial conditions during the years leading up to the Depression. For instance, there were four cooperating companies in 1928 and 1929. This number jumped briefly to twelve in 1931, before the market worsened and industry franchise dwindled to nine participating corporations in 1932.[17] By the TFS's own account, the protection effort "was greatly slowed up during the past biennial due to the depressed financial conditions of the private cooperators."[18] In the years since the first lookout was erected on State Forest Number 1, only eight other lookouts had been secured with industry sponsorship. Until additional funding became available, the most efficient manner with which to extend the fire control network and improve the forest fire record was to place the "eyes of the forest" in the forest—on top of the tallest trees.

The most economical way to effectively improvise a protection system in the various patrol districts under blanket protection was to construct a series of temporary observation points high above the ground by either spiking trees or building tree cabs like the ones in Milam and Bronson. The simplest technique, spiking trees, relied on driving 9- x 9/16-inch galvanized telephone steps into the trunks of large trees. The spikes would end at a seat, or crow's nest, where one might "stay on watch for long periods without discomfort."[19] Over fifty trees had been spiked by 1926, allowing patrolling officers to "see over many miles of their district" and reduce time spent in the saddle.[20] With a few strategically placed trees, "the men are spending the greater part of their time, during dry weather, on lookout points and in fighting fires," then in traveling by horseback, the *News* concluded.[21]

Figure 9. Some spiked trees have survived, like this shortleaf pine near the junction of FM 2426 and Forest Road 114 in Moore Plantation, Sabine National Forest. Photograph by the author.

As late as 1996, Johnny Columbus and Columbus Lacey recalled one such tree near Bannister during an interview with USFS archaeologist Faye Green. "There was a big, uh, black gum tree out here with a lookout tower . . . and the Texas Forest Service man sat up there on top of it, and filled the tower," Columbus remembered before Lacey interrupted. "William Wood did that didn't he? . . . I bet that thing's still there."[22] In fact, some are, such as the more accessible shortleaf pine used as a lookout from 1935 until about 1955 on the south side of FM 2426 in the Moore Plantation (fig. 9).

Forest Protection and Fire Control   **37**

Spiked trees evolved into experiments with tree cabs beginning in 1929.[23] While they were still viewed as "a more or less temporary structure," the cabs were built at one-sixth the cost of a steel lookout tower and were designed to be more comfortable than the crow's nests built atop a spiked tree.[24] Each platform was about 6 feet square and constructed on the side of a tall tree or extended above the crown of one of four upright supports.[25] Along with the extra space, the cubicles were equipped with a canvas roof covering and curtains that could be shuttered during "disagreeable weather conditions" (fig. 10).[26] Accessing the cabs was still dangerous, however, as observers were required to climb steel ladders 80 to 112 feet high that were secured to the side of the hosting tree. These hazards were demonstrated by an article written by Larry J. Fisher in 1937:[27]

> Perry Moye was in a jam. A few moments before, the wind had slammed down the trap door in the floor of the tiny tree cab and there was Perry Moye, forest Patrolman, locked in atop a swaying, tall, pine tree, 106 feet above the ground. A steel ladder connected the cab to the earth. But the ladder wasn't doing the fire-watcher any good just now—he couldn't reach it. Finally, crawling cautiously through a window, he managed to swing himself into a position where he could free the hasp. He swung below like a circus acrobat but, unlike the acrobat, there was no net to break a sheer fall of a hundred feet. But he didn't fall; and that, Perry believes, is the most exciting moment he has had in his 12 years of work with the Texas Forest Service.[28]

Fisher eventually went to work for the TFS in 1943 as a visual aids specialist in the Division of Information and Education, and Moye continued as a lookout in the new steel tower that replaced the Votaw Tree Cab throughout the 1930s. He would be highlighted in the news again by Fisher in 1939 for hosting the Perricone quadruplets, Anthony, Bernard, Carl, and Donald, at his tower. "Usually called 'A,' 'B,' 'C,' and 'D' for short," they were the only known male quadruplets in the world at the time, and they stopped for a photograph with Perry at his tower.[29]

Longtime Forest Service employee Knox Ivie, who helped construct the majority of the tree cabs, relates that there was incentive among the three- or four-member crews to build the platforms quickly. The group, often camping until the job was done, wanted to make their beds aloft where the mosquitoes and ticks were fewer and the wolves that "would wake us up with their howling" were less menacing.[30]

Early-detection structures were not exclusively in trees, and a curious lease exists in the TFS archives. In a manila folder, which is marked as a "locked file," the papers preserve an agreement with the state for one acre of land in Anderson County. The purpose of the lease, details the contract, is for erecting "a fire detection cab held in place by a sufficient number of guy wires to make the fire detection cab rigid and stable."[31] The interest arises, not only because of the language of the arrangement but also because of the price paid for the privilege. In an era when dozens of similar leases were signed with "a dollar in hand," Alton and Mary Langston received $5 when they penned their names on the document on May 20, a pretty penny in 1932.

By August, the *News* was broadcasting that "a unique fire detection structure" had been erected on Walker Mountain.[32] The broad hillside had been cleared of timber, the article related, and it was impossible to find a suitable natural position for observation. Compounding the problem, the division could not afford a steel lookout and initially settled on the use of three trees on the edges of the hillside to provide protection, but blind spots made them ineffective. The answer, engineered by Ivie, was to construct a 5-foot-square cab on top of a 78-foot-tall creosoted piling donated to the state by the Southern Pine Lumber Company. The pole was 19 inches at the base and tapered to a 10.5-inch diameter at the top. Unbelievably, the pole was raised and positioned into a 4-foot-deep hole using a stump puller, steel cables and pulleys, a 35-foot-tall gin pole, and a team of mules.

Ivie recalled building the pole cab during a 1970 interview, saying that it took two weeks of preparation to secure the ladders and cross-members to the pole before it was raised. Not surprisingly, there were doubts among area residents that the project could be completed without spectacle, and Ivie mused, "It was generally known when the raising would take place, and people from all over the country came to see it go up. A lot of them never believed that it could be done, but I proved them wrong" (fig. 11).[33] The design caught on, according to a TFS

Figure 10. The tree cab between Moscow and Corrigan in 1930. Temporary structures like this were erected to create a rudimentary fixed point detection system. Many were replaced by steel observation towers when ECW funds became available during the Depression. Image courtesy of Texas A&M Forest Service.

Figure 11. The Langston, or Walker Mountain, pole cab was engineered by TFS employee Knox Ivie. Image courtesy of Texas A&M Forest Service.

activities summary, which boasted "similar pole cabs are being constructed in other southern states according to our specifications."[34] It seems likely that it also inspired the USFS to include an "observation mast for emergency lookout" purposes in the 1938 USFS *Standard Lookout Structure Plans.*

Benchmark descriptions capture the evolution of several fixed point observation stations from spiked trees or tree cabs into steel lookouts. The Smith Ferry Lookout, for example, which still stands in Tyler County, was preceded by a "hickory tree used as a forestry lookout." This 1931 description was later revised in 1949 to note that "the Smith Ferry Lookout Tower has been erected over the station."[35] Likewise, the Fails, Peach Tree, Rocky Hill, Soda, Central, and Weches Lookouts replaced the Cromeens, Peach Tree, Ewing, Williams, Pollok, and Cutler Tree Cabs.

Improvised lookout stations were gradually replaced by more functional, permanent posts. Pamela Conners points out that one method for determining the significance of a lookout is to group the various types into broad categories defined by the intended use of the structure.[36] This philosophy is similar to the functionally specific architectural style described by Peter Steer and Keith Miller,[37] who followed Mark Thornton.[38] In this scheme, towers are segregated based on how and when they were staffed and by the location of the rangefinder or alidade. Primary lookout structures, they suggest, are occupied continuously by an observer throughout the fire season. These differ from secondary lookouts that are activated only for seasonal use during periods of moderate or high fire danger. In turn, both of these designations are different from "emergency" or "project" towers that are used intermittently during high danger periods or for specific, short-term construction or clearing operations. Finally, observatories can be classified based on the rangefinder vis-à-vis any accompanying facilities, and a 144-square-foot floor plan is often used to distinguish between "live-in" and "observation only" lookouts.

Texas had a fair share of emergency and secondary lookouts, some live-in towers, and at least one "unclassified" station atop the New Boston water tower (fig. 12), which was obviously built for another purpose. Depending on the rigidity of the definition, certain project towers also existed, like those deliberately moved to "Critical Areas" to provide fire protection to defense plants operating at capacity during eras of conflict. For the most part, however, the Texas system was built of primary, observation-only lookout structures. As the system matured, the state favored smaller floor plans and galvanized-steel construction.

Discounting the homegrown lookout structures that were constructed using limited funds at the beginning of the fire control campaign, at least ten manufacturers of prefabricated towers contributed to the detection network in Texas. Many of the brittle, folded construction plans for these towers are preserved in the archives of the Texas A&M Forest Service. The collection includes hand-drafted sheets detailing cab and pier specifications and the erection procedures for popular Depression-era builders like the Aermotor Company, IDECO, and the EMSCO Derrick and Equipment Company. There are others from regional manufacturers lesser known outside the oil patch, along with one wooden tower model designed by the Southern Pine Association.

Tower procurement during the Depression advanced through federal contracts, which are detailed in correspondence between the Division of Forest Protection and Aermotor.[39] In response to an inquiry from the TFS in 1944, Aermotor's president reminded the state that "we furnished you towers thru the United States Department of Agriculture, Forest Service, of the type known as our MC-39."[40] Another letter, written by Aermotor in 1948, reinforces the use of government contracts: "It appears that most of the towers you refer to must have been purchased by the U.S. Forest Service or other Federal agency, and allotted to you."[41] A final letter written in 1958 is the most specific (fig. 13), indicating that "in the year 1933, Texas Forest Service acquired at least 17 of these [99.75-foot platform ladder MC-40s], which were furnished by us [Aermotor] on a contract from the U.S. Forest Service."[42]

Likewise, EMSCO fabricated towers for the TFS that were designed in accordance with US Department of Agriculture, Forest Service, specifications.[43] An inquiry from the Division of Forest Protection in 1944 reveals that thirty-one outside-ladder, 99.75-foot towers were purchased during 1934 and 1935. These towers were originally quoted as "TW" forest towers but conformed to drawing "H-2283-D."[44] The latter designation was carried in Texas A&M Forest Service property inventories, though an examination of standing examples reveals "TW" stamps in the metal components.

In a state with "big oil," purchasing refurbished oil-field derricks from faltering oil companies or rig contractors such as Lee C. Moore, B. R. Parkersburg, the Nashville Bridge Company, or the Dunlap Manufacturing Company became more economical, especially after the infusion of federal money ended with the close of the Depression-era "make-work" campaigns. As an example, when the TFS solicited quotations for towers in 1958, a new 100-foot Aermotor lookout with an inside stairway was priced at $6,460,[45] whereas a used, 122-foot galvanized-steel Nashville

Figure 12. TFS built a plywood shelter atop the New Boston water tower. The structure was part of an improvised ("emergency") detection system used in the vicinity of the Red River Arsenal. This photograph was taken in December 1942 during the early period of World War II. Image courtesy of Texas A&M Forest Service.

F.C.D. REC'D. JUN 13 1958 T

AERMOTOR COMPANY

(FOUNDED 1888 BY LAVERNE NOYES)

MANUFACTURERS OF

ELECTRIC PUMPS, WINDMILLS, WATER-SUPPLY GOODS
SPECIAL STEEL TOWERS

2500 ROOSEVELT ROAD

WILLIAM H. HAMILTON, PRESIDENT
MELVIN T. JENSEN, SECRETARY & TREASURER
JAMES R. WALSH, 2ND VICE PRESIDENT

CABLE ADDRESS "AERMOTOR"

CHICAGO 8, ILLINOIS, U. S. A.

June 11, 1958

Texas Forest Service
P. O. Box 460
Lufkin, Texas

Attention: Mr. J. O. Burnside, Head
Fire Control Department

Gentlemen:

We have your letter of June 5, 1958, File 2.461, requesting descriptive literature and prices on 7' x 7' metal cabs for 100-foot outside ladder type Forest Fire Lookout Towers.

Before we can quote prices on 7' x 7' cabs only, we must know on what type of tower the new cab will be used and the number required.

We list below the towers that seem to be the ones you would most likely have:

Drawing Number	Height	Description	Remarks
MC-40	99'-9"	Platform Ladder Tower	In the year 1933, Texas Forest Service acquired at least 17 of these towers, which were furnished by us on a contract from U. S. Forest Service.
MC-41	99'-9"	Straight Ladder Tower	
LX-24	100'-0"	Outside Ladder Tower	
LS-40	100'-0"	Inside Stairway Tower	This is the most commonly used of our own towers.
MC-39	99'-9"	Inside Stairway Tower	A commonly used tower based on Government specifications.

170' stairway

CHICAGO AMARILLO CORDELE, GA. DALLAS DES MOINES HARRISBURG KANSAS CITY MINNEAPOLIS OAKLAND OMAHA

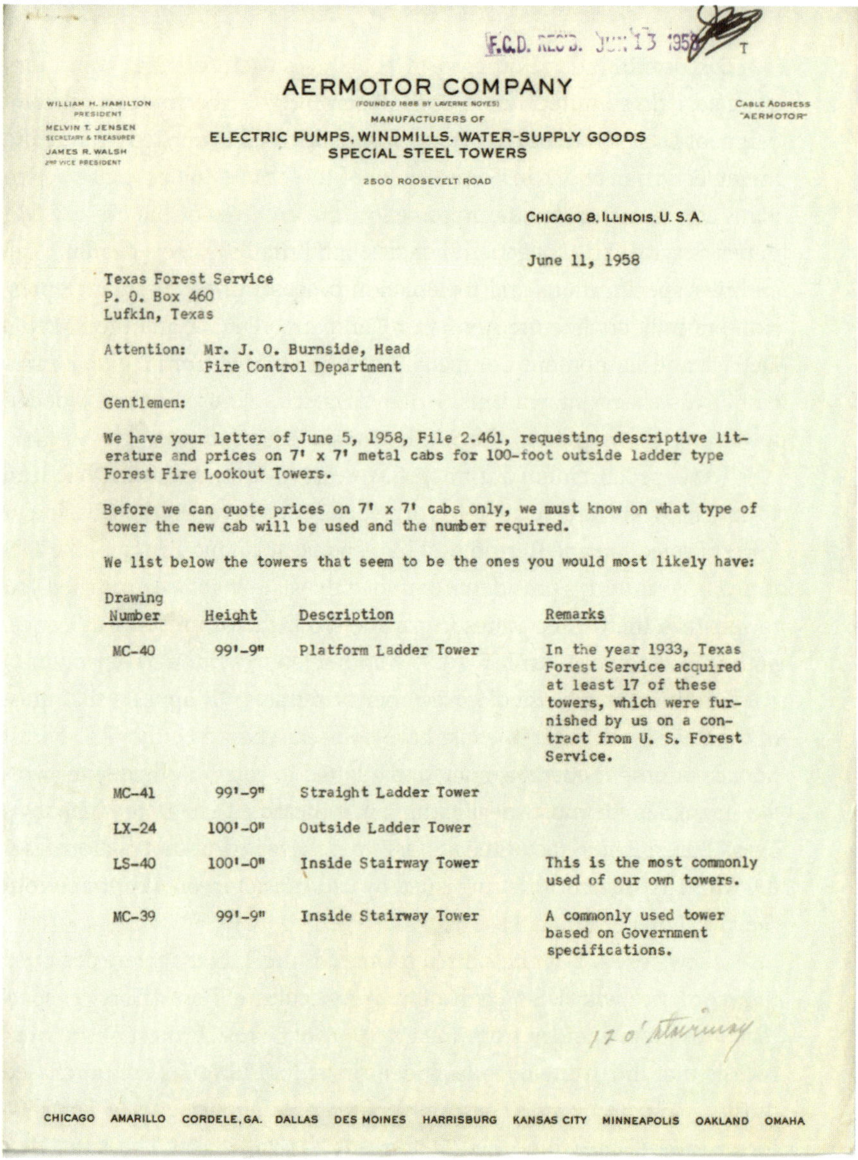

Figure 13. Correspondence between the Aermotor Company and the TFS demonstrates how contracts were drawn during the ECW era. The letter also describes the manufacturer's tower models and details the quantity of lookouts delivered to the TFS. Note the seventeen MC-40s inventoried here are in agreement with White's budgetary letter to Governor Ferguson written in October 1933. Image courtesy of Texas A&M Forest Service.

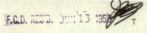

Bridge Company derrick cost $200. The latter could be retrofitted with an identical Aermotor MJ-66 cabin for an additional $800.[46]

Less abundant were towers of wooden construction. Both Siecke and White communicated with the National and International Creosoting Companies, the Texas Creosoting Company, and the Kirby Lumber Company in 1932 regarding the "erection of a lookout tower constructed of creosoted lumber," which could be used "for the purpose of discovering forest fires in this vicinity."[47] Initially, these could not compete with steel lookouts, however, because of engineering challenges made clear by a 1932 letter from the American Wood Preservers Association to the Texas Creosoting Company. Especially in the low-relief Piney Woods, where towers had to be built skyward to afford clear views, wooden towers lacked the required strength: "We have tried using various designs.... The biggest trouble we have in designing wood towers on this order is with the fittings, and usually we have to use larger sizes of material than may seem necessary to provide holding power for the bolts. In fact, it is absolutely impossible to develop the member strength by bolts at joint details."[48]

However, during the Depression, there was a strong industrial lobby to provide a wooden alternative that could compete with the steel industry. For that reason, the 1938 US Forest Service's *Standard Lookout Structure Plans* included a number of timber tower designs. Regionally, these were supplemented by drawings engineered by the Southern Pine Association. With the evolution of durable wooden connectors, such as Teco split rings, it became possible to transfer beam loads and raise wooden towers to the required heights. Beginning in the early months of 1938, the first prefabricated, 100-foot lookout tower built of creosoted lumber was erected at the recently completed headquarters of the Division of Forest Protection in Lufkin.

Announcement of the success was made quickly in the *News*. Boasting that it was not only the first such tower in Texas and that it "could be seen for miles," it also highlighted that the "exact lengths were easily handled by machinery at the National Lumber and Creosoting Company's plant." No material, it went on, "had to be rebored or recut in erecting the tower." Rounding out the article with a Texas-sized claim, the article concluded that "the wooden structure has many advantages, chiefly that of low cost of materials, low construction cost, permanence and less maintenance expense."[49]

Additional wooden structures followed. Absurdly, a 99.75-foot steel tower built in Weches by enrollees from the Lufkin CCC Camp was replaced by a wooden tower of identical height four years later in 1938. Others were completed at USFS Piney (1939), USFS Nogalus (1939), USFS Tenaha (ca. 1939), Bon Wier (1940), Emilee (1941), and Wolf Hill (1949). Interestingly, the latter was described as a "45' creosoted pole tower, outside ladder, steel cab."[50] Most were torn down or replaced during the 1950s as they weathered and became unsafe in the humid climate, though USFS Nogalus—actually tied in height with TFS Bon Wier—once held the record as the tallest all-wood forest fire tower on any national forest in the United States.[51]

Finally, one 102-foot Bilby tower was purchased secondhand from the US Coast and Geodetic Survey (USC&GS) and erected at Saratoga in 1950. Designed of prefabricated steel, the portable tower was light enough to be transported by truck and could be erected in a day over benchmarks as mapping efforts progressed across the country. A veteran USC&GS

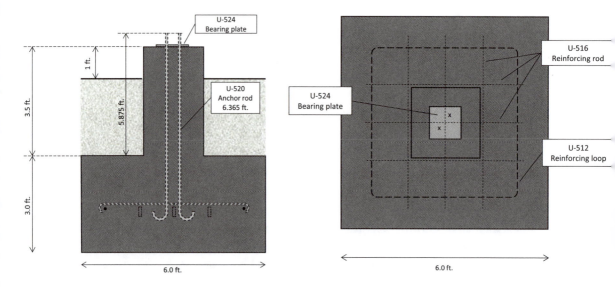

Figure 14. Cross-sectional (*left*) and plan views (*right*) of the pier for a 99.75-foot Aermotor MC-40 lookout tower.

employee, Jasper S. Bilby, designed the reusable tower with help from the Aermotor Company in 1926. It includes an inner instrument tower that was free from vibrations caused by surveyors working on the platforms of a detached outside tower. The tower gained popularity and became "one of the most enduring and widely used" types of triangulation towers because it was lower in cost and more durable than the wooden towers it replaced.[52] The three-legged structure was shaky as a lookout, though, and the straight ascent up a corner post to the 6-foot hexagonal cab was unpopular with crews. Even O. C. Braly, the FCD's trusted rig contractor, wrote in his assessment, "My inspection shows that it can be modified at a lot of expense, and you still wouldn't have much of a tower. It is made of light material and is weak."[53]

Work progressed rapidly once an observation site had been selected. Anchor bolts were shipped in advance of the lookout kits so laborers could pour the tower's foundation piers before the remaining components arrived. Without a rocky substrate, excavations were made to the appropriate depth and dimension of each pier, as workers were careful not to exceed the bottom grade and disturb the underlying soil. Each of the piers looks like an upside-down "T" in profile, with a wide flat base and narrower rectangular column that extends just above the elevation of the ground. Depending on the tower's height, the concrete was reinforced, but, regardless, each pier was meant to support the two precisely positioned J-shaped anchoring rods. For a 99.75-foot Aermotor MC-39, the wide, basal portion of the pier measured 6 feet 9 inches wide and 7 feet 3 inches tall. The anchor rods protruding about 5 inches above the base, meanwhile, descend nearly 7 feet into the pier (fig. 14).

A crew of CCC laborers supervised by a competent foreman could complete a lookout tower in as few as ten days once the foundations were ready.[54] Texas Forest Service correspondence during the same period supports this, and W. E. White once estimated that the Smart School Lookout could be finished within three weeks' time.[55] Likewise, the Wakefield Lookout was surveyed on November 30, 1937,[56] and the lease for the property was signed on December 18.[57] Records show that the tower was then completed by January 10, 1938.[58] The building crew at the Honey Island Camp, P-55, likely held the record though. They reported that the "actual time" spent constructing an EMSCO lookout in Hardin County was only three days.[59]

After the piers were poured and the bearing plates and angle clips secured, the corner posts were raised using a gin pole and tackle.[60] Utilizing finished portions of the tower, laborers progressed up the structure in a series of sections that became narrower as the tower's elevation increased. A horizontal strut, secured to the vertical legs of the tower, formed the base of each section. Heavy wooden planking, often with dimensions of 2 inches x 12 feet x 10 feet, could then be rested on top of each horizontal strut to provide a convenient resting platform for workers while building two intersecting, diagonal braces upward within each partition. To strengthen the tower, horizontal stiffening struts were bolted together near the center of each "X" that formed when the two diagonal braces crossed. Additionally, a series of vertical hangers were installed between the center of every other main strut and stiffener strut (fig. 15).[61]

Stamped or painted part numbers appeared on every component and corresponded to the manufacturer's specifications to guide the assembly crews. With this modular construction forming the essence of the design, towers could be erected to a series of predetermined heights by providing additional sections to the base. When finished, the inward-leaning, battered tower was composed of four walls with a braced but open frame.[62] Because of the standard cab dimensions and the consistent tapering angle of the corner posts, any additional height, as measured from the floor of the tower cab to the base plates, meant increasing the dimensions of the footprint.

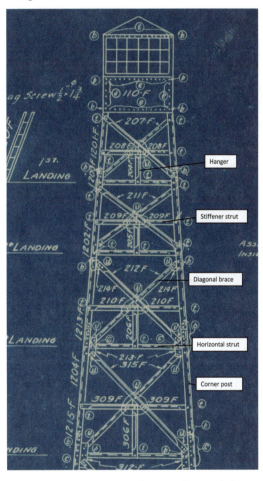

Figure 15. Structural elements on a typical fire lookout tower.

The cabins at the top of each lookout were designed to house the equipment and support the personnel required for fire detection. Nevertheless, they were engineered apart from the towers and were sometimes interchanged between them. The FCD often played hopscotch with surplus cabs or built their own to reduce project costs. One complicated series of cab transfers that serves as an example occurred near Livingston in 1957.

Polk County, in the north-central portion of the Beaumont quadrangle, once maintained a disproportionately high number of lookouts with overlapping "seen areas." As operating costs increased, pressure amplified to consolidate the number of towers in the district. To meet this objective, the FCD decided to demolish the 99.75-foot MC-40 south of Livingston and simultaneously relocate a 120-foot Aermotor MC-39 from the Soda community, located 12 miles east of Livingston. As part of the reorganization, the original Livingston Tower site south of the city would be abandoned in favor of a new location north of the city, close to the historic position of the Buck Tree Cab.

The TFS contacted contractor O. C. Braly during June 1957, soliciting an estimate for the demolition, transportation, and construction of the towers in question. An additional complication required Braly's attention, however, since the Soda Lookout had been standing without a cab after it had been cannibalized to replace the deteriorated cubicle on top of the Cairo Springs Lookout in Jasper County. Braly was asked, therefore, to carefully save the cab from the original Livingston structure and transfer it north of the city before placing it on top of the taller, recently reconstructed Aermotor from Soda.

After the move, the *Silsbee Bee* reported that "L. D. Clifton, crewleader in the Livingston crew area, will be at this new location to assume his duties."[63] Only then was the shorter Livingston Tower transported to the district headquarters in Woodville, where it was stored before ultimately being re-erected at Saratoga. Meanwhile, the tower north of Livingston survived through the 1980s because of its use as a communications facility. Vandalism increasingly became a problem, though, and in 1986 a General Electric repeater high-band radio valued at $3,500 was stolen from the cabin when someone pried the tower door off its hinges.[64] The station was decommissioned and sold for $50 in March 1993.[65]

O. C. (Oley Cecil) Braly

Perhaps no one sweated or cussed more on top of Texas' lookouts than O. C. Braly, a rig and tower builder who did contract work for the TFS for at least twenty-two years. Records are scanty before 1952, but those that are preserved suggest Braly had an immense knowledge of the trade and a trusted relationship with John Oliver "Joseph" Burnside and the Fire Control Department. One internal memorandum written in 1956 notes that "this man did some tower movement and erection for us in some of the other districts and can give us the approximate cost that you requested."[66] By no means is it an exhaustive list, but the number of towers he was personally involved with is documented to be over forty (table 1).

Table 1. Lookout towers constructed, reconstructed (2), or dismantled (D) by O. C. Braly for the TFS or USFS. Some towers sported a cabin of his design (BR cab). This summary is based on available memoranda and is not exhaustive.

Arp (BR cab)	Latch (BR cab)
Barnum (2)	Leesburg (BR cab)
Bastrop	Liberty Hill (2)
Bath (as bidder 1958)	Lindale
Bird Mt. (2)	Livingston
Cairo Springs (BR cab)	Magnolia (BR cab)
Candy Hill	Mauriceville (BR cab), (2)
Chandler	Meadows (BR cab)
Chita (2)	Moscow (2)
Church Hill (BR cab)	Moss Hill (D)
Conroe (2)	Onalaska (2)
Cushing (2)	Pine Mills (BR cab)
Daingerfield (BR cab)	Pool (2)
Devils (D)	Poynor
East Mt. (BR cab)	Redwater (?)
East River (Texas Forestry Museum)	Salem (BR cab)
Elkhart (2)	Saratoga
Elysian Fields	Smithland
Etoile (2)	Smithville
Evadale (R)	Soda
Gilmont (BR cab)	Starrville (BR cab)
Grayburg (BR cab)	Vidor
Hi Point (D)	Weches
Karnack 1 (D)	Wilhite
	Winchester

His ability and expertise are obvious in a series of communications leading up to the replacement of the Weches Tower in 1958. As discussed, the original steel lookout constructed at the site in 1934 was quickly substituted by a wooden design in 1938. In Texas' humid environment, the creosoted structure deteriorated and was on the list to be replaced—for the third time. A number of discussions ensued between the FCD, the Aermotor Company, and TFS director Alfred D. Folweiler to determine the cost of a replacement. Folweiler also began planning for the changes in a September 1957 "Report on Programs and Needed Facilities, 1960–1975" and solicited $18,100 to cover the capital expenses necessary for erecting the new infrastructure.[67]

Separately, Braly was busy erecting the Vidor Lookout for the TFS in 1958, an EMSCO TW-1 tower that had been purchased secondhand from the USFS and dismantled at Cyclone Hill.[68] During the process, Braly informed the FCD that eight 122-foot derricks, "in nearly new condition," were available for purchase in the Talco Oil Field at considerable savings to the state.[69] By September, the former Pan American Life Insurance No. 3 derrick had been secured for use at Weches, and Braly advised the FCD that reducing the tower's height by 18 feet would allow it to fit on "the tower bases now in place from the steel tower once located at the Weches tower site."[70] The existing foundations were examined by state engineers and found to be larger and with greater bearing surfaces than required, and Braly's recommendation was approved. His ingenuity saved the Forest Service $125, and the tower was completed as planned.

Braly also designed custom cabs for a handful of other repurposed towers, resulting in additional cost savings for the state. In fact, his cabs appeared on so many lookouts that 1960s-era property inventories even included a special "BR" notation identifying towers that had a cab constructed using his design.[71] The Pine Mills Station, a standing example with a cabin of his construction, was one of six oil-field derricks purchased for use as fire lookout structures in 1960 to expand the detection network into portions of Wood, Camp, Upshur, Franklin, and Titus Counties (fig. 16). The TFS still maintains the tower for communications, and while closed to the public, it is listed on the National Historic Lookout Register. The derrick preserves period artifacts, including an L. C. Moore placard and an original sign stenciled with orange paint, warning visitors to "climb at your own risk." A careful eye also reveals subtle design differences in the shape and number of window lights between this and Aermotor MJ-66 cabin fabrications.

The tower represents an unusual design with tubular legs and semicircular brackets that are clamped together to secure the horizontal cross-members to the corner posts. The corner posts, in turn, are anchored to the piers with reinforced base plates that have wider-spaced anchoring bolts. As on most oil derricks, a ladder with rest platforms ascends the outside of the structure, and the base of the frame allows space for moving drill pipe. Also, the interior of the structure has been left empty to accommodate the traveling block that would be necessary for drilling operations.

Braly was also called upon by Crewleader Robert Redding at Leesburg in 1962 when he wanted to assess the feasibility of adding an electric elevator to the repurposed Parkersburg derrick he had to climb daily. Braly was familiar with the tower, having just erected it and constructed

Figure 16. The Pine Mills Lookout (A) is a surplus oil derrick with tubular legs (B) that was manufactured by L. C. Moore. O. C. Braly constructed the custom cabin on top of the derrick. Note the difference between this anchor plate and the Aermotor and EMSCO designs (C).

Forest Protection and Fire Control **49**

one of his custom cabins at the top. Braly's plan called for mounting a motor-driven drum on top of the tower that could spool a wire cable connected to a metal basket. Redding was so tired of climbing the ladder to the top of the tower several times each day that he was willing to make the modification at his own expense. Discussions for installing the device escalated to Director Folweiler, who asked, "Precisely what motivates Redding into desiring an elevator? Is it fear of using the ladder, or is it fear of affect [sic] on his heart if he climbs the ladder, or is it something else?"[72] Evidently the lift was never built, and Jeff Hagen was promoted from crewman to crewleader at Leesburg in 1963.

Braly's craftsmanship is on display at the Texas Forestry Museum in Lufkin, along with a restored fire Jeep and the cab from the second tower at Cyclone Hill. The combination of an obsolete detection system with mounting repair costs, urbanization, and landowner friction at the East River Lookout site convinced the TFS to remove the East River Tower from Montgomery County in 1972. At the same time, Ed Wagoner, the executive secretary at the museum, petitioned the TFS for a surplus lookout that could be used as an outdoor exhibit. Because Texas law prohibited selling or donating state property, Patrick Ebarb and the director of the TFS worked out an arrangement whereby the Aermotor MC-39 at East River could use museum funds to be moved from storage at Jones State Forest and re-erected as a fire observation and educational piece in Lufkin. Legally, though, the tower remained state property for use in the prevention of wildfires whenever it was needed. Agreeing to these arrangements, Wagoner informed Braly in March 1974 that the board of trustees had accepted his $3,300 bid for transporting the tower from Conroe and rebuilding it on the grounds of the museum. Perhaps not as spryly as he once had, Braly completed the lookout at the museum. He passed away shortly after finishing the project on November 21, 1975, and received short mention in the TFS publication *Have You Heard*.[73]

Reconstructing the Network

Most often, a trip to a lookout tower culminates with upward glances and, sometimes, an adventurous trip up narrow, rattling risers to an observation cabin at the top of the structure. In the excitement, it is easy to forget to look down. Not down *from* the tower, but down *at* the tower. Admittedly, it was years before I had. It is only now, after hours in the Piney Woods and time spent reconstructing partial records from the state and federal forest programs in Texas, that looking down has become the focus of every visit to a fire tower. In looking down, I hope to discover the details of the lookout's piers—those four concrete footings that support, or did support, the legs of the tower. The recognition that the piers were key to understanding the height, and often the type, of the lookout became obvious on a humid morning as I stared down at two sets of them on the northern boundary of the Jones State Forest. Simply, after years of looking *up* at towers, the destruction of much of Texas' forest fire protection system forced me to look down at the only remnants of these historic structures.

As a geologist, I attempt to understand natural systems that are, superficially, filled with random variability. The profession relies on observations and data trends that are, collectively, coupled with inferences and experiences that help us generate a series of plausible scenarios. These will attempt to explain how a place, a system, or an environment changed through time. Often, there is lateral thinking and unconscious biasing in our interpretation of the data.

Lookouts, in contrast, offer a refreshing set of predictable engineered conditions with measurable details down to a fraction of an inch. In Jones State Forest, this was my eureka moment as I inspected the piers left by two generations of lookout structures. By the middle of the 1930s, when the fire detection system in Texas experienced the most rapid period of change, convergent evolution, standardization, and engineering principles directed the best practices for keeping a protected, 49-square-foot observation platform aloft. The requirements were well understood. The lookout structure had to be durable enough to withstand exposure, while still accommodating convenient ways to raise people and equipment to a protected seat above the elevation of local obstructions for the sole purpose of detecting and positioning wildfires. As Suiter pragmatically wrote, "Fire lookout cabins were mostly just housing for fire finders."[74]

Standardized cab dimensions and standardized tower heights meant one thing to me that morning—repeatability. Any changes in pier design or pier dimensions, like those observable in the northern section of the Jones Forest, could be interpreted to reflect variations in lookout height or subtle preferences among manufacturers. Analyzed alongside photographs, available records, and the position and orientation of any associated concrete slabs, it should be possible, I reasoned, to compare the overgrown footprints of destroyed towers with the design attributes of still-standing examples to predict the type of structures used at lookout sites with scanty documentation.

Windmills, transmission towers, cellular towers, water towers, oil derricks, and lookouts all became interesting waypoints as an atlas of

Figure 17. Details of the "cross pattern." Each manufacturer and/or lookout model employed distinctive anchoring systems. These observable attributes, combined with measurements of length and hypotenuse, are useful in determining the type and height of a tower at a lookout site when the tower has been removed and records are unavailable. In combination with observations from associated structures (relay houses, riser landings, or ladder platforms) it is possible to make inferences on the model or manufacturer of a lookout. Though drawings for towers with this design called for keeping the pier's edges parallel to the clamp, and thus offset 45 degrees with reference to the frame of the tower, the tendency in Texas was to square the piers so that they would be parallel with the tower frame. Pier example (*center right*) from the E. T. Usher Forestry Station, Florida.

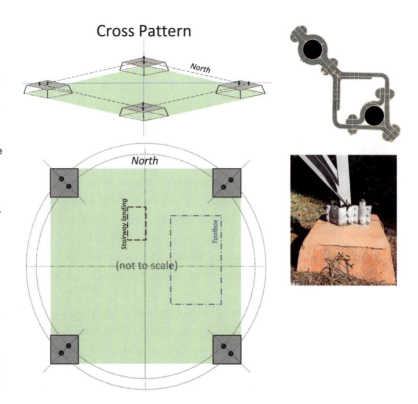

towers and anchoring styles slowly grew. Among steel lookouts, two distinctive classes emerged, which I informally dubbed the "cross" and "diamond" categories. Based on a time-saving procedure used while working in the field, I have the habit of sketching the dashed outline of a square in my field notebook before heading out to investigate a tower site (figs. 17 and 18). Onto this square are added four additional, smaller squares that share corners with the larger square where it forms right angles. Together, the smaller squares define the lookout's concrete piers, which when connected by the larger perimeter square, represent the basal area, or footprint, underneath the tower.

Because of the relationship between a tower's height and the dimensions of the base (appendix 6), comparing the field measurements (the perimeter's length and hypotenuse) of discarded towers with the dimensions documented on manufacturer's drawings provides a useful estimate of the forgotten tower's elevation (fig. 19). With the perimeter sketch already drawn in a notebook, any field labor is reduced to recording the length of the sides and the hypotenuse of the triangle created by half of the square. The position and dimensions of associated features, such as landing platforms or relay slabs, can also be easily inserted on the drawing. Importantly, as it relates to the "diamond" and "cross" classes, this process also includes adding two pencil points on each of the four smaller squares that represent the location of the steel anchor bolts embedded in each concrete pier.

The base map constructed at Jones State Forest contained the dimensions for two tower footprints, complete with pencil points indicating the position of the anchor bolts. The westerly piers belonged to "Tower No. 7," a shaky-looking, 87-foot derrick with a 9- x 9-foot cabin that extended beyond the derrick's corner posts. Manufactured by IDECO and erected in 1931, a ladder, with resting platforms, ascended the southern face. A more modern replacement, a 100-foot Aermotor LS-40 with stairway risers and landing platforms, was built just to the east to succeed the original tower in 1962 when the folded angle braces of the first structure deteriorated beyond repair. Both lookouts were engineered with the walls of the cabin facing cardinal directions; by default, the tower footprint and piers were also oriented in this fashion.

A striking difference between the anchor bolts became obvious on the map. Considering just the northwestern pier of each of the towers as an example, the International lookout's anchor rods were positioned diagonally across the pier in a line oriented from northeast to southwest. In contrast, the anchor bolts on the northwestern pier of the Aermotor were positioned in an opposite fashion diagonally across the pier from the northwest to the southeast. When viewed from above on the map, connecting the vectors created by the anchor bolts formed either a diamond-shaped pattern or a cross-pattern (an X) between the four piers. It took time to understand why, but the answer emerged as the atlas of tower observations grew.

It turns out that the various patterns occur because of the fashion in which the corner posts, or legs, are fastened to the anchoring bolts embedded in the piers. Two systems were common for steel lookouts. In the first, two symmetrical components are placed vertically on either side of the anchoring bolts to create a ridged bracket that snugly clamps around the threaded anchor rods protruding from the pier. Individually, each bracket is rectangular in profile and has holes milled into it to accept nuts that secure it in place with its mirrored counter-component. In cross section, each piece has been bent in several places to create two separate structures. At one end, the steel has been shaped to create a half circle with a small radius. On the opposite side, a series of 45-degree folds work together to form two sides of a square. When the two halves are tightly secured, the finished clamp creates a circle that nestles around one anchor bolt and a square. Within the square is space to accommodate two features. At one corner, the second anchoring rod can be secured to the clamp with help from a winged bracket. At the other, the angle iron from the corner post can be nested and locked into the brace.

This arrangement results in the tower leg being seated within one corner of the rectangular space formed by the bolted clamp; it is in line with anchor rods. If the angle iron forming the leg is to be parallel to the sides of the pier, though, then the entire clamp and the anchoring rods need to be rotated. In this configuration, the corner posts are squared with respect to the piers, and the tower has four clean, but open, planar faces. The anchor rods, however, lie both inside and outside the face. In this formation, the vectors drawn connecting the anchor rods and viewed from above result in a pattern that resembles a cross (or X). One advantage of this design is that the distance between the anchoring rods can be easily surveyed. Once the centerpoint of the tower is staked, an engineer can trace two circles with different radii and intersect the midpoints for each anchor rod. This was likely a handy, time-saving trick in the field. The design was perfected by Aermotor and used on a variety of its towers. It was also popularized by William B. Greeley,[75] as well as the USFS.

The other pattern is more easily visualized. Again, we begin by assuming that the corner posts are built parallel to the sides of the pier to create a clean, planar tower face. To simplify the attachment of the corner post in this design, an angled steel clip is milled to accept the anchor rod. This angle clip is slipped through the anchor rod on the outside of the tower leg and secured to the pier with a square washer and a nut. The longer, vertical portion of the anchor clip has also been milled to accommodate a series of bolts, which are then tightened to the corner post when construction of the tower begins. An angle clip is secured to each side of the leg, necessitating that two are required per pier to firmly hold the corner post in place. Because these attachments are both outside the frame of the tower and parallel to the angle iron of the leg, the configuration of the anchoring rods is opposite to the pattern created by towers held steady with clamps. For towers of this design, connecting the vectors created by the anchor bolts results in a diamond-shaped pattern when viewed from above. There are subtleties between manufacturers and models with "diamond" designs, too. Notably, the number of bolts securing the angle clip to the corner post varies depending on tower design and height. Some manufacturers also elected to change the linear arrangement of the bolts into an off-centered, "back and forth" configuration (see fig. 18).

Piers for wooden towers contrast with both of these designs (fig. 20), and the dimensions of the base are shorter than those for steel towers of comparable height. The USFS Tadmor Lookout, presumed to have been

Diamond Pattern

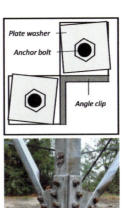

Figure 18. Details of the "diamond pattern." Each manufacturer and/or lookout model employed distinctive anchoring systems. These observable attributes, combined with measurements of length and hypotenuse, are useful in determining the type and height of a tower at a lookout site when the tower has been removed and records are unavailable. In combination with observations from associated structures (relay houses, riser landings, or ladder platforms) it is possible to make inferences on the model or manufacturer of a lookout. Pier example (*center right*) from the Mayflower Lookout Station, Texas.

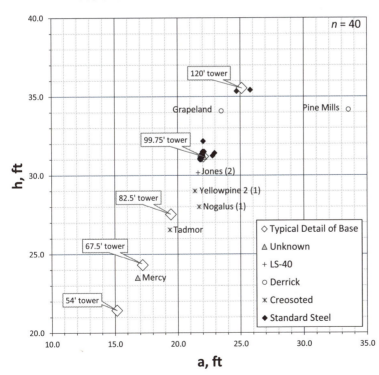

Figure 19. Field measurements can be compared to standard lookout tower dimensions to help determine the height, and often the manufacturer or model, of a lookout tower. With a library of dimensions built from existing construction plans, this methodology can be duplicated throughout the country. When sites are overgrown, minor bows in the tape may influence the length of either *a* or *h*, resulting in measurements that are less accurate in comparison to the typical details of the base.

Forest Protection and Fire Control 55

Figure 20. More variability occurs in the anchoring systems for creosoted wood lookouts. Note how the asymmetric example from a CT-1 tower differs from both the "diamond" and "cross" patterns typical of steel towers. The Tadmor site demonstrates these design elements. Other wooden towers, like USFS Piney Woods, employed three-rod anchoring solutions. Pier example (*lower right*) from the Tadmor site.

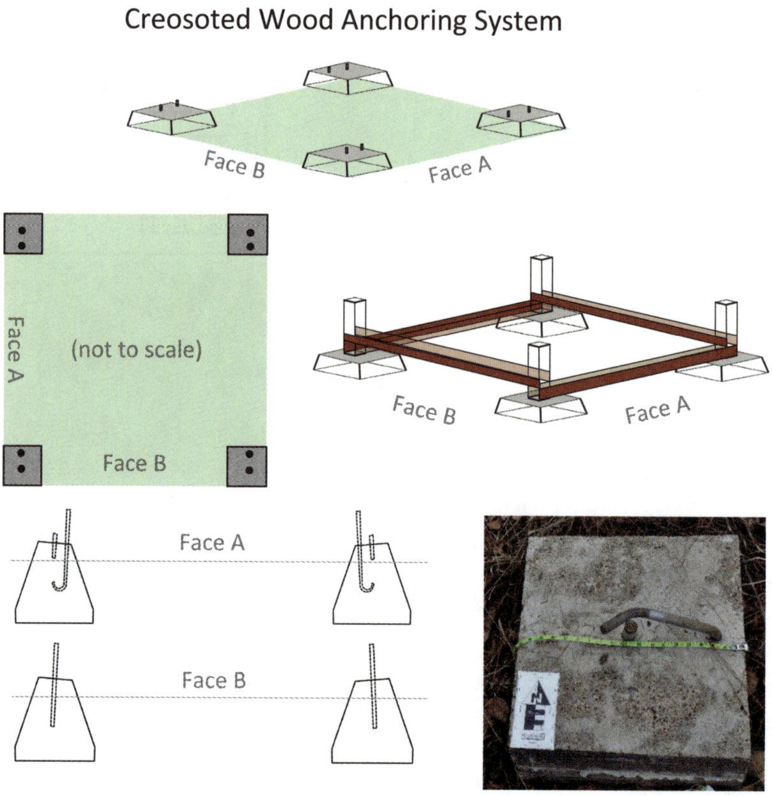

modeled after the CT-1 design, offers a convenient place to study the differences. Beginning again with four squares neatly drawn to represent the piers, each contains pencil marks for two steel rods that would have anchored the tower. The position and design of these rods do not conform to either the "cross" or "diamond" configurations, however. In fact, one rod—actually a dowel—sits centered in the pier and has not been threaded to accommodate a nut. The other steel protrusion, an anchoring rod sensu stricto, has been placed at the very margin of the inside wall of the pier. This rod protrudes higher above the concrete than the dowel and is threaded. At Tadmor, one even preserves an anchoring nut.

Viewed directly from the sides of the square, two anchors protrude from the piers in only one direction. This, in period blueprints, is termed "Face A." Orthogonal to this face is "Face B," on which only the centered dowel is visible. The asymmetrical design is the result of the position of the first horizontal struts, which are connected nearer the elevation of the pier on Face A. The two beams forming the horizontal strut on this face are anchored to the pier via the anchor rod and an angle clip. The four dowels, in turn, pierce each of the corner posts. In the opposite direction,

56 Chapter 3

the horizontal struts have been staggered above those on Face A and do not require a footing rod.

It is possible to understand more about the lookout by mapping the distribution of other objects associated with the tower piers. For example, it becomes obvious if an observer ascended the tower by way of a staircase or ladderway based on the location of the concrete landing. An internal concrete step parallel to the piers suggests the risers ascended from side to side up the inside of the frame, whereas a landing angled with respect to the piers indicates the risers were positioned near the corner posts. These differences, in combination with the position of the anchor bolts, offer additional clues about the model and manufacturer of each tower.

Multiple sets of piers indicate tower replacements, which were sometimes made as tower maintenance costs increased or safety standards evolved, usually from models with ladderways to those with internal staircases. These differ from other types of piers left by water towers at some compounds, such as those at USFS Ratcliff and the Hollis, Arkansas, CCC encampment (fig. 21), which display nonconforming basal dimensions. Relay houses and toolboxes also leave their mark and had to be fitted in the remaining space below the tower. On lucky occasions, names, initials, and camp or project numbers may have been pressed into the curing cement.

Figure 21. Off-center rectangular steel anchor at the Mercy Lookout site (A). The 1938 construction date is visible on the pier. Rectangular anchors for the water tower at the Hollis, Arkansas, CCC camp (B and C). These have been identified from camp drawings and surveys published by TAKAHIK River Valley Hikers. Photographs by the author.

The piers at the Onalaska Tower site are heavily inscribed and rich with history. Placards indicate that the tower was constructed by laborers from the Livingston CCC camp under the supervision of Knox Ivie, Superintendent S. W. Coats, and Foreman George Grimshaw (fig. 22). Other notations celebrate Siecke and White's roles as the director and chief, and

Forest Protection and Fire Control 57

Figure 22. This Onalaska Tower pier recognizes Supervisors Ivie, Coats, and Grimshaw. Stenciling on other piers at the original tower site acknowledges builders from CCC Camp P-61-T Livingston, Director Siecke, and Chief White. Photograph by the author.

Benjamin D. Hawkins's position as an inspector. Hawkins was also present during the construction of the Shepherd (Location 1) Lookout.

The piers at other towers such as Denning, USFS Moss Hill, Woodville, Kirbyville, and USFS Liberty Hill, serve as monuments to construction crews. The most richly ornate, such as USFS Moss Hill, lie quietly in a desolate forested tract north of Zavalla and east of FM 2109. Before the thoroughfare was straightened to run directly north-south at the base of Moss Hill, the main road ascended the hillside and passed just west of the lookout compound. A walk along that route provides pleasant access to the station, rich with the ghosts of CCC enrollees such as O. D. Page, who, along with others, left their names in the curing concrete. Adjacent to their names appear simple illustrations like those that might be made by a child. Prominently, a depiction near Page's name portrays a figure with an elongated, heart-shaped head, circular eyes, and a circular nose.

In the most spectacular instances, absolute dates left in the concrete can solve riddles, as at USFS Nogalus. The isolated guard compound on the Houston-Trinity County line once boasted an L-2400 tower, garage and storeroom, and a dwelling for the observer. The 120-foot wooden lookout, at one time the USFS's highest such tower, was eventually replaced by a conventional steel tower from the Ozark National Forest in Arkansas. Field observations supplement the record left by historical memoranda, and the March 30, 1956, inscription on the landing platform for the steel tower captures when the wooden structure was condemned.

Nonproductive Time

Productivity is measurable in many ways. Sometimes, employers measure the portion of the workday not associated with the performance of a task for which one is paid as nonproductive time. More than likely, this is familiar to everyone reading. For lookouts, the constant hours of tower sitting meant disproportionately high rates of nonproductive time. In fact, this luxury encouraged many artists and poets to accept summertime employment in fire towers. Gary Snyder, Philip Whalen, and Jack Kerouac all took their turn as towermen in the North Cascades during the Beat years. Later, Edward Abbey was also inspired to write his semi-fictional novel *Black Sun* based on his experience as an observer. The freedom afforded to lookouts is best described by the experience of Philip Conners at Apache Peak. He mused, "The life of the lookout, then, is a blend of monotony, geometry, and poetry, with healthy dollops of frivolity and sloth."[76]

During a 2015 interview, Mervis Lowery reminisced about his time on Texas' Etoile Lookout during the mid-1960s. When asked, "What did you do when you were up there?," Mervis replied, "Do you really want to know. . . . Lay on the door and sleep. . . . Reason why we laid on the door was to keep anyone from coming in on us."[77] As time passed and emphasis was placed on cost efficiency, the era "when fire crews simply had to stand by and wait for a forest fire to become visible" disappeared, and employers demanded more productivity from fire personnel.[78] In Texas, these changes began with Director Folweiler in 1949.

The annual reports of the TFS provide yearly summaries of patrolmen activities. Categories include duties such as firefighting, maintenance, outreach, and lookout observation. It is clear why more effective forms of fire spotting evolved after collating the annual number of hours devoted to tower observation. During a Depression-era peak in Texas during 1936, the official tally of observation hours stood at 104,838. Assuming an 8,760-hour-long year, these statistics indicate that nearly 12 years of continuous tower staffing were necessary to protect the forest from fire during one calendar period. This was a tremendous investment in comparison to the 14,492 hours (or 1.7 years) expended for fighting fires that year. Though 1936 was the maximum, and the number of lookouts in the state network had risen, the decade-long average shows that almost 8 continuous years of tower sitting was required for each calendar period. Of course, this type of system became increasingly unsustainable, especially when subsidies and labor waned.

Seen Area

On a clear day, a person visiting a lookout tower in the Lone Star State may have been rewarded with views extending nearly 35 miles. More commonly, midcentury descriptions suggest that the effective perimeter of each tower in the generally flat Texas Piney Woods was between 180,000 and 200,000 acres.[79] Certainly, this impacted the minimum 100,000-acre requirement for designating protection units and later influenced the configuration of each of the towers in the system. This acreage, or "seen area," translated to a territory with a 9.5- to 10-mile radius encircling every tower the lookout was responsible for monitoring. To put this into perspective, if you are one of the millions of Texans residing in a megaplex, a circle with a similar diameter could be drawn from Discovery Green—on the east side of downtown Houston—past the 610 Loop and only slightly shy of the Beltway; or another circle with a diameter larger than the distance between the Alamo and the 410 Loop. Otherwise, it would allow you to see the steeple of the Juarez Cathedral from the Wyler Aerial Tramway in El Paso, or the 635 Loop from the Cotton Bowl Stadium at the Dallas Fairgrounds. While these distances seem large, securing appropriations and constructing a complete protective network of lookouts and communications across the Piney Woods was a monumental task and one that ultimately took federal support to complete.

4 Cussed and Discussed
The 1920s

JEFF O'QUINN WAS THE TALK OF LUFKIN on February 23, 1923, when he stopped farming for employment as a patrolman with the eight-year-old Department of Forestry. Eighteen years later a *Pineywoods Pickups* article offered a glimpse at his career, which appeared below a hastily sketched stick figure meant to be Jeff.[1] The character was bent, but smiling, and holding a cane. As he looked back, the editor explained, the sixty-five-year-old remembered how busy it was "fighting fires all day and the biggest part of the night, day after day." In his own words, O'Quinn was "cussed and discussed" at a time when residents in the depths of East Texas still relied on fire to improve grazing, kill ticks, and open range. "People wanted to know if I had gone crazy," he said.

O'Quinn's employment came during a time of increasing responsibility and expansion for the Forest Protection Department, thanks largely to Siecke's effective leadership. Only a decade had passed since Foster employed the first forest patrolman, but their ranks had swelled to forty-four by 1926.[2] The growing field force began impacting fire statistics too, and there was a 50 percent reduction in the number of forest fires between 1924 and 1925.[3] The following year, 1926 statistics saw an additional 28 percent decrease in fire occurrence and an 80 percent reduction in the area burned.[4] Cautiously, the agency reported that while weather conditions were favorable, the average fire size had been reduced from 202 acres in 1925 to 59 acres in 1926. With pride, the change was attributed to the fact that "fires have been more promptly detected and extinguished."[5]

The field force still teetered during especially dry years, as was the case during 1929 when a protracted drought caused incipient fires to spread rapidly throughout the fall months. Patrol officers worked between fourteen and sixteen hours on successive days to keep the fires in check, but as the *Center Daily News* reported, the state's thin blanket patrol was simply "confronted with more fires than could be handled." Outside of the recently established protection units, the article continued, "many of the fires burned over considerable areas before they were extinguished or before the general rain checked them." Conditions were somewhat different in the protection units, and an increase in personnel and funding meant that fires "were extinguished before they reached any considerable size."[6]

Increasingly, fire losses were also being minimized through enforcement, and many of the decisions addressing the willful or negligent ignition of fires originated during the Thirty-Eighth Legislature in 1923.[7] While a resolution had been enacted in 1884 making it unlawful to deliberately set grass fires, the older legislation primarily focused on protecting the stock industry in West Texas. After 1923, however, additional regulations required spark arrestors on locomotives, logging engines, and other wood- or coal-burning engines operating near forests.[8] Perhaps more important, the new laws created penalties for those setting unauthorized forest and grass fires. Thereafter, when sufficient evidence could be secured that a fire was ignited with malicious intent or through carelessness, the defendant was prosecuted or fined.

The *Ninth Annual Report of the State Forester* published in 1924 announced details of the first two prosecutions under the fire laws of 1923.[9] In a Polk County case, the offender ignited a fire on his own property but was punished when it escaped and damaged an adjacent owner's land. The second conviction occurred in Sabine County when an individual maliciously set fire to another person's property. As the *Eleventh Annual Report of the State Forester* pointed out two years later, "A few convictions of the persistent violators would undoubtedly create a more wholesome respect for the fire laws and would have a far reaching effect."[10]

By 1938, O'Quinn had received a special ranger's commission and was investigating forest fire law violations as a peace officer. A *Corrigan Press* article reported these types of "public contacts" and noted that investigations into fifty-one instances of woods burning led to convictions in thirty-five cases. Director Siecke was quoted in the article saying "that where fire law cases have been investigated, many of the old offenders have ceased to fire the woods. We firmly believe that this enforcement work is very necessary in connection with our educational work in fire prevention."[11] Unfortunately, it's not clear how many were true convictions, and one colleague recalled years later that he had met Jeff and his partner in the woods with a suspected firebug: "One day I came across them when they had a man tied to a tree in the woods. 'Jeff, what are you going to do?' I asked. 'We are going to make him confess,' they said. 'You can't do that,' I said. 'Why not?' they answered. 'He's guilty.'"[12]

The Thirty-Eighth Legislature in 1923 also increased the Department of Forestry's budget and approved state purchase of lands for use as state forests.[13] Momentum continued into the Thirty-Ninth Legislature in 1925, when a joint committee of legislators and citizens was created to make recommendations on how to manage the state's timber resources. At the time, the *News* responded to rumors that the state or federal government contemplated large purchases of cutover lands for reforestation purposes, thus removing them from the tax rolls.[14] Readers were reminded that it was impossible for the federal government to purchase even "an acre of forest land in Texas" without consent from the state legislature. But, the *News* continued, the recently authorized Committee on Forestry took a firm stand against the state's purchase of large cutover acreage. Instead, its recommendation was that legislation be enacted making it possible for private capital and private initiative to improve forest cover on cutover properties.[15]

Simultaneously, forestry agendas were advancing throughout the South. In 1926, the Brown Paper Company, operating in Monroe, Louisiana, purchased two lookout towers to protect its timberlands. Likewise, the Industrial Lumber Company in Elizabeth, Louisiana, purchased its own lookout.[16] That same year, the Crossett Lumber Company erected five towers in southern Arkansas to surveil its recently reforested holdings.[17] In the latter case, the company even erected guard dwellings below each tower and connected telephone lines between the stations. A year later, the Alabama Commission of Forestry completed a tower on land donated "by a public spirited citizen,"[18] and in 1928, the Federal Forest Reserve Commission approved three national forest purchase areas in Louisiana.

Back in Texas, a pivot was occurring in patrol activities during the mid-1920s. Instead of crisscrossing their districts on horseback, patrolmen began spending a "greater part of their time, during dry weather, on lookout points and in fighting fires."[19] By 1924, approximately half of the "meager funds" made available to the Department of Forestry were devoted to protection work,[20] and a fixed point detection system was being improvised as trees were spiked and watchful eyes sat waiting for signs of smoke. But the patrol districts were still large, and the agency still lacked necessary equipment.

Massachusetts had just erected its forty-third fire tower in December 1925 when the *News* announced plans to inaugurate a lookout system in Texas with help from matching federal funds.[21] The next year, the first steel fire tower was constructed and equipped on State Forest Number 1 in Kirbyville. Slowly, other towers followed within the decade as protection units were established outside state forests (table 2).

The Kirbyville Tower, as reported in Bulletin 18, was assembled in a region of high hazard and immediately "increased the efficiency of the fire detection and suppression work."[22] W. Hickman Smith, otherwise known as "Uncle Hick," and James V. Sheffield had expansive views from the cabin and could take in a 30-mile-wide panorama that included smoke billowing from mills in Deweyville, Wiergate, and Jasper.[23] Critically, they also had a telephone connection with auto patrolmen in Kirbyville and Jasper, along with horse-mounted patrol officers adjacent to the forest.[24] In the first fall fire season alone, the position reported 135 fires and hosted 931 visitors.

Table 2. Lookout towers constructed between 1920 and 1929.

1920	1921	1922	1923	1924	1925	1926	1927	1928	1929
						Siecke SF to Evadale		Bath *to Elkhart*	Liberty Hill (State)
								Bird Mt.	
								Ratcliff 1 *to Cushing (?)*	
								Yellowpine 1 *to Elysian Fields*	

The tower was also a splendid opportunity for Uncle Hick to demonstrate the value of a lookout system, as proven on October 17, 1927. Spotting a fire, he and his helper telephoned Kirbyville's scoutmaster Dowling. Before long, twelve Boy Scouts "were going liggidy split for the reserve" to help suppress the flames.[25] The results, said the *Kirbyville Banner*, were "serious fire averted; Smith well pleased; scouts happy; Dowling—tickled to death."[26] To build momentum, the *News* suggested, "it is impossible for horse patrolmen to learn of all fires when they first start, unless discovered and reported by a lookout watch man. It is, therefore, of importance that all of East Texas be covered with a system of lookout towers. Such a system would greatly reduce the acreage burned yearly by forest and grass fires and save the State thousands of dollars damage."[27]

The Department of Forestry continued to use the *News* as a platform to advocate for a statewide system of lookouts in 1928 and envisioned that this "complete system of lookout towers . . . should be connected with telephone lines" to guard East Texas forests.[28] The publicity, and Siecke's lobbying efforts, began to sway industry representatives, and the number of protection units quickly grew after the first one was approved in 1927.

State Forest Number 1, on which the Kirbyville Lookout sat, was a new component in the Department of Forestry's effort to demonstrate reforestation techniques and sustainable forest harvests. The cutover land was established as a state forest in 1924 and was quickly followed by additional acreage near Conroe in 1925. At that time, the Conroe tract was creatively called State Forest Number 2, though both forests would be renamed in honor of significant forestry figures by the middle of the century.[29] The 1,700-acre Kirbyville tract was quickly divided into a series of educational and experimental plots to begin studying the impact of razorback hogs, fire, and thinning on longleaf pine regeneration.[30] Other research programs focused on direct seeding of southern yellow pines, the care of pine seedlings, and the control of brown-spot needle blight in longleaf pines. There were also trials meant to determine how certain species such as mulberry, locust, poplar, and bamboo adapted to the Piney Woods.

At Conroe, where the forest cover was dominantly loblolly, the grounds were managed to "demonstrate to the farmer and other timber land owners, just how timber can be grown as a crop. Only simple, practical measures will be followed that can be duplicated by the private owners."[31] Modest cottages were constructed on both forests, and superintendents were appointed to protect the state's holdings and monitor the experiments at each facility. While plans had been made to erect a fire tower on State Forest Number 2 as part of the building campaign, funding was perpetually short,[32] and the Conroe courthouse cupola had to suffice as a lookout until 1931.

Senator I. D. Fairchild introduced Senate Bill No. 351, transferring 2,150 acres covered with second-growth shortleaf pine in Cherokee County from the state prison system to the Department of Forestry during 1925.[33] This legislation created State Forest Number 3, where the 5- to 10-inch-diameter pines were meant to demonstrate scientific forest management techniques in the shortleaf pine country typical of Northeast Texas. In short order, a fourth state forest was inaugurated when lumberman John Henry Kirby donated 600 acres of his Tyler County holdings to the state in 1928. The acreage for State Forest Number 4 was donated to Texas A&M University by the "Prince of the Pines."[34] In announcing the gift, Kirby stipulated that the 600-acre bequest near Warren, Texas, be used for scientific investigations focused on reforestation and the rejuvenation of forestlands in East Texas. He also required that a loan fund be established using revenue generated from the sale of forest products harvested on the tract "for worthy boys" enrolled in forestry studies.[35] Quickly, an 80-foot-tall tree cab was erected to serve as a vantage point in Protection Unit 8.

Gradually, Jones's vision was reaching fruition. Public perceptions were changing, forests were regenerating, and the influence of the TFS was increasing. Importantly, lumber companies were expressing interest in forest conservation and began to recognize the necessity of being at the table when state forest policies were being adapted.

O'Quinn must have felt somewhat disheartened by the artist's sketch that accompanied the 1941 *Pineywoods Pickups* article about his career, because a correction titled "With Apologies to Jeff O'Quinn" was offered the following quarter. "Jeff says he's not as old as the picture showed him to be," the editor wrote below a new illustration featuring a caricature donning boxing gloves and knocking out a bull. Whether it was legislation, enforcement and educational campaigns, favorable weather, or luck, the decade of the 1920s provided the Piney Woods with their first chance to heal. Perhaps it was also the dedication and perseverance of a small number of visionary conservationists like Siecke, Fairchild, and O'Quinn. "Give him a bowl of squirrel mulligan stew," the O'Quinn's apology concluded, "and he's set for anything."[36]

Texas' First Steel Lookout: Siecke State Forest

The State Forest Number 1 lookout was the first fire observation tower constructed in Texas and was completed in June 1926. The tower was erected soon after the cutover longleaf acreage was purchased for use as a state forest and was meant to demonstrate scientific land management practices. As the first reforestation experiments took place, however, the tower loomed above the nearly level, sandy, stump-dotted prairie.

A dozen men with recognizable names from the earliest TFS rosters participated in building the tower. Included among them were E. E. Jones, J. M. Turner, V. V. Bean, H. J. Eberly, H. F. Munson, R. A. Wilson, and Jeff West.[37] Along with Siecke, the state forester at the time, many left their mark on the landing platform at the base of the tower, an artifact that is still visible today (fig. 23).

Immediately, the tower became a showpiece that was widely featured in newspapers across the state, and it was described as having "considerable educational value in addition to its main usefulness."[38] By 1935, "the old steel tower" overlooked the barracks, mess, and assembly hall of CCC Company 845's encampment. The company was tasked with state forest improvement projects, and the 1934 work plan written by C. B. Webster, the Division of Forest Management chief, indicates much time was spent landscaping, constructing fire lanes, building hog-proof fences, and completing timber-stand improvement projects in the vicinity of the tower. Later, research plots were established to monitor longleaf pine growth variables and experiment with species hardiness in Texas climates. By the end of 1935, work at the forest was finished, and the company moved to Beaumont to begin developing Tyrrell Park.

The station saw continued service during World War II as a fire lookout tower and Aircraft Warning Service (AWS) observation platform. As at other towers with military roles, strict protocols prevented visitors from

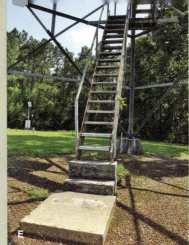

climbing the lookout or being close enough to overhear conversations for fear of espionage. The Kirbyville Tower was elevated and repositioned after the war when the TFS announced changes to the detection system in 1948. At that time, the original lookout was moved to Evadale. Meanwhile, a 40-foot-taller Aermotor MC-39 from Sheffield's Ferry was shuttled from Jasper County to Kirbyville and re-erected to see above the skyward-reaching crowns of the flourishing secondary stand. The piers from this larger tower, still bearing Kelley's name and a December 7, 1948, inscription, can be observed today.

Symbolically, the inscribed landing platform from the original lookout was incorporated into the footprint of the replacement tower, greeting modern visitors and encouraging them to reflect. Ten flights of stairs lead guests through the trap door of the cab and into the olive-drab, clapboard-sided, and ceilinged walls of the space. A map table still adorns the room, and a sea of greenery extends outside the window to the horizon. Now over eighty-five years old, the Depression-era lookout, formerly at Sheffield's Ferry, stands at the Siecke State Forest. The site is accessible on state property with oversight from district forest personnel and is safe to climb. Today, the tower is in need of maintenance and painting to stabilize the structure and extend its life.

Even though the tower was moved to the site from a nearby duty station, one can argue that these actions occurred three-quarters of a century ago and do not significantly interfere with aspects of the tower's original function. In the author's judgment, it is worthy of inclusion on the National Register for its role in the development of Texas forestry, the scientific study of southern pine forests, and the contributions of the CCC during the Depression. Advocates should increase awareness for this significant piece of Texas history more than any other lookout structure and solicit state protection, calling volunteers, tradespeople, carpenters, and steelworkers into action.

Figure 23. Work progresses on Texas' first lookout tower, an 80-foot Aermotor LS-40. After the site was prepared, piers were poured (A), and construction advanced section by section up the structure (B). The 80-foot LS-40 was replaced by an Aermotor MC-39 moved from Sheffield's Ferry in 1948 and is still accessible today. The cab interior maintains the original woodwork and map table (C). The riser landing from the original tower was incorporated in the construction of the replacement lookout (D and E). The tower at the district headquarters in 2016 (F). Images A and B courtesy of Texas A&M Forest Service. Photographs C–F by the author.

5 Come Up Sometime
The 1930s

Come Up Sometime

Some people think the CCC'S
Are just a bunch of bums,
Who do not have a place to sleep,
Or live out in the slums.
Well, now, dear friends, you've got us wrong,
Why not come up sometime?
I'm sure you'll find more neat men,
Than in your town or mine.
Say, every cent they are paid
I'm sure they deserve,
And if you dare to call them names,
You've got more than nerve.
Men from colleges and schools,
Were sent here to work.
They've got intelligence and friends,
You'll never see them shirk.
Now why not give the boys a hand?
I'll say "they do their part"
Look upon them as your friends,
And open up your heart.
So boys, why not join the CCC,
And live with the happy throngs,
And send your dear mother the monthly check,
To fill her heart with song.
When you join the CCC,
You'll have a burden to bear,
But don't pay attention to what people say,
If they had any gumption they would be there.
Some people think of the CCC,
As a place for CHARITY to dine,
So if you don't believe it is the proper place,
Why don't you come up sometime.[1]

THE GLOBAL ECONOMIC COLLAPSE that began in 1929 had lasting impacts on the nation's landscape and population. As one 1936 *Lufkin Daily News* article surmised, the country had fallen from a "bright fantastic dream of an unending prosperity built from eternal stock market pyramidings" to catastrophic unemployment that signaled the arrival of the Great Depression.[2] Nationally, fourteen million were jobless.[3] Many were vagrant, and drought, fire, and industrialization had taken a heavy toll on agricultural productivity and the natural environment. Unemployment was especially acute for "hundreds of thousands of boys in their late 'teens and early twenties," the *Lufkin Daily* continued, and "brought the stark tragedy of futility, of not being wanted, of a vicious idleness which are into the character of the greatest resource any nation has, its strong young men."[4]

Locally, sawmills were closing in East Texas. Robert Maxwell and Robert Baker's *Sawdust Empire* portrays a state that was severely impacted by the "cut-and-get-out attitude" that prevailed at the time,[5] while naturalist Roy Bedichek described "a weary land devastated by unscientific cultivation which followed in the wake of the insatiable sawmills."[6] Randolph Campbell also portrayed pre–World War II Texas as a "largely impoverished, rural, agricultural state,"[7] while Francis Abernethy recalled in the foreword to *Land of Bears and Honey* that many Piney Woods families were "living on subsistence farms, planting every inch they could plow, and eating anything they could shoot or catch and put into a pot of stew."[8]

Political changes in 1933 demonstrated the level of discontent for incumbent state and national leaders in response to the Depression. Miriam Ferguson took her second, nonconsecutive oath of office on January 17, 1933, and "resurrected," as Mark A. Wellborn described, "a progressive voice" from the state's "political past."[9] Together with her husband-adviser James Ferguson, an impeached former Texas governor himself, the "Fergusonism" pulpit advanced antibusiness reforms aligned with labor.[10]

The inauguration of President Franklin Delano Roosevelt two months later on March 4, 1933, offered a coordinated relief effort at the national scale and promised a New Deal for Americans. During his inaugural address, Roosevelt emphasized that Washington's "greatest primary task" was putting people to work on projects that helped "stimulate and reorganize the use of our natural resources."[11] Wasting little time, Roosevelt outlined the Emergency Conservation Work (ECW) program, of which the CCC was a part, during his second week in office. His strategy for the CCC leaned on existing governmental departments to provide oversight and temporary work for up to 250,000 young men—a staggering figure 43 percent larger than the ranks of the regular enlistment of the army.[12]

The Emergency Conservation Act was passed by Congress less than four weeks after Roosevelt had taken office;[13] it authorized the president to hire the unemployed in "construction, maintenance and carrying on of works of a public nature in connection with the forestation of lands belonging to the United States which are suitable for timber production, the prevention of forest fires, floods and soil erosion."[14] These provisions were also extended to lands in private ownership, but only under cooperative conditions for controlling forest fires, preventing disease and outbreaks of forest pests, and mitigating erosion or flooding.

Robert Fechner, a union vice president, was appointed director of the ECW program on April 5, 1933, and became the chairman of an advisory council with delegates from four governmental departments.[15] The complicated administrative structure the president sketched out on a table napkin meant that the Labor Department oversaw enrollee recruitment, the War Department controlled the camps, and the Departments of Agriculture and Interior approved and supervised work programs, which were sometimes designed and administered by technical agencies like the TFS or the Texas State Parks Board.[16] Local experienced men (LEM), who were qualified by the technical agencies, accounted for approximately 10 percent of the strength of the CCC and executed the various projects. "No matter how unwieldy" these arrangements may have appeared, Wellborn argued that the budding CCC structure "provided the operational dynamics that were necessary for administrative proficiency."[17]

Initially, the president personally approved each conservation project, an overwhelming assignment that curbed progress and distracted his office from other relief measures. To expedite things, Fechner recommended that the army petition the president to authorize its "wartime powers," which would allow for the transport of enrollees and the procurement of campsites and supplies without open-market procedures.[18] Roosevelt approved.

With this structure in place, Texas enrollees were governed in camp by a base commander and military assistants who were reserve officers in the Eighth Corps Area.[19] In the field, however, the workers were directed by LEM or project leaders, who were officials from the administrating state or federal technical agencies. While the organizational chart was effective despite its complexity, the enrollees found themselves responsible "to various governmental agencies depending on the time of day."[20]

The ECW program immediately provided sufficient labor to initiate "many of the planned, but never executed conservation projects" sought out by state and national agencies,[21] while also benefiting a pool of the nation's youth and their families. In this effort, the Labor Department coordinated with the State Relief Board, which worked with county and municipal agencies to identify qualified young men and their families relying on public relief. After enlisting and reporting to camp, a $30 monthly salary was provided to each enrollee. The income was divided between the laborer and his family, with the larger component, just over 83 percent, being sent directly to the enrollee's home. By the end of April

1934, the War Department announced that it had already dispersed nearly $2.8 million in wages to Texas enrollees and their dependents.[22] By the end of August 1936, that amount had grown to $10.8 million.[23] In addition to the stipend, a second provision of the Emergency Conservation Act allowed the president to guarantee housing and provide "subsidence, clothing, medical attention and hospitalization" to each laborer during their period of service.[24]

Recruitment stations began to appear across East Texas in April 1933 to satisfy the state's 11,250-person enrollment allocation. Roosevelt's expectation that the program be operational by July 1, 1933,[25] was met when enrollment reached its authorized strength nationally.[26] A *Tyler Journal* piece explained the enlistment process to readers in Smith County, indicating that an applicant must first submit the appropriate eligibility form to a local board for review.[27] If qualified, the individual's application was forwarded to the army, where regular selections continued prior to the cadet being sent to Dallas for conditioning and, later, camp assignment. Only one month had passed since congressional approval, but the *Tyler Journal* article went on to indicate that the army's Dallas District would furnish 850 enrollees for "reforestation duty." The *Montgomery County News* also reported on June 1, 1933, that the Montgomery County quota of 30 boys all left for the forestry camps.[28]

The Texas Forest Service's annual reports began praising the program in early 1934, declaring that the relief effort had taken "hundreds of thousands of boys out of home units that were on relief and many thousands of young men off the highways, and has given them constructive work under wholesome supervision and discipline."[29] Unquestionably, Siecke's leadership and the availability of aid allowed the TFS to benefit handsomely, and many of these "boys" were heading to projects administered by the TFS.

Work programs were organized into six-month periods that began in April 1933 after enrollment quotas had been met and the work plans were approved (fig. 24).[30] As the system evolved, each six-month project was numbered in succession and informally designated within either a summer or winter period. In this scheme, the first six months of the CCC during the spring of 1933 were labeled as the "First Period." With the completion of the project's agenda in the fall, the Second Period, a winter enrollment, began during October 1933. As James Steely noted, dates surrounding the establishment of new camp assignments had a "nationwide tendency" to be flexible to avoid transportation bottlenecks and allow for a few additional weeks of labor on unfinished work items.[31]

As each project was authorized, it received an alphanumeric identifier that assisted administrators. The scheme utilized a letter prefix,

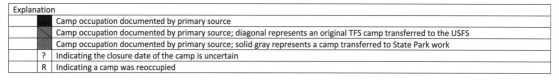

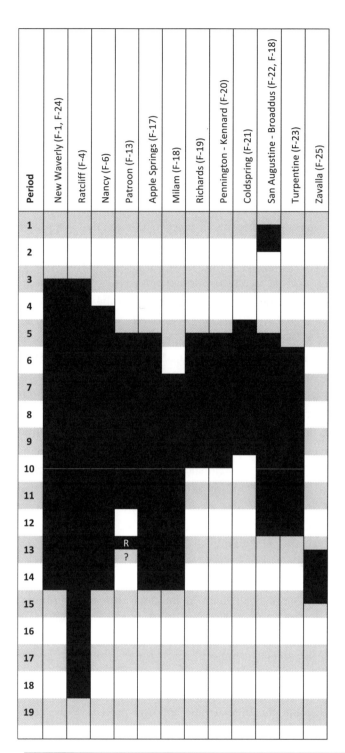

Figure 24. CCC projects overseen by the TFS (*left*) and USDA Forest Service (*right*). The diagram represents the length of time each camp was operational and displays the periods in which it worked. The dates for federal camps are less well defined (e.g., F-25, Zavalla), with information coming from multiple, discontinuous primary sources.

Explanation		
		Camp occupation documented by primary source
		Camp occupation documented by primary source; diagonal represents an original TFS camp transferred to the USFS
		Camp occupation documented by primary source; solid gray represents a camp transferred to State Park work
	?	Indicating the closure date of the camp is uncertain
	R	Indicating a camp was reoccupied

which referred to the technical agency sponsoring the work or the type of landownership. In this fashion, "F," "P," "SP," or "S" designations were made to identify Forest Service, private forest, state park, or state forest projects. In Texas, all camps designated by the letter "P" or "S" were the responsibility of the TFS, one of the most prominent technical agencies in the state.[32] The numbers following this designation then referenced the order in which the work plan was approved.[33]

Because the War Department preferred dispatching standard military companies of two hundred soldiers, each project was completed using the labor from one company. In practice, though, the strength of each company fluctuated because of changing enrollment quotas and enrollee desertions. As Texas sat within the army's Eighth Corps Area, each deployment received a company designation in which the value in the "hundreds-place" identified the unit as one from the Eighth Corps Area.[34] The TFS project near Lufkin, for example, was referred to as P-57-T, or private forest project, Camp 57, Texas. To realize the goals of project P-57, the army deployed and administered two hundred tree soldiers from Company 838.

Lufkin district commander Captain C. W. Hanna related that the assignment of regular army officers to CCC camps during this early phase disrupted the routine work of the army. Candidly, he wrote that "few understood exactly what was expected of them," as orders and memoranda "followed each other in rapid succession." The first few months were, he said, "a time of frantic hurry and confusion."[35] To better administer the camps, the Eighth Corps Area organized each state under its umbrella into a CCC district. But Texas geography still made this model overwhelming, and the state was divided into sub-districts during June 1933, with Lufkin becoming the headquarters for Sub-District No. 1.[36]

Mandates for rapid implementation and state participation in emergency conservation work resulted in "a multi-governmental administrative alliance involving state and federal efforts."[37] Director Siecke and Chief White were immediately effective in their proposals, and the early authorization of forest-related projects was "strongly influenced by the Texas Forest Service and other well-organized agencies."[38] Certainly, the time both men had spent in the USFS had given them insights into the agency's priorities and the bureaucratic approval process. This, combined with White's military service record and Siecke's responsibility of reporting to the Texas legislature and Texas governor since 1918, prepared them thoroughly for advancing their agenda.

Siecke submitted reforestation camp proposals to the president in early May for state forest developments at Kirbyville and fire protection work in three private forest camps at Lufkin, Livingston, and Groveton.

The Corrigan *Plain Dealer* made clear the complications of the new, untested conservation alliance: "Robert Fechner, director of emergency conservation work in Washington, certified the four Texas camp sites to the President and when approved by the chief executive agreements must be negotiated between the State and Federal Governments fixing the amount of compensation that landowners will pay for the improvement of their lands. Likewise, the Governor must agree to pay part of the cost of improving the State forest lands." The article went on to note that Texans B. F. Williams, the state reclamation engineer, and Lawrence Westbrook, director of the State Relief Commission, were traveling to Washington to discuss financing and submit plans for the location of additional camps. Progress was predicted to be slow, though, because the negotiations involved coordinating government work on private properties. "Texas," the article observed, was "one of the first to propose reforestation work on privately owned land. The emergency conservation work committee in Washington to date has devoted practically all of its attention to establishment of camps in national parks."[39]

The TFS established the Division of Forest Protection during 1922 and quickly moved the office from College Station to Lufkin in 1925 so that Fire Chief Howard J. Eberly could more easily direct resources across the 8-million-acre Piney Woods protection area. By 1933, the Lufkin office was staffed by Chief White; Isaac C. Burroughs, the assistant chief; and Ivan H. Jones, the assistant forester. When word arrived that the TFS was to become a technical agency overseeing twelve of the forty-one camps allocated to Texas during the First Period (see fig. 24), Lufkin naturally became the TFS's CCC administrative center. Headquarters were established in a brick bungalow at 312 Shepherd Avenue by June 28, 1933.[40]

Initially the small TFS staff responsible for "selecting the supervisory personnel and planning and directing the field work" for the technical program was overwhelmed.[41] An early, but temporary casualty was the publication of the *News*,[42] which suspended publication in April as the first conservation work plans were drafted. The organization's continued visibility during this period was largely due to the dedication of the Forest Service's staff, who "cheerfully worked long hours" to carry out their responsibilities.[43] Even Siecke and Fechner contributed a widely circulated article indicating that the twelve Forest Service camps were occupied by twenty-four hundred enlisted men and explained how Texans were participating in the national reforestation effort. The pair discussed enrollee eligibility requirements, causes for delay in establishing campsites, and the forest conditions that necessitated the work. The article also introduced aspects of the fire protection program and highlighted how forest improvements and the construction of fire lanes benefited the interests of private landowners.[44]

In August 1933, the *Lufkin Daily News* reported that President Roosevelt had given orders to extend all CCC camps for another month. "Perhaps," Lufkin's *Camp Chatter* related, "by that time we will have found that certain corner made famous by Hoover prosperity."[45] News of a second recruitment round was broadcast across East Texas, and the *Plain Dealer* advised that boys aged eighteen to twenty-five should be in Livingston to register with the County Board of Welfare and Employment if they were interested in joining the CCC.[46] Many enrollees also chose to reenlist for the Second Period that fall. At the Trinity Camp, for instance, the company maintained 92 percent of its First Period ranks.[47]

The efficiency with which Siecke and White worked, especially during the early days of the ECW program, is captured by a budgetary letter White wrote to Governor Ferguson on October 28, 1933, after the legislature made a special forestry fund available during the biennial period between September 1, 1933, and August 31, 1935. The funding was earmarked for cooperative projects between the state, the federal government, and the seventeen CCC forestry camps administered by the TFS. In his letter, White explained that Siecke had authorized him to submit the Forest Service budget to Ferguson before leaving for Washington, DC.

The letter announced the true beginnings of the statewide fire control network, and White indicated that the "original Fire Protection plan called for forty-four lookout towers." Explaining the budgetary items, White continued: "Up to the present time the Government has supplied us with . . . 17-one-hundred-foot steel towers." With these lookouts already delivered, the TFS was requesting the governor's approval to spend $20,466 to purchase twenty-seven additional steel 99.75-foot fire detection towers, built "per U.S. Forest Service specifications at Federal Government contract price."[48] Ferguson approved the original $35,000 budget on November 1, 1933, but it was later revised in April 1934. The update was made to reflect changes in the estimated cost of tower materials, alidades, and telephone equipment. When the revision was submitted, the tower price had decreased by approximately $86 apiece so that the necessary infrastructure could be purchased for $18,165.48.

Forty-two CCC camps were approved for Texas in June 1934, nine of which were on private forests and three on state forests.[49] During July 1934 the USFS entered Texas as a technical agency, and four CCC camps were transferred to this agency for work on the newly created TNFs.[50] At the same time, new federal camps were also established, such as F-1, New Waverly, and Camp F-4, Ratcliff. As Hanna pointed out, "Up to this time, the Texas Forest Service and, for a few months the State Park Division of the National Park Service, were the only Technical Services with whom the Sub-District had come in contact."[51]

A later announcement from Fort Sam Houston in October 1934 indicated that 7,600 vacancies would be filled within the Eighth Corps Area. The jurisdiction anticipated 21,000 junior enrollees and 3,600 LEM would be necessary for Fourth Period projects and estimated recruiting about 3,480 Texans.[52] Among the changes, the new period relocated an Austin-area company to Groveton and one at Meridian to Huntsville. Meanwhile, companies from Colorado migrated southeastward to Cleveland, Jasper, and Crockett (Pennington) to escape the cold, while three Oklahoma companies moved to Lindale, Zavalla, and New Waverly.[53] In all, sixteen of the twenty-two East Texas camps were engaged in forestry work, providing employment for 3,800 enrollees.[54]

November 1934 saw plans for enlarging the Lufkin headquarters for Sub-District No. 1.[55] These changes were probably the result of the sub-district's ballooning responsibilities, which had grown in the Fourth Period to include supervision of over 5,000 enrollees.[56] The expansion occurred within a month, and the warehouse on East Dozer Avenue was moved to a larger facility across the city at the Leach Building on the corner of Angelina and Burke.[57]

Congressional impasses threatened appropriations for the work-relief bill during the early months of 1935 and brought uncertainty to ECW programs.[58] As C. W. Hanna observed, "The official and legal life of the CCC came to an end on March 31" without passage of the bill. Eventually the two houses agreed, however, "pay came through as usual," and enrollment for the Fifth Period concluded by April.[59] The district braced for changes afterward, though, brought about by surging enrollment numbers and a nationwide effort to enlist 520,000 men in more than two thousand camps throughout the country.[60] Other national changes had local implications, and racially segregated camps were also implemented during the Fifth Period.

Texas was reorganized into fifteen districts to better administer the growing number of work programs across the state,[61] and the Lufkin headquarters split shop to create new administrative offices in Tyler.[62] The headquarters officially began operations on July 1 at a building formerly occupied by the Southwestern Transportation Company near the Cotton Belt shops on Social Street, and additional warehouses were located on North Fannin and East Oakwood.[63] The first camp under administration by the new district was nearing completion at Tyler State Park by the end of July, and seven others were established by late August.[64]

Together, the new headquarters complex oversaw eighteen camps in Northeast Texas during its first period of operation,[65] housing nearly

4,000 enrollees in what was called the "North Texas headquarters."[66] By the fall of 1935, eighty-five CCC camps were assigned to Texas, enrolling approximately 17,500 men. Of these CCC camps, twenty-one were engaged in forest protection and improvement projects in national, state, or private forests.[67] The Lufkin District alone administered twenty-five, housing 4,822 enrollees.[68]

March 1936 marked the third anniversary of the CCC, and celebratory articles appeared throughout East Texas to recognize the occasion. The *Huntsville Item* praised the efforts of Camp P-74,[69] and it acknowledged the laborers who erected telephone lines, built roads, fought fires, and assisted in the construction of the six new Walker County lookout towers.[70] The Maydelle Camp celebrated its third anniversary "under the shade of tall pines and beside a lake created by their own hands" that June.[71] The day's activities included a tour of the state forest, a picnic lunch, a camp inspection, and a baseball game against the Jacksonville Soil Conservation Service team.

The TFS had adapted to its supervisory and planning responsibilities. With a reduction in the number of projects administered by the service, the agency again began to distribute the *News*. Proudly, TFS proclaimed in July that during the Sixth Period, Texas forestry camps ranked sixth in the number of accidents among the other twenty-three units on national forest, Tennessee Valley Authority, state, private, and levee camps throughout the southern states in Region 8. During this period, the *News* explained, Texas forestry camps had a total enrollment of 8,222 men, or an average of 1,370 men per month.[72] Twenty-eight accidents occurred during this period, averaging 4.6 per month, or 3.1 accidents per thousand men. The majority of the reportable incidents, the summary continued, were related to the improper use of hand tools and falling objects. Surprisingly, the *Tyler Journal* noted four months later that "many enrollees have been trained in the hazardous work of fire tower construction, and up to the present time no accidents have occurred in completing the erection of 55 towers."[73]

The *Lufkin Daily News*, meanwhile, devoted an entire section of its Centennial Edition to forest products and the Forest Service during August 1936. The series of articles celebrated the CCC on its anniversary, and chaplain Captain William P. Hardegree emphasized that the program was "a moral and physical re-education to the physically, morally, and mentally undernourished and misshapen human wreckage of a changing social and economic world."[74] Other articles demonstrated how Lufkin bustled as the district's hub, and H. H. Buckles described the seventy-five "happy, well-paid" administrators working at headquarters.[75] Another story featured the motor pool and the thirteen mechanics and drivers who serviced fifty-four army vehicles, nine headquarters trucks, five ambulances, six passenger vehicles, thirty-five utility units, generators, saws, and compressors.[76] Many of these vehicles were used to transport the enormous quantities of food from the city's warehouse to the various camps within the district three times per month, and a complementary article made clear that the enrollees "march on their stomachs."[77]

While a nationwide poll conducted by the American Institute of Public Opinion suggested 82 percent of the country supported the continuation of the CCC,[78] enrollment figures suggest some measure of dissatisfaction among the tree soldiers. Within the Lufkin District, for example, the *Lufkin Daily News* reported in August 1936 that nearly half (45 percent) of the recent 162 discharges were made for reasons of physical disability, discipline, or desertion. W. C. Norman, the personnel clerk offering the information, went on record saying that a large number of desertions follow each new enrollment. Some boys just "can't take it," he said.[79] Even among the brass, Eighth Area commander Major General Johnson Hagood was also removed from duty earlier in the year for contemptuous "reference to CCC activities as 'hobbies,' 'collecting postage stamps,' and 'taking an interest in butterflies.'"[80]

Nevertheless, the program was successful in supporting Texans and restoring the state's natural resources. Statistics compiled in October 1936 observed 76,804 Texans had been employed by the CCC since the first enrollment period in April 1933.[81] Through August 1936, the CCC had invested $44.9 million in Texas, and over 101,000 acres of forest stand improvement work had been completed. More staggering, enrollees had planted 26.6 million trees. As the Eighth Period began that fall, the *Tyler Journal* commented on the busy season awaiting 2,800 enrollees in seventeen national, state, and private forest camps. Chief among that session's objectives were the continued development and protection of acreage within the "four newly created National Forests," including the replanting of approximately 15 million pine seedlings on 15,000 acres. Importantly, the summary also hinted at the expansion of the lookout tower network into the northern section of the Piney Woods, saying "eighteen additional towers will be erected on state and private lands and additional towers and telephone lines will be constructed on national forest lands between now and March 31, 1937."[82]

E. O. Siecke reported on forestry progress in March, noting that sixty-two lookout towers had been constructed during the period between June 1, 1933, and January 1, 1937.[83] These had been connected through a communication system spanning 1,300 miles of telephone line that were strung over fifty-nine thousand telephone poles cut and creosoted

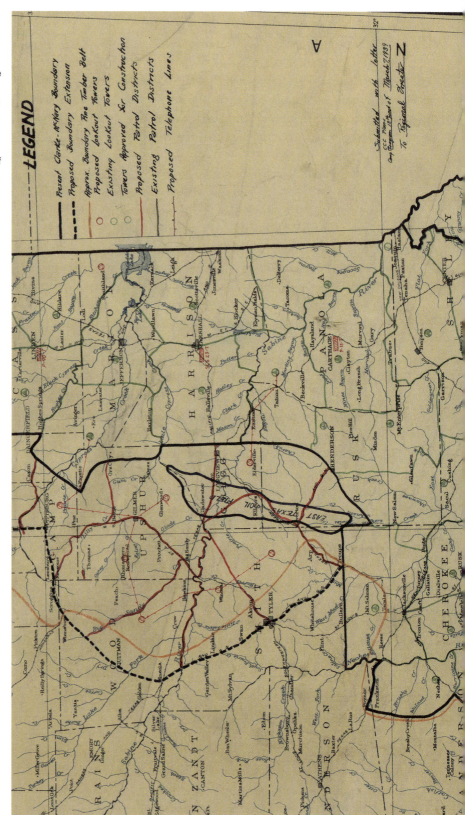

Figure 25. A portion of a map prepared and submitted to the USFS before the Thirteenth Period program was fully defined in March 1939. The draftsman proposed an expansion of the lookout network into the fringes of the Piney Woods through the extension of the Clarke-McNary boundary. Tower construction in the counties suggested was not realized until the 1950s and 1960s. Image courtesy of the National Archives, Atlanta.

at State Forest Number 3.[84] Meanwhile, the forestry camps administered by the TFS "maintained their seemingly invincible hold on first place in Region 8" with the fewest number of accidents to enrollees during the last half of 1937, when only three were reported.[85]

Livingston enrollees eased their schedule to celebrate the camp's fourth anniversary in the spring of 1937. Reporter Lillie Childs, a stenographer for the East Texas CCC headquarters and the "only girl CCC reporter," attended that celebration along with members of the Coushatta tribe, noting that the event was "practically" the fourth birthday of the CCC. The company was organized on May 20, 1933, and was the first "to be sent into the depths of East Texas" under the technical direction of the TFS.[86]

Only fourteen fire towers (twelve steel lookouts and two creosoted towers) were constructed in the two-year period between 1937 and 1938.[87] With the lookout infrastructure largely completed, TFS camps continued to expand the network of roads used for fire protection and local transportation. Bridges and rock fords were built,[88] and water wells were drilled at fifteen tower sites.[89] With fewer projects, the number of TFS camps declined so that only six were active by the end of 1938.[90]

Fifty-seven CCC camps were operational in Texas during the Thirteenth Period between April and September 1939. The TFS and USFS each operated six camps,[91] and a March 1939 map submitted to the regional forester proposed expanding lookout coverage into portions of Smith, Gregg, Upshur, Wood, Camp, Morris, and Red River Counties (fig. 25).[92] Westward expansion was also envisioned for portions of Houston, Leon, and Madison Counties. Seemingly, nine additional towers were approved for construction near Pinehurst, Lovelady, Latexo, Elkhart, Bryans Mill, Henderson, and Saratoga, but these were never built. Finally, the map also proposed that towers be built north of Bon Wier, south of Burkeville, and at Smithland. Prominently on the map are the positions of the "present Clarke-McNary Boundary" and the "Proposed Boundary Extension." Perhaps, because of a failure to extend the position of the boundary or inadequate matching state funds, plans for westward expansion were not realized, and the desired changes did not occur until the 1950s and early 1960s.

Camps at Jasper, Linden, Livingston, and Trinity were disbanded when the Fourteenth Period began in October 1939. At the time, the *News* reported that the four camps had cumulatively constructed fourteen fire towers during their tenure.[93] With these reductions, only four TFS programs were authorized for the Fourteenth Period. Flagship camps in Lufkin and Woodville were continued, while the remnants of

Figure 26. The 120-foot creosoted wood lookout tower at Bon Wier was constructed by enrollees from Newton Company 839 during December 1940. It was tied in height with a similar structure at USFS Nogalus (L-2400). The Bon Wier Lookout was replaced by a surplus oil-field derrick in 1958. Image courtesy of Texas A&M Forest Service.

the Livingston Camp were moved to Humble and a new camp was established at Newton.[94]

W. E. White wrote to the vice president and general manager of the Kirby Lumber Corporation regarding Protection Unit No. 9 in August 1939. In his letter, the Division of Forest Protection chief noted that the agency "spent considerable time" picking the location for two new lookouts "to handle the situation" in north-central Newton County. In his letter, White also acknowledged that "the Army had begun to erect the buildings for a CCC camp site [at Newton]." When Camp P-92-T was operational and the enclosed lease agreement was executed and returned, White assured the general manager, work would commence on the fire towers at Bon Wier and Burkeville.[95] During October, another letter from White to the vice president indicated that TFS "plans call[ed] for the erection of a 120' lookout tower on this site [Bon Wier] as soon as we receive same from the government."[96]

By the spring of 1940, the Newton Camp was busily completing a lookout at Burkeville, Number 71 in the state system, and a 120-foot wooden tower was to be started at Bon Wier. The wooden tower, named after the community of Bon Wier, was unusual not only because of its height but also because it was to be constructed with "a catwalk with railings built around the outside of the cab for observation purposes" (fig. 26).[97] Evidently these additions, which appeared on the Thirteenth Period work plan, had been approved.

More interesting than tower construction was the TFS pivot from laborious forest improvement projects to the development of a tree nursery at Camp P-93, Alto. The nursery was envisioned to meet the "growing demands in East Texas for trees to plant on worn-out agricultural lands and cut-over forest lands" and was to replace two small facilities operating at the state forests near Conroe and Kirbyville.[98]

Sites in nine counties were inspected, but David Anderson later recalled that the slope of the land, soil type, drainage, and dormancy period during the winter ultimately led to the selection of the 73-acre site near Alto, which embraced several Caddo culture ceremonial and burial mounds dating to AD 1000.[99] Enrollee efforts assured that over one million pine seedlings were raised for distribution across the state by the end of the first season.[100] Incrementally, the operation was forecast to expand from six million to ten million seedlings, with continued support from the cadre of CCC enrollees. The nursery continued production for decades, only closing in 2008.

Texas National Forests

In a clever, if not ultimately damning, move Texas did not relinquish the republic's unappropriated lands when the treaty of annexation was signed with the US government in 1845.[101] Quickly, the state's domain disappeared through veteran and settler land grants, leaving few public spaces for recreation or demonstration. Maxwell and Baker pointed out that devoid of federal lands, the federal Bureau of Forestry had only a "secondary interest in the region west of the Sabine" River and did not station any forestry personnel in Texas.[102]

Provisions of the Weeks Law offered a mechanism for the federal government to purchase certain nonagricultural forestlands for watershed protection and national forests, but only when states passed enabling acts authorizing such purchases.[103] Candidly, the Texas Department of Forestry recognized in Bulletin 4 that "public spirit has not as yet been sufficiently aroused among private owners to lead them to turn over any part of their holdings." Likewise, there was "no immediate likelihood" that the government would enter new regions under the land purchase provisions of the Weeks Law unless sanctioned by the state.[104] Despite reluctance in Texas, other southern states such as North Carolina, South Carolina, Georgia, and Tennessee had authorized federal land purchases as early as 1901. Others quickly approved enabling legislation: Alabama in 1907, then Arkansas (1917), Florida (1921), Louisiana (1924), Oklahoma (1925), and Mississippi (1926).[105] The Lone Star State remained slow to react, though, and Bulletin 4 continued: "It would be desirable for the next legislative session to pass such enabling act as expressing to Congress approval of this work."[106]

Lufkin representative John W. Laird made an attempt, proposing an awkwardly titled consenting act during the Thirty-Eighth Legislature in 1923.[107] But it failed to gain momentum even though the State Parks Board was approved that same year.[108] A decade later, while Roosevelt was emphasizing nationwide conservation measures to generate employment, another Lufkin senator, John S. Redditt, understood the potential economic impact of the new administration's vision. His district comprised about half of the commercial forest regions in Texas and stood to benefit immensely from the new federal programs if a bill inviting the government to purchase property in Texas could be negotiated. Fortunately for Texans, Senate Concurrent Resolution (S.C.R.) No. 73 passed the Senate on May 22, 1933, and the House the following day. It was signed into law by Governor Ferguson on May 26.[109]

The original legislation consented to federal acquisition of state lands for three purposes, including use for national parks, national forests, and the purchase and eventual resale of property to Texans on long-term payments. The latter was meant to assist the citizenry in owning "their own homes with sufficient land for the support and maintenance of their families." A key provision of the law was that each of these activities should "assist the unemployment situation in Texas." Importantly, S.C.R. No. 73 also stipulated that the state forester, along with the board of directors of A&M College, "first approve the purchase of any land for National Forests."

By September 1933, newspaper announcements demonstrate that USFS representatives were beginning to organize reconnaissance surveys in East Texas to identify suitable tracts of land for inclusion in future national forests. These expeditions were coordinated by the chief land examiner, C. E. Beaumont, and included a team of nearly a dozen men who set up a field office in Lufkin.[110] Within the group was Sherman L. Frost, a Yale graduate who would later serve as an interim director of the TFS. Along with students and graduates from forestry programs in Michigan, Louisiana, New York, Pennsylvania, Washington, and Iowa,[111] the crews set out to evaluate the region and define tentative park boundaries.

Not surprisingly, the organized groups of surveying teams with strange equipment evaluating the woods caused some suspicion, and Carolyn Frances Hyman related that some residents were openly hostile to the crews, believing them to be "revenuers." It is obvious from her writing in the late 1940s that she was fortunate to have had armchair discussions with members of the original crew, who recalled the scarcity of drafting tables in the sawmill towns of East Texas and the necessity of using pool tables in Groveton during the evening hours to compile their field data.

Her conversations clarify some aspects of the modern-day forest boundaries. For those who have studied the thick, boldface proclamation boundaries on a map and wondered how they were delineated, we have some clues. In addition to boundaries drawn out of convenience along the margins of Texas' original land surveys, it seems others are as arbitrary as they appear. It must have made for good conversation as members of the original party reminisced with Hyman. With the continuation of Prohibition into the early 1930s, some "impassable" objects—whiskey stills—were encountered during transects through the woods, necessitating "right-angled offsets" to avoid what may have become nasty confrontations.[112]

Reports produced at the conclusion of the examination mirror the intent of the Clarke-McNary Act and S.C.R. No. 73. In them, land acquisitions were justified where relief employment was needed, in the context of proximity to navigable waterways, on lands with low agricultural value, and in stands that were suitable for demonstrating sound forestry principles. Three examples highlight these objectives. In the first, a report concerning the Crockett Purchase Unit noted, "The establishment of this unit and the purchase of the Houston County Timber tract is of economic importance, in that it will aid in perpetuating two large lumber concerns, of this section, which employ hundreds of sawmill workers."[113]

A second, written by William Willard Ashe after a reconnaissance trip during October 1923 observed:

> Erosion of soil is extremely active on naked soils on hillside sites; and while the damage which is at present being done is not so extensive as formerly . . . the hillside lands in the old field stands are channeled by many gullies, erosion in which, however, has been greatly reduced through the reestablishment of the forest and the soil covering of litter. In places the alluvial lands have been seriously damaged by sand bars deposited thereon, the sand being the soil eroded from the hills.[114]

Finally, in a third example written as a supplement to Beaumont's work, Joseph E. Kircher, the USFS regional forester, commented:

> This unit is characterized by extensive second growth stands of timber of a size and density which allow early or immediate operation. One of the chief objectives in establishing National Forests and one which applies with special emphasis to the Southern Pine Region, is that of demonstrating the proper methods of handling timber lands. If purchases are restricted to lands carrying little or no timber large enough to warrant profitable operation, it is obvious that demonstration will necessarily be confined to planting and protection against fire with possibly a few thinnings in the older stands. And the all important phase of timber harvesting will be postponed for a long period. If, on the other hand, a moderate proportion of the acreage is made up of the older age classes it will be possible immediately to start demonstration, illustrating all phases of forest management.[115]

When Beaumont submitted his short reports in October and November 1933, over 1.5 million acres were eyed for federal acquisition in five purchase units, including the Angelina, Crockett, Sabine, Sam Houston, and

Tejas Units.[116] Upon review, four proclamation boundaries were established by the National Forest Reservation Commission during its March 26, 1934, meeting. Within these proclamation boundaries, which form the mapped margins of each forest, the government was then authorized to acquire suitable timberlands.[117] Interestingly, Peter Stark points out that the original plans called for uniting the various purchase units into a single forest called the Sam Houston National Forest.[118] This intent is made clear in Joseph Kircher's supplemental report, where he anticipated "that all of the Texas units if acquired will for administrative purposes be grouped as one National Forest."[119] For this reason, the name of the Sam Houston Purchase Unit was changed to "San Jacinto" in November 1934 to avoid confusion. When these efforts did not materialize as planned, the San Jacinto Purchase Unit's name reverted back to "Sam Houston" during June 1936.[120]

The attempt to establish a single national forest would have had implications for revenue sharing within *all* of the counties encompassed by the envisioned Sam Houston. Certainly this was Kircher's understanding when he wrote that "the benefits from receipts will be distributed to all the counties in the four units and not merely those embracing lands within the Crockett division."[121] Stark notes that this arrangement could have created barriers to the land acquisition program. For instance, counties dominated by cutover forests would benefit unfairly by collecting a percentage of the revenue generated by other counties in which merchantable timber was harvested. To avoid opposition to government progress, each proclamation boundary was therefore established as a unique national forest that would be administered under an organizational "invention" called TNFs to achieve unity.[122]

Kircher's supplemental report implies that there were open communication channels between the government and Director Siecke. In fact, Siecke even made "a number of suggestions as to areas which he believed [would be] most suitable for federal acquisitions." His first, apparently, was 85,000 acres of Delta Land and Timber Company holdings in the future Sam Houston National Forest. Despite contact between the two offices, Kircher's February 15, 1934, report indicates Siecke was requested "to make an explicit statement as to his attitude and that of the Board of Directors of A&M College" toward the selected units.[123]

The letter from Kircher to Siecke, written three days before the report, is preserved in the Texas A&M Forest Service archives. In it, Kircher seeks purchase approval for the lands delineated by Beaumont's team as the law required:

> You are no doubt somewhat familiar with the investigation carried on by Mr. Beaumont during the past sixty days looking towards the selection of areas in your State to be recommended to the National Forest Reservation Commission for establishment as National Forest Purchase units. Mr. Beaumont has tentatively selected the following four areas. . . . My object in writing you at this time is to ascertain whether or not these areas meet with your approval and that of the Board of Directors of A&M College as provided in the State Enabling Law, being Senate Concurrent Resolution No. 73.[124]

Though subsequent communications are not accessible, Siecke's letter to Kircher in April 1934 summarizes the outcome of the conversations between the TFS and the Board of Directors:

> [Siecke] tried to give them a picture of the National Forest Acquisition Program since its inauguration in 1911. The present expansion and probable further future expansion of this program in connection with a National Land Utilization Program was emphasized. The Board finally adopted the enclosed resolution.

Seemingly, approval was not without controversy, however, and the letter continued:

> From the standpoint of policy, I suggest that you do not give publicity to the recent action of our Board of Directors. It seems desirable for the Board and also for the State Forester to drop out of the picture to the fullest possible extent. You, of course, realize there are apt to be protests from county courts and other groups in case a considerable acreage is purchased in any one county because of the fact the National Forest land will not be subject to state and local taxes. If vigorous protests are made to the Board, it may embarrass the Board members and also the project. Therefore, I believe it will be distinctly advantageous to keep the Board and the State Forester in the background.[125]

While freshly penned legislation authorized TNFs, the USFS was undergoing administrative changes. When the Department of Agriculture's Bureau of Forestry became the Forest Service in 1905, the nation's sixty forest reserves were predominantly west of the Mississippi River, and six administrative districts were created in the western United States to administer the lands. After the passage of the Weeks Law in 1911, however, forests were established in the eastern and southern United States.[126] As the number of forests grew, regional administration became necessary, and an Eastern Region was established in 1914.

With continued expansion brought about by emergency conservation measures, the Eastern Region's acreage swelled and was divided in 1934 to form a Southern Region, or Region 8, with headquarters in Atlanta, Georgia.[127] Despite these growing pains, the work of the USFS continued unabated in Texas, and an extensive land acquisition program commenced within the boundaries approved by the state and president. Loren L. Bishop, the first supervisor of Texas forests, established an office in the Federal Building in downtown Houston to oversee and realize plans for acquiring between 1.4 and 1.7 million acres of land within the proclamation boundaries.[128] The first federal land purchase was completed on March 26, 1934.

By November 1934, 500,000 acres of land had been acquired in the newly formed Angelina, Davy Crockett, Sam Houston, and Sabine purchase units. Four CCC camps were quickly organized under USFS command during the winter of 1934 to begin improving the forests.[129] Additional acreage was approved by the National Forest Reservation Commission in March 1935,[130] and the number of USFS-administered camps swelled to fourteen by the winter of 1935. Loren Bishop outlined enrollee accomplishments in 1937, including the planting of fifteen million seedlings, the construction of 500 miles of earthen forest roads, and the development of six lakes for public recreation. He also highlighted the "large number of forest fire lookout towers, cabins, and other administrative improvements" completed by the men.[131] By the time President Roosevelt officially proclaimed the lands as national forests on October 15, 1936, about 631,000 acres were under federal administration.[132]

The majority of the purchases, accounting for 83 percent of the forest area, were made from just seven lumber interests.[133] Intriguingly, a June 1937 analysis comparing the sale price and assessed value of the lands approved for purchase recognized that the results were "in direct contradiction to the statement frequently advanced, that owners do not pay taxes on a reasonable valuation of their cut-over and second-growth timberlands" (table 3). With the exception of the Houston County Timber Company, "the sale price was definitely less than the assessed valuation" of the property, and the average government purchase price (excluding the arrangements made with the Houston County Timber Company) was $1.67 per acre less than the assessed value of the property.[134]

Consolidating the acreage under federal protection was meant to restore forest cover on "vast areas of bare, gaunt, East Texas cut-over

Table 3. Comparison of the sale price and the assessed value of timberlands purchased by the US government during TNF acquisitions. The information was compiled by Paul W. Schoen, who later became the forest management chief in June 1937, using data obtained from L. L. Bishop and the county tax assessors for Angelina, Trinity, Sabine, Houston, Montgomery, Shelby, and San Augustine Counties. The tabulation was completed twice, including and omitting lands sold by the Houston County Timber Company.

Vendor	Acres	Sales price per acre	Assessed price per acre	Sales price % of assessment
Houston County Timber Co.	94,126.40	$12.50	$6.00	208.3
Pickering Lumber Co.	90,076.10	$3.00	$5.50	54.5
Temple Lumber Co.	84,559.60	$2.50	$4.00	62.5
Delta Land and Timber Co.	82,774.30	$2.60	$4.00	65.0
Long Bell Lumber Co.	73,349.60	$3.50	$4.03	86.8
Trinity County Lumber Co.	61,418.70	$3.75	$5.00	75.0
Kirby Lumber Co.	57,024.70	$1.75	$4.56	38.3
Total	543,329.50	$4.23	$4.73	84.3
Total excluding Houston County Timber Co.	449,203.10	$2.85	$4.53	63.7

lands,"[135] while simultaneously providing watershed protection, erosion control, and recreation. Chief among the government's ambitions were producing repeatable crops of forest products and having the forests serve as demonstration areas that showcased the long-term value of an educated land-use policy.[136] But provision that the land would be used to "ease unemployment" or that it could be transferred back to the citizens of Texas has intermittently caused controversy.

Camp Segregation

The professionally captured photographic panorama of Company 893 in Pineland taken during August 1933 suggests the strong military influence at the camp, even during the early periods of ECW progress. Orderly rows of surplus, high-roofed military tents appear between dirt avenues dividing the sparsely, but pleasantly treed campsite. The entire company appears in the foreground. Some members wear civilian clothes, many don military uniforms, and others wear kitchen attire. There are at least 174 men in the photograph. What is most striking about the image, however, is the physical separation of 16 African Americans, who stand more than shoulder-width apart from the majority of the other enrollees.

Together, they represent only about 10 percent of the company. The image is not uncommon, and a similar photograph of Company 888 taken in Weches during October 1934 shows an even smaller minority. There, only 9 of the 192 men, a mere 5 percent, are African American.

While these meager percentages were hardly inclusive, that the units were integrated was considered a progressive win. As Roosevelt's ECW policy sprinted for approval through Washington, an important amendment was attached by Illinois Republican Oscar Stanton DePriest, the sole African American member of Congress. In the draft legislation, DePriest added a statement that "no discrimination shall be made on account of race, color, or creed." After the conservation act passed, Steely noted, the army adhered to this policy, largely by leaving in place its own "established racial arrangements of generally assigning African American recruits to kitchen or bivouac duties."[137]

By December 1933, however, the Pineland Camp was about to change history, in both Texas and the country. Staring at the image taken of Company 893 in August, one can only wonder how many of the "boys" stayed on for the Second Period and which, if any, played a role in shaping the future policy. Wellborn provides some of the most complete published accounts that allude to a series of race-related incidents that occurred in East Texas at the forestry camps. Some had become violent. It was at the Pineland Camp in December, though, where twelve African American enrollees allegedly deserted camp. A change in camp command had taken place after the First Period, and the new commander stated that he would not tolerate "laziness, disrespect or insubordination." In hateful language, a liaison officer's report of the incident concluded that the deserters "were too well treated. . . . They believed that they were just as good as whites, [and] were their equals in every respect."[138] The rest is difficult to read, but Wellborn's treatment of the event is important in that it offers an unfiltered version of history that could have been easily brushed over. The remaining excerpts of the inspector's report are equally bigoted: "Most of the Negroes in the camp were a poor type of the lower class. Those that were good were influenced by the bad. Never having been shown such consideration before, they forgot their place. The kind treatment they had been receiving from Lt. Berrett [the previous camp commander] inculcated in them a feeling of overconfidence and superiority."[139]

Connie Ford McCann was an enrollee at the Pineland Camp in 1933 and was unusual in that he kept a diary of his experiences. These, and the photographs he took, have been digitized by the University of North Texas Libraries "Portal to Texas History" collection and are available online in one of the most comprehensive databases for Texas historians. Fortunately, his narrative—an alternative narrative—survives about the reported incident:

> Dec 15—Several of the "tougher" fellas in camp got drunk Sunday the 17th [November] and mistreated several of the Negros. O. Le Blanc was one of them. Resulted in his being discharged + all the Negros deserting at one time. What a mess! The Capt. went to Sub-dist hdq + asked for an investigation so that the mess wouldn't mar his record. The investigation came on the 21st—but didn't satisfy himself as to the cause + result of the whole business so I understand.[140]

Relations only deteriorated in East Texas camps during early 1935. An unsigned letter to Major Carl S. McKinney, the commanding officer of the East Texas District, written jointly by the US and Texas Forest Services suggested that African American work projects be consolidated into one camp:

> The assignment of Negroes in various size groups to the East Texas Forestry camps constitutes a marked disadvantage to the technical service and no doubt also complicates the administration of such camps by the War Department. It is not practical to work Negros in the woods unless a sufficient number are available to make a truck load, which comprises from 20 to 25 individuals.[141]

The issues escalated to ECW director Fechner, who then wrote to Texas governor James Allred that the integration of camps "was done without my knowledge or consent in the early months of this work, but I corrected it as quickly as possible after it came to my attention."[142]

Fechner's decision had immediate effects in the East Texas camps. After May 1, 1935, segregated "white" and "colored" companies appeared, and a special suffix, "C," was added to each company's identification number to indicate which units were composed of "colored" enrollees. As Lufkin district commander Hanna explained, "There were too many colored enrollees for one camp and not enough for two, [so] all possible colored men were assigned to Co. 897 at the Oakhurst Camp and the overflow was assigned to Co. 838, Lufkin, which became 'half white and half colored.'"[143]

Pioneering Units

The gravel road is nearly level as I drive with the windows down through the forest. Occasionally, I'm jolted from my mental wanderings by a pothole or a muddy splash as I make my way toward the site of the Richards Camp. This corner of the Sam Houston National Forest seems isolated today, and it's hard to imagine how a young enrollee—or a minority enrollee—must have felt as they approached their campsite for a new job away from their trusted support networks at home.

Lufkin (Company 838; P-57-T)

The Battle of '33

> We're down here in East Texas (They say to save the old pine tree)
> And we're doing six months of battle in the war of '33
> Sometimes we get mighty lonesome for ma, pa, and sis to see;
> Many nights we've wished for home in this battle of '33
> Often I dream of chicken fried, cakes and pies ma cooked for me;
> But now I've learned to eat my spuds in this battle of '33.
> Oft times at work I'm weary and tired and wish that I'd never seen a tree,
> For "doodling" stumps ain't no fun in this battle of '33.
> Often I wish that I were home again so my "sweetie" could be with me—
> And again we'd stroll down lover's lane as we did before the battle of '33.
> But in Lufkin as the twilights begin to fall, our sweethearts we go to see—
> A few happy hours we spend with them, the girls in the battle of '33.
> So after the war clouds are lifted away—but should other depressions be
> May the sons serve the purpose as their fathers did in. '33.[144]

Company 838 was organized at Fort Sam Houston in May 1933 and sent to Lufkin under the command of Lieutenant F. R. Undritz. Two army sergeants and fifty-five junior enrollees were initially dispatched to clear underbrush and establish the campsite. *Happy Days*, the CCC's nationally distributed newspaper, ran a story on July 29 describing the conditions in Lufkin. The new site sat at the edge of the County Fair Grounds, W. A. "Hank" Atkinson explained, and was "gorgeous, beautiful, plentiful." The city donated use of a baseball field, a volleyball court, tennis courts, and even "indoor baseball, wrestling, and boxing" arenas on the grounds. The only complaint at the time was that the shower facilities were "of those open-air" type "with two kinds of running water—cool and cooler."[145]

By July 1933, Project P-57 was well under way with 190 enrollees working to reforest the area around their base and actively building roads and bridges and fighting fires. Atkinson even commented in *Happy Days* that it was "surprising how quickly the boys have turned from soda jerkers, cake eaters, and what have you, to men who know their axes and saws in the woods."[146] While some crews were working in the forests, others were busy crisscrossing the Piney Woods in trucks from the motor pool to supply other district camps with food and equipment.[147]

Surveyor and writer Isaac Burroughs was now in charge of handling requisitions for the TFS, and there were "orders, vouchers, shipping records galore" on his desk. But things appeared to be running smoothly, and the *Lufkin Daily News* suggested that readers "drop in his office some time and marvel how one man so efficiently can perfect so many major details. We should think the Texas Forest Service realizes the fact—incidentally employing and keeping a real man for a real man's job."[148] The organizational and network capabilities of the TFS are clear from this, as well as other descriptions, many of which were written by Atkinson.

One month later Siecke and White were both dinner guests at the Lufkin Camp, Atkinson reported in a "Camp Chatter" story. Also in attendance was Major Guthrie from Fort Sam Houston, "who co-ordinates his time between the Civilian Conservation Corps and the regular army. The major was making an inspection tour of all Texas CCC camps." With genuine respect for his superiors, or perhaps even some fawning, Atkinson continued: "It has often crossed the writer's mind just why the Texas Forest Service stands out in the eyes of firms, organizations, and even government individuals as an outstanding harmonious organization. After knowing these two fine gentlemen, it is easy to understand why the forest service functions so smoothly."[149]

A community open house followed the next week to "acquaint the people of Lufkin with the forestry recruits and explain their work," complete with a baseball game, boxing and wrestling matches, and a camp tour. Sandwiches, punch, and ice cream were served to approximately 130 guests while camp supervisor R. E. "Bob" Erwin and his commanding officers made welcoming speeches.[150] Most likely Erwin, having personally supervised the construction of a 190-foot-long bridge across the Angelina River near Ewing, touted the camp's early successes.[151]

Educational statistics offered by the *Lufkin News* in December 1933 painted an uncharacteristically high level of academic success among Lufkin enrollees, breaking somewhat with the notion that many of the "boys" had incomplete rural schooling opportunities. The publication of the camp newspaper, which often featured simple but thoughtful poetry, reinforces the group's educational achievements. Within camp, eleven men had completed as many as three years of college work, and thirty-seven others were high school graduates. An additional eighty-seven had finished at least two and a half years of high school.[152]

By October 1933, *Happy Days* reported that the "steel for fire lookout towers" had arrived at camp and that "work on the construction of three towers will begin in a few days."[153] By June 1934, the camp had built thirteen lookouts in portions of the Palestine quadrangle,[154] and *Happy Days* reported on the near completion of a fourteenth by July 21.[155] There had been no accidents in the construction, the newspaper reported that October, and the piece commended the "demon tower-building crew of Co. 838."[156] The names of many within this "demon crew" are still visible on the ladderway slab below the Central School Tower, just where they left them in the curing concrete on February 21, 1934.

The order to segregate enrollees had immediate impacts on diversity at the Lufkin Camp. In May 1935, the camp "became half white and half colored,"[157] and by July 3 the remaining white enrollees began transferring to other camps in the district. By October, the camp was "wholly colored," and the unit was reassigned as Company 838 (C). The redesignated company continued to operate from Lufkin until the autumn of 1939,[158] when the enrollees were transferred to a new camp in Newton.[159] The vacated Lufkin Camp was backfilled by tree soldiers from the camp at Trinity, which was dissolved at the beginning of the Fourteenth Period.

Evidently, Company 838 (C) was moved from Lufkin after an evening of unrest initiated by social disparities. As one foreman remembered, the enrollees often walked into Lufkin during the evening to socialize, in what was then the segregated section of town. Handsome uniforms and extra pocket money attracted the attention of local bachelorettes, and resentment built until three enrollees were badly beaten by competing area suiters. In response, the enrollees quietly organized a retaliatory offensive. One Saturday, the men congregated along a railway grade before storming into town "and avenged themselves . . . on males of proper age." A swift retreat through the woods on the outskirts of town had been planned to return to the barracks. But the strategy quickly went astray when gunshots "began to ring out in the woods between the camp and the town" from skittish residents in the wooded fringes of Lufkin who feared an "insurrection" when they noticed groups of enrollees running in the dark back to the nearly deserted camp.[160]

The buildings at the Lufkin Camp found a second life during World War II, when Southland Paper Mills leased the compound from the

USFS. After officially opening on February 15, 1944, German prisoners of war were barracked there and sent into the Piney Woods to cut pulpwood in support of the American war effort. The camp remained operational until 1946.[161]

Ratcliff (Company 1803; F-4-T)

The *Crockett Courier* documented construction of the CCC camp at Ratcliff during August 1934. The announcement observed that the camp would "be operated much as the other CCC camps scattered over the state" but was explicit in that "this particular camp will be operated by the National Forestry Service instead of the State Forestry Service as is the case with all of the present camps."[162] By October 15, two hundred men from Company 1803 transferred to Ratcliff from Eagletown, Oklahoma, and immediately began to construct telephone lines, grub rights-of-way, grade roads, and build pile bridges.[163] During this early period, enrollees also dismantled the 87-foot IDECO tower constructed by the TFS in 1928 and replaced it with a 99.75-foot Aermotor MC-39 (fig. 27).

Near camp, enrollees also began moving 9,500 cubic yards of earthen fill to repair the dam impounding what was once the 4-Cs millpond, transforming the lake and its surroundings into a regionally important public recreation area. When finished, the shores of the 70-acre lake were encircled by a 3.5-mile scenic drive, punctuated by campsites and planted with shortleaf pine. A beach, a bathhouse, picnic facilities, and a play area were also provided.[164]

For some fortunate enrollees like Hubert Doss, the combination of the swimming beach and his lifeguard certificate meant an assignment as a lifeguard for three consecutive summers. Hubert Doss's memories are captured in an interview conducted by the USFS in 1991. Doss arrived at Camp F-4 soon after it was finished in 1933 and remembered the lookout tower and weekend fire guard duty. He recalled "some good fires" that sent crews rushing out to battle the flames. As might be expected, fire duty was not

Figure 27. The Aermotor MC-39 at Ratcliff is a Depression-era structure located at the margins of Camp F-4-T (A). The MC-39 was built by Company 1803 enrollees to replace an 87-foot IDECO lookout that had previously been erected by the TFS in 1928. The remains of the older tower still appear south of the standing tower (B). While disturbed, piers for a water tower also remain at the site (C). Photographs by the author.

popular, and Doss reminisced that when "the fire bells rang the second time . . . we'd as many as we could we'd take off into the woods and hide if we could." Interviewing him, Bobby Johnson laughed and asked, "You mean to get out of the fighting?" Doss replied, "Yes sir, get out of going to the fire. But most of the time a sergeant, he'd round us up."[165]

Other oral interviews and newspaper articles provide insights into work and leisure at the camp. Fount Kelley was interviewed in 1991 and remembered working with the KP, or kitchen police, and as a telephone operator connecting the camp's fire towers with the Crockett ranger office. Kelley remembered three towers in particular, one at camp, another in Weches, and a third "between here [Ratcliff Lake] and Highway 287 towards Pennington," referencing what was likely the Nogalus Lookout. "I could talk to those towers," he told Johnson. When Johnson asked, "Who'd they have in those towers? Forest Rangers?" Kelley replied that they were staffed by CCC boys. Later, he helped establish national forest boundary markers and was on fire detail.[166]

Despite favorable enrollee recollections and enthusiastic printed material, tension and conflict within East Texas camps appear to have been somewhat widespread based on documented events at Pineland, Lufkin, and Ratcliff. In one serious event at Ratcliff, the *Mexia Weekly Herald* reported an incident involving an F-4 enrollee who was being detained at the Houston County jail for assaulting a woman in 1937. An altercation occurred when the jailer entered the cell shared by the enrollee and another inmate being held on seven counts of burglary. After the enrollee's cellmate shot the jailer in the face with a .23-caliber automatic, the pair escaped the prison in the jailer's vehicle.[167]

The Ratcliff Camp continued operating until the end of 1941.[168] Many TFS employees cycled through the facility during the seven years it was in operation, providing them with important field and management experiences that helped make them successful later in their careers. The longtime head of the FCD, Joseph Burnside, benefited from time at camp, while Don Young, then a district ranger with the USFS, oversaw development of Ratcliff Lake.[169] Andy Anderson worked as a junior forester at Ratcliff,[170] and Knox Ivie was superintendent of the camp, once directing the field activities for some two hundred men.[171] The impact of their combined experiences was best expressed by Anderson, who said, "I have the feeling that modern day foresters lack the crusading spirt of years past. . . . We set our dreams pretty high. But didn't we do pretty good with our dreams?"[172]

National Forests and Grasslands of Texas's Only Depression-Era Lookout: Ratcliff Tower

The *Crockett Courier* announced completion of the Ratcliff Tower in April 1928, part of a protection unit that had just been organized between the TFS, the Southern Pine Lumber Company, and the Houston County Timber Company.[173] Ratcliff, and another 87-foot IDECO lookout constructed at Bird Mountain in Anderson County, were critical pieces of infrastructure necessary for providing intensive fire protection to the contributing landowners. As some of the state's first fire control structures, the lookouts were still a curiosity to the public and a showpiece for the agency. A year later, a Forest Guard meeting was held at Ratcliff, where W. E. White, Walter Bond, Isaac Burroughs, and William O. Durham were photographed at the base of the structure (see fig. 6). Soon afterward, Isaac Burroughs surveyed the location for the state on February 20, 1931, by marking the center of the tower with "a 1-inch square buggy axle driven at point where diagonals from center of piers intersect."[174]

The state-owned lookout was transferred to the federal government when Houston County Timber Company lands were acquired for part of the Davy Crockett National Forest. At the time, Chief White wrote to Director Siecke from Lufkin explaining that the TFS could not find any leases related to the Ratcliff or Yellowpine Towers.[175] "It is possible," he noted, "that we have never secured leases for these two tower sites, and we may have only obtained verbal agreement to erect the towers." Siecke responded the following week from College Station, confirming that they "made a thorough search" of files there and were "unable to locate executed tower site leases for either tower."[176]

The USFS swiftly removed the 87-foot International following the land acquisition, and Company 1803, Ratcliff, erected a new 99.75-foot Aermotor MC-39 in its place "at a point closely adjacent to the old site" (see fig. 27).[177] Two primary documents indicate that this change occurred in 1936. The first, the *Twentieth and Twenty-First Annual Reports* recorded that "one 87 foot tower transferred to the U.S. Forest Service" during 1936 was torn down.[178] The second, a summary of activities for Company 1803 written in 1936, noted that "a fire tower 100 feet high has been erected and another tower 90 feet high has been taken down."[179] While documentation is scarce, the author believes the original 87-foot IDECO may have been returned to the TFS and re-erected at Cushing in 1936.

Ratcliff was an important station linking the state and federal fire control networks, and the observer could speak directly with the Weches observer or the USFS district office in Crockett. In fact, a 1949 memorandum of understanding between the two agencies indicated that the Ratcliff Tower would be staffed from February 1 through April 10 each spring, and from July 20 through

November 10 whenever the fire danger rose above Class III.[180] This active schedule meant that Lookout Charles McCurdy was at the tower in October 1947 when he received a report from a local farmer about smoke he had just discovered. Fortunately, his quickly responding crew and the cache of equipment at the tower meant that the firefighters were able to control the blaze and save the farm buildings.[181]

The TNFs made plans to disconnect telephone and electrical connections to the towers at the beginning of the 1972 fiscal year when they decommissioned the lookout network. Only six lookouts would remain on the grid, one memorandum reported, because the facilities served as radio repeater stations. Power stayed on at the Ratcliff Tower because of its location at the edge of the Work Center and the need for continued radio communications across the district.[182]

The Ratcliff Lookout still stands in the Davy Crockett National Forest at the margin of the Ratcliff Work Center. The tower and the collection of nearby buildings were all built by the CCC. It is one of two federal towers in the state and the only one that dates to the Depression. In addition to the integrity of age and location, the structure's design is characteristic of New Deal construction and represents one of America's best-branded lookout manufacturers. The tower is also significant to the history of fire control in Texas and demonstrates the shared responsibilities and common mission of both the state and federal fire control programs. It represents a tangible symbol of one of the first protection units and earliest lookout sites. Along with the existing tower, piers from the 1928 structure are preserved at the site. With these factors in mind, the lookout should be prioritized for state and federal protection. It should be open and accessible to all citizens for education and demonstration.

Conroe (Company 1804; S-62-T)

Many who have entered the CCC came into camp knowing no one. We were in a strange place far from our friends. It was hard to accustom ourselves to the rough life of a work camp and a strange city to spend spare time in. A compensation for the roughness that made it endurable is the friendships we formed, not only with each other in camp but outsiders. We met fellows from all parts of the state. Fellows who bring new customs, new modes of speech, and actions. These friendships in the main are not lasting but a few are made which will endure for the rest of our lives. We met those who gain our admiration and continue to hold it thru close association, which is one of the greatest tests. We gain in brea[d]th of mind, in consideration of our fellowman. The close association necessitated by the type of life we are living makes a test of our abilities to give thought to their needs, their wants, and their rights. Friendship[s] in the CCC are formed more solidly than concrete. If the CCC has given you a friend, you have gained something all the world can't buy. Make friends and make them lasting.[183]

The *Montgomery County News* reported on June 1, 1933, that an army captain was in Conroe in conjunction with the Forestry Department scouting for the location of a CCC camp. During his visit, he indicated to reporters that fifty men would be sent within a week to make arrangements for the arrival of a full quota of men.[184] True to the captain's word, the June 15 issue ran the headline "Tree Army Unloaded Today" from a train carrying 170 men. Once they had disembarked from the Conroe train station at about one o'clock, the story reported, "The men were lined up and following the Conroe Boy Scout troop" as they marched toward the encampment just west of the fairgrounds.[185] A follow-up story appeared on June 22, explaining that camp construction was nearly complete and that enrollees were already busily engaged with reforestation work. The tree soldiers were a "fine bunch," the paper reassured area residents, and the men were often seen on the streets after work, "happy and carefree."[186]

Project S-62-T was one of the first TFS projects to be approved and was designed to transform State Forest Number 2 into a properly managed experimental forest where visitors could see the results of scientific forestry. Enrollees worked chiefly on improvement cutting within the forest boundaries and graded and graveled 20 miles of all-weather state forest fire lanes. These not only served fire control purposes but also provided "the many citizens confined in office or shop" with an opportunity to "enjoy the many benefits of a day in the woods."[187]

Road construction projects also progressed outside the state forest boundaries during the first two years Company 1804 was stationed in Conroe. In May 1934, for instance, seventy enrollees were working on a 27-mile-long roadway "through some of the richest timber lands" in two counties. The 20-foot-wide thoroughfare gave laborers practical knowledge to tackle "different drainage problems" through the creation of box culverts and pile bridges, one *Happy Days* story explained.[188] The roadway itself made the adjacent wooded area accessible to firefighters while it also served as a much-needed firebreak. By February 1, 1935, a staggering

31,402 working days of labor had been exerted in Montgomery County on this and other such projects.[189]

The enrollees were compelled to stop the routine work of camp whenever it was necessary to respond to forest fires. These were so prevalent during the first fall at camp that during November 1933, the tree soldiers responded to sixty-four fires and expended 3,126 workdays firefighting.[190] Sixteen fires alone were extinguished on November 25.

Camp happenings published in the *Montgomery County News* and the *Conroe Courier* focused almost entirely on the social and welfare activities within camp and offered only glimpses of the company's work schedule. Dances supervised by camp commander Captain George Magnusson and chaperoned by local mothers were reported, while sports scores indicated an active recreation program and an opportunity to compete against other area camps. The stories also demonstrate that enrollees had access to educational programs, which commenced as early as October 1933. Later announcements in January 1935 explained that an educational program was also arranged "for the colored enrollees of the company."[191] Notwithstanding the delay, the following week's news reported that "school was progressing rapidly with the colored boys, everybody doing their best."[192]

That same month, the *Montgomery County News* ran the headline "Educational Aim at Camp Proves Benefit to All" and offered a glimpse of the scholastic attainments among the enrollees, where only approximately 20 percent of the 235 enrollees had completed seventh grade.[193] Four camp instructors taught courses in English, arithmetic, spelling, writing, reading, algebra, business arithmetic, history, typing, business English, and civics. There were also classes in general law, bookkeeping, shorthand, first aid, and dramatics. Nearly 90 percent of the enrollees participated in the various classes, and attendance stood at 94 percent during January 1935. The program was successful enough that a two-room school was constructed at the campground, and the mess hall and recreation hall were also used as classrooms.

Company 1804 was disbanded on January 1, 1936, and the majority of the enrollees were transferred to other districts, chiefly in Arizona and Colorado.[194] These enrollees contributed to the fire protection network before leaving Texas through road construction projects and the completion of four Montgomery County lookout towers at Splendora, Keenan, Willis, and Farris.[195] Already, by April 1934, *Happy Days* was reporting that the camp enrollees were traveling "far afield" to complete fire tower construction projects.[196]

Maydelle (Company 833; S-54-T)

The Maydelle Camp was established during the First Period in June 1933 to support State Forest Number 3. Despite pleas to maintain the camp into 1938, completion of projects within the work area forced the closure of the camp four and a half years later in December 1937. During its lifetime, the project was supervised by both the Lufkin and Tyler administrative districts and had completed stand improvement cuttings on the state forest. It provided enough timber to manufacture twenty-four thousand 18-foot telephone poles, three thousand fence posts, and three thousand cords of fuelwood. The telephone poles were treated with creosote at an on-site demonstration plant and were used to link communication lines between fire towers.[197] Camp enrollees also mapped timber, built bridges, fought fires, and on at least two occasions in 1935 and 1937 responded to airplane crashes.

They also built fire lookouts. "Under the fire protection of this camp," the *Rusk Cherokeeean* announced, "comes 300,000 acres of richly timbered land. Overlooking these acres are our fire lookout structures one at camp, one at Bird Mountain, one at Rusk, and one at Langston's Polo [sic] cab. . . . Two new towers are near completion, one at Neches and one at Love's Lookout near Jacksonville."[198]

F. L. Carroll, "a widower," signed the lease for the Neches Lookout during the summer of 1935. The author speculates that before the tower was erected on Daly Mountain, a prominent wooded hill of considerable relief north of Neches, it had been located elsewhere in the state's detection network. This hypothesis is based on the fact that nearly all of the new towers constructed in northeastern Texas during 1936 conformed to Aermotor MC-39 specifications, while the structure at Neches is an EMSCO TW-1 design. Though a number of these lookouts were purchased under contract for the TFS, most had been erected by 1934. Moreover, of the three non-Aermotor towers erected during 1936, each was located at the periphery of the detection system where it would have extended coverage into new acreage. Unfortunately, documents have not been discovered detailing these moves, but a surplus tower freed from Geneva, Pinckney (Location 1), or even White City, Gum Springs, or Forse Mountain may have been recycled at Neches.

The Neches Lookout still sits quietly in northeastern Anderson County, only accessible up an arcing, gated track off FM 2267. Rounding the final turn, the tower becomes visible through the trees at the crest of the hill. While the structure is shaky and in need of stabilization, with 1960-era yellow paint marks still indicating bolts in need of replacement, the steelwork appears strong with minimal signs of swelling. The tower components are unaltered, and each is stamped with original part numbers. An EMSCO placard remains bolted to the northern face of the tower, and the angle iron is marked "Tennessee USA." Though access is restricted, the tower is owned and maintained by the TFS for fire communication purposes. The tower grounds have retained sufficient integrity of setting, materials, and association with Depression-era fire control activities to warrant its inclusion on the National Register.

Motion pictures filmed by the TFS bring Company 833 to life, and the lookout at camp is captured in a series of short four- to thirty-second segments. In the longer clip, the viewer watches as a forest patrolman descends from below the cab of the Aermotor platform-ladder MC-40. Unbelievably, the individual hardly pauses as he alternates between risers, descending one set as you would a ladder before sidestepping on the platform and climbing down the next like a series of stairs with his hands behind him on the rungs stabilizing his torso. There are no handrails, just the steeply angled ladders. Once the patrolman is on the ground, the silent film flashes an informational card: "When leaving for a fire, the patrolman calls his emergency patrolman who comes and takes his place in the tower." Then a pickup truck arrives at the base of the tower, and the emergency man ascends the ladder.

In the film, a large-diameter circular saw blade hangs from the angle iron below the first platform adjacent to the ladder. Judging by the size of the saw blade in comparison to the patrolman, it must be nearly 3 feet in diameter. At first puzzling, a shorter, four-second film clip provides an explanation. The camp, built near the site of the Mewshaw sawmill and at the site of the creosoting plant, used the saw blade as a fire alarm bell. In this second clip, a patrolman leans heavily on the ladder, laboriously banging the saw blade with a mallet. The sharp-sounding alarm bell would have, no doubt, called the camp's standby fire crew into action.

The piers of the Maydelle Tower are visible at the I. D. Fairchild State Forest, as the 2,375-acre area is known today. The tower, which was moved just north of Maydelle in 1960, still stands beautifully, albeit without its circular saw blade. It was donated to the Maydelle Water Supply Corporation in 1989 for use in ambulatory and water supply communication capacities.[199]

There are no captivating videos, but Camp S-54's handiwork is also displayed at Love's Lookout. The tower, situated alongside State Highway 69 in Jacksonville, is a local landmark that has become part of a highway department rest area. The lease for the tower was signed by the city's mayor on August 31, 1935, and William M. Smith was appointed patrolman soon after the tower was constructed during the summer of 1936. In just thirty days between August 19 and September 23, at least 1,380 visitors registered in the ledger that was kept inside the cab. The *News* estimated that there were an additional eight hundred "would be climbers who have started the long journey to the top but on reaching the second or third landing have thought better of the idea." Men and boys, the article related, accounted for the "greater number of registrants, but the fairer sex is by no means shy of climbing" and contributed to 35 percent of the visitations. Smith, at watch in the cabin, playfully said that many of his guests were children, but he classed them "as men and women after they have made the climb to the cab."[200] While the article was written during the midst of the Depression, visitors registering with Smith had come from across Texas, Oklahoma, Arkansas, Louisiana, Kansas, New Mexico, Alabama, Florida, Missouri, Iowa, Indiana, Ohio, Michigan, Illinois, California, and New York.

The tower sat next to a 63-acre recreation area constructed by the Civil Works Administration, which boasted a swimming pool, picnic area, and rock amphitheater.[201] The combination of the lookout and the park's facilities, which could be reached from Tyler by traveling along a narrow, winding highway past rolling hills, kept visitors coming to the park into the 1940s. It was so popular, in fact, that it was even mentioned in *A Guide to the Lone Star State* when it was published by the workers of the Writers' Program. During the decade, visitors might have met Clyde Tidwell at the tower, if he was not busy extinguishing up to a half-dozen fires he spotted from the lookout each month.

Tidwell became a patrolman in 1944 and worked at the station until his death in 1959.[202] A reporter once asked Tidwell when he was sixty years old, "What do you think about day in and day out, sitting comfortably in your cane bottom chair high over the Pineywoods?" After scanning the "magnificent panorama," the same "richly timbered land" that the Maydelle Camp once protected, Tidwell said, "I think about a lot of things." Then, with a riddle he added with a smile, "How right the world is in lots of ways, and how wrong [it is] in others." The author, Tim Parker, concluded his article by observing that "lovers had written their names" all over the tower. The whispering "plans of marriage and pledges of long life together drift up to the fire-watcher on warm summer evenings," Parker added, prompting him to ask Tidwell what he says when he encounters such youngsters. Tidwell's only reply was that "I just look in the other direction and forgive what I know."[203]

The tower received regular maintenance through the 1950s. One curious summary detailing work completed in 1957 noted that crews "painted [the] inside and top of cab and [the] base of wind mill tower."[204] This suggests that, at one point, the structure was retrofitted with an antenna similar to the ones attached to the cabs at Siecke State Forest and Woodville.

A problem was discovered with the tower lease during 1961 when the Texas Highway Department planned a Highway 69 expansion. Simultaneously, the TFS was reviewing plats for an isolation fence around the tower, and a comparison of maps indicated that the new highway right-of-way would overlap the property leased to the TFS. After researching historical property records, the Highway Department produced a deed in their favor for 20.22 acres of land signed by Texana Love on May 3, 1929, years prior to the agreement with the Forest Service. Thereafter, the TFS continued to operate the duty station but under permit from the Highway Department.

In 1974, a radio agreement allowed the Cherokee County emergency coordinator to position a radio transmitter at the top of the tower. This contract, along with Patrick Ebarb's recommendation to Bruce Miles seven years later suggesting that the lookout served some "historical value," placed the tower on the list of structures to be retained by the TFS while others were being auctioned for sale.

In 1992, Area Forester Joe Fox wrote the Texas Department of Transportation seeking permission to transfer the tower site to the City of Jacksonville. Under the revisionary provisions of the deed between the state of Texas and the Love estate, though, the Transportation Department's district engineer recommended returning the tower grounds to the Love estate. Through a series of formalities, the Texas Department of Transportation then released the property to the Love estate, which then deeded the parcel to the City of Jacksonville for use in developing a roadside park.

Joe Fox wrote to James Hull informing him that they were processing the application for the Love's Lookout tower to be added to the National Historic Lookout Register in 1993. At the same time, Fox advised, ownership of the tower was being transferred to the city as part of a restoration project for the area. The park now offers motorists traveling on Highway 69 a safe, convenient place to stop. The facilities include well-managed gardens and a patio area overlooking the scenic, narrow valley. On the patio, miniature, decorative towers complement the Aermotor that looms over the highway.

The tower's future seems secure, and the structure has been well maintained by the city. Though not open to the public, an original Aermotor placard is visible through the fence, and a 1959 USC&GS marker persists on the lawn. A high number of visitations at the side of the highway make the educational opportunities for this tower superb, and developing proper signage and interpretive programs at the park would help convey the significance of the fixed point detection system and history of the CCC in the Piney Woods.

Towering pines shade the remains of Camp S-54-T today, but there is little physical evidence of the barracks or courtyards that occupied the site. Firefighters continued to use the facilities for training and education into the early 1950s, and crewmen were taught telephone line maintenance, the use and care of fire tools, firefighting methods, and lookout protocol during the short courses.[205] Field days and demonstrations were also held at the CCC-constructed pond, and high school students learned about selective cutting and seed tree selection from foresters.[206] But as the cost of maintaining the infrastructure gradually rose, the structures—and eventually the tower—were sold or removed.

University of the Woods

What Is Success?

It's doing your work
With a bright happy smile
And doing the things
That are worth the while.
It's giving your time,
Your money and brawn,
It's forgetting and forgiving
Mistakes that are gone.
It's loving and hoping
for your neighbors good cheer—
It's doing honorable deeds
Every day of the year.
And as the years roll bye [sic]
Just give your best;
And, my friend, I'm sure
You'll find that's success.[207]

The CCC offered enrollees a second chance at education. Many had dropped out of school in an effort to provide additional labor on family farms as they struggled to eke out a marginal existence, while others left to find work in forestry-related industries. For these enrollees, the camps provided defined work schedules and sufficient nutrition for the first time in their lives and presented them with leisure time that could be filled with the academic and vocational opportunities that were offered to them by the camp's educational program. In fact, by June 1934, the *Tyler Journal* was reporting that upward of five thousand CCC members in the 129 Eighth Corps Area camps were seeking a better education. At that time, more than one hundred subjects were being taught by 299 instructors in offerings ranging from music to public speaking, mining to aviation, and mixing dough to handling explosives.[208]

A year later, Major General Hagood announced that of the ten thousand men enrolled in the forty-two Texas camps, seventy-five hundred had enrolled in the educational program.[209] Some enrollees, such as those at Livingston, were able to attend Livingston High School while performing their camp duties,[210] while others at Company 839 in Trinity enrolled at Sam Houston State Teachers College.[211] The TFS supplemented the efforts of the camp educational adviser and, during June 1936, offered courses in conservation, road and bridge construction, mechanics, leadership, woodcraft, forestry surveying, welding, and saw filing to any of the nine hundred men in the eight forestry camps. As the *News* pointed out,

> The principal objectives of the vocational program are to instill in the enrollees the principles of conservation; to develop in the men an appreciation of their work, their civic duties and responsibilities; to offer instruction designed to prepare the boys for definite lines of employment after leaving the camps; to increase the efficiency of groups of enrollees engaged on the various work projects; and finally to meet the needs of those who will, after their discharge from the CCC, resume high school and college work.[212]

These were largely the qualities that defined America's "Greatest Generation," of which Dwight T. Smith Sr. was part. Smith joined the CCC in 1939 and was stationed at Apple Springs, where he served as a lookoutman in the Davy Crockett National Forest. During 1943, he enlisted in the US Army Air Corps and flew thirty-five combat missions over Europe as a gunner and flight engineer. Later, he returned to forestry as a counselor at the Job Corps Center in New Waverly before he transferred back to the USFS center in Apple Springs, where he had started as an enrollee thirty years before.[213]

The *Lufkin Daily News* went as far as to report that the "CCC educational program has been called the greatest experiment in adult education ever carried on."[214] Within the seventeen camps administered by the Lufkin District in 1936, fifteen had educational advisers.[215] Certainly, the "College of the Pines" led many Texans to academic and vocational achievement, teaching a generation the fundamentals of construction, forestry, civics, communication, and collaboration. Many are, in fact, the faltering skills that we, as Americans, struggle to possess in the twenty-first century.

Building the Core Network: 1934

"News and Gossip" reported by the *Lufkin Daily News* indicates that Lufkin Company 838 was nearly finished with the Fern Lake Tower in March 1934.[216] Supervised by construction foreman Joe E. Barnes, the lookout was one of thirteen completed by camp laborers in Angelina, Houston, Trinity, and Shelby Counties by June 15.[217] In other camps throughout the Palestine and Beaumont quadrangles, tower construction and the completion of roadways, bridges, and telephone lines were being prioritized to secure access into isolated districts for the sole purpose of fire prevention and detection.

The speed with which the protection network developed during the first few periods of Depression-era relief programs was striking. Before 1934, the TFS had incrementally added to the lookout network but had managed to secure funding for only nine lookouts and twenty-six tree cabs,[218] despite efforts to lobby the legislature for additional funding and petition industry representatives for expanding cooperatives. While cost-sharing protection alliances between industry and the TFS declined during the Depression, the poor economic conditions unexpectedly benefited the agency with an infusion of ECW expenditures and a special state emergency appropriation that accompanied the "make-work" campaign. With these resources and the "reservoir of manpower" available after 1933, the protection system expanded to forty-four lookouts in 1934, fifty-nine by the end of 1936, and seventy at the close of 1938 (table 4). The advantages of the president's program were not lost on the Forest Service, and Bulletin 25 surmised, "It is conservative to say that many objectives have been fully realized ten years in advance of the normal growth of forestry in Texas."[219]

Table 4. Lookout towers constructed between 1930 and 1939.

1930	1931	1932	1933	1934	1935	1936	1937	1938	1939
Horsense *to Camden*	Conroe		Willow Springs	Alto		Appleby	Lufkin	Humble *to East River*	Moores Grove
				Apple Springs *to Mt. Zion*		Beckville 1	Pinckney 2	Liberty Hill (USFS) *from Shepherd 1*	Mt. Zion *from Apple Springs*
	Etoile								
								Mercy	Nogalus
				Bronson		Cushing 1	Pool *from Farris*		
				Central School		Deadwood	Rocky Hill *to Paxton*	Smart School	Piney Woods (Piney)
				Chireno *to Shepherd 2*		DeKalb	Sheffield's Ferry *to Siecke SF*	Wakefield	Rusk 2 *from Rusk 1*
				Chita 1		Devils	Soda *to Livingston 2*	Weches 2	
				Cleveland *to Kenefick*		Douglassville			
				Cyclone Hill		Fred *to Ariola*			
				Cypress Lake		Hallsville			
				Denning *to San Augustine*		Hooks *to Redwater*			
				Dodge		Jefferson 1			
				Fails		Kildare			
				Farris *to Pool*		Love's Lookout			
				Fern Lake		Mims Chapel			
				Geneva		Mt. Enterprise			
				Horton Hill		Neches			
				Huntington		Negley			
				Jackson Hill		Scottsville *to Karnack 2*			
				Keenan		Snap			
				Kirby SF		Timpson			
				Kountze		Union Hill			

Based on relative dating:

Chambers Hill (1936)
Dreka (1934)
Forse (1937)
Four Notch (1937)
Gum Springs (1936)
Hi Point (1937)
Mt. Vernon (1939)
Neblett (1938)
Tadmor (1939)
Tenaha (1938)
White City (1934)

Table 4. Lookout towers constructed between 1930 and 1939. (continued)

1930	1931	1932	1933	1934	1935	1936	1937	1938	1939
				Livingston 1 to Saratoga 2					
				Maydelle 1 to Maydelle 2					
				Mayflower					
				Moscow 1 to Moscow 2					
				Moss Hill					
				Newton to Karnack 1					
				Onalaska 1 to Onalaska 2					
				Peach Tree Village to Barnum 1					
				Piney Woods (Weldon)					
				Pinckney 1 to ?					
				Rusk 1 to Rusk 2					
				Shady Grove					
				Shepherd 1 to USFS Liberty Hill					
				Smith Ferry					
				Splendora					
				Spurger					
				Tabernacle					
				Votaw					
				Weches 1					
				Willis					
				Woodville					
				Zion Hill					

Cradle of Scientific Forestry in Texas: The Woodville Lookout

Tyler County was at the heart of the Texas lumber industry at the turn of the twentieth century, visited by famous names in forestry such as Gifford Pinchot, Henry Solon Graves, and Herman Chapman. Along with these distinguished guests, Yale University forestry students, including Aldo Leopold, also converged in the county for field camps and the study of southern forest conditions. The excitement of these encounters and the impact they had on American forestry has been described by Dan Utley, who wrote about the "Yalies in the Deep Woods."[220]

The EMSCO tower at Woodville (fig. 28) commemorates the importance of scientific forestry and forest fire protection, goals advocated by these early visionaries. The lookout was built in 1934 by Company 891 enrollees and can still be visited on the grounds of the TFS District Office in Woodville, just north of a Texas Historical Commission marker recognizing the serendipitous meeting between some of the nation's leaders in conservation. The structure maintains an original EMSCO placard on the north wall of the tower, as well as pier inscriptions, and preserves an inscribed toolbox once used to cache firefighting equipment. It is listed on the National Historic Lookout Register because of its ties to early southern forestry, the CCC, and the TFS.

The patrol district around the tower was officially designated a shortleaf protection zone. Patrolman Allen Riley questioned this designation in 1937, writing to his superiors in an attempt to renegotiate his title to one of a "Longleaf Patrolman." White responded to Riley and acknowledged that a small portion of his patrol area was, in fact, longleaf pine. "The greater portion, 75 per cent or better," White continued, was "regular shortleaf" country.[221] The reason for Riley's inquiry was that the higher number of fires and greater acreage burned in the state's eight longleaf patrol districts resulted in larger salaries for patrolmen protecting longleaf forests.

During the late 1940s, a relay building was erected below the lookout, and a "windmill type" radio mast was mounted on top of the tower cab. This addition raised the height of the tower to 160 feet and permitted better communication throughout the district. The system was replaced in 1959, however, when a 300-foot radio tower was installed to improve communication on the outer fringes of the district and eliminate "dead-spots" and "high ridges."[222]

"Operation Sage Brush," a series of military war games, created excitement at the tower during 1955. Upon request from a commander in the 728th Aircraft Control and Warning Squadron, towers at Woodville, Timpson, Soda, and Chita were used by detachments of troops during their training. The TFS established firm limitations on use, requiring that no borings be made into the structure during the installation of military radio equipment. The Forest Service also required that state personnel could continue to access the tower or cabin whenever their duty demanded. Finally, the TFS also insisted that military personnel would not be permitted to enter the cabin and that military occupation would not be allowed to interfere with the "normal movement of Texas Forest Service personnel or vehicles." This was deemed particularly important at the Woodville Tower, as it was a "district headquarters area where there is occasionally a considerable flow of traffic by civilian and Texas Forest Service personnel."[223]

The tower and district office also served TFS training needs, with short courses on fire control and law enforcement. During these events, crews from duty stations throughout District 4 converged in Woodville to receive instruction and focus on community fire prevention problems.

During 1962, towers across East Texas were inspected for signs of wear or deterioration. This resulted in a number of structures being strengthened or replaced. When the Woodville Tower was examined on March 21, thirty joints showed minimum signs of swelling. Following the examination, the director approved repair of the structure, and it was refurbished in 1962.[224]

Today, the tower is in need of maintenance and painting to ensure that it is stabilized for the future. Informational signs and regular interpretive events on the district grounds would increase awareness of the state's fire control legacy and the mission of the Texas A&M Forest Service.

Figure 28. Aerial view of the Woodville Tower and District 4 headquarters in February 1954 (A). Note the contoured agricultural fields and "windmill-type" radio antenna secured to the top of the lookout cab. Similar antennas were located at the Love's Lookout and Siecke SF Towers. Concrete toolboxes constructed by the CCC were filled with firefighting tools (B). An example, albeit without a roof, is preserved below the Woodville Tower today and is complete with CCC-era inscriptions. Images courtesy of Texas A&M Forest Service.

Samuel Gerald summarized the initial wave of tower construction in a July 1935 *Claude News* article.[225] Fifty-two towers, he reported, had already been built in twenty-two counties throughout the Beaumont and Palestine quadrangles. This number reflected the forty-four lookouts built during the first year of the CCC program as well as the nine others that had been constructed between 1926 and 1933. The suspicion that one lookout was accidentally omitted from Gerald's count is partly confirmed by statistics from the 1936 *Texas Almanac and State Industrial Guide*, which included fifty-three towers. Gerald's discrepancy may have arisen because of the rapid evolution of the network and the fact that the early tower system was dynamic. By September 1936, the *Hereford Brand* had reported that it was "desirable to check and recheck [the system] to improve lookout locations or find places where new ones are needed." In an era before elevated surveys were convenient, it was difficult "to tell from the ground what may be seen from a hundred feet in the air."[226]

Other lookouts moved as landowners changed or leases were canceled. This was the case for a number of state-managed towers that were absorbed by the USFS. For a variety of reasons, understanding changes in the network during this period is difficult, and progress can occur only by stitching together information from annual reports of the state forester, observations from maps and notations in surveyor's logs, and firsthand field observations. For instance, the 1936 *Annual Report of the Texas Forest Service* details that eight 99.75-foot lookouts were transferred to the USFS and that one was torn down.[227] Two years earlier, the annual report also indicated that two 87-foot lookouts were transferred to the federal government.[228]

One of the most complete collections of documents concerning this opaque period is safely tucked away at College Station, and it was by chance that it was located among files concerning the Willow Springs Lookout. This tower has always hinted at a complicated past. Before the TFS began using a grid system to locate fires and label towers, lookouts were simply identified by their name and a number. This number coincided with the order in which the towers were erected, and the first tip-off that something was unusual about Willow Springs lay in the fact that it was built quite early (late 1933) but maintained a high identification number (No. 74). That it lay in a corner of the Sam Houston National Forest where there were a lot of early changes in the tower network reinforced suspicions that the structure may have a storied past.

After a decade of assumptions concerning the early Texas National Forests lookout network, and during an effort to chronicle the Willow

Springs Tower history, this incredible bundle of papers was discovered. Informally dubbed the "Willow Springs Collection," the most important five pages describe the fates of nine enigmatic lookout towers after they were acquired by the federal government. The correspondence was initiated by W. E. White in a 1938 letter to L. L. Bishop. In his typically precise style, White asked for an update on the nine lookouts that were transferred to the USFS. The two-page enclosure attached to White's letter described the status, as known by the TFS, for each of the mystery nine. B. B. Whipple replied to the inquiry on USFS stationery, still with a downtown Houston address.[229] The letter's fallout plugs the holes: Farris Lookout was moved several miles south to a sandy hillside near Pool Chapel; Shepherd (Location 1) was transplanted at USFS Liberty Hill; Geneva was torn down by the TFS; Ratcliff was acquired and replaced; Yellowpine, Moss Hill, Apple Springs, Cyclone Hill, Dreka, White City, and Willow Springs were all standing on federal landholdings.[230] Other correspondence detailed complicated landownership changes and short-term lease agreements at Willow Springs. These, no doubt, convinced the USFS to return the Willow Springs Lookout to the TFS in 1943.[231]

Land surveys were required before the tree cabs or towers could be erected, and engineers such as Millard Lawrence, Benjamin D. Hawkins, Ivan H. Jones, Knox B. Ivie, William O. Durham, Calvin F. Ballard, Winston Kirkland, and Isaac C. Burroughs crisscrossed the state identifying suitable locations. Often, they were joined by local patrolmen such as Alfred L. Chatham in Protection Unit 5 south of Huntsville or John M. Turner at State Forest Number 1. On occasion, White would also accompany the parties afield. With compass, plumb bob, and tapes, the surveyors had the difficult job of identifying highpoints with broad viewing areas suitable for lookout construction during an era when detailed maps were scarce. Once potential sites were selected, the teams pivoted, acting as land agents to determine ownership and negotiate leases.

Landowners displayed a range of sentiments when approached about potential agreements, especially in time. The leases were short documents, written in favor of the TFS, which granted the state "the possession, occupancy and use" of each property. The Weches Lookout lease, signed in early September 1933, was typical in that one square acre of land was acquired from Mr. and Mrs. Stanton for the sum of one dollar and remained binding to the heirs or assigns of the lessor for the time that a tower was located on the premises and "used for fire protection purposes." When Lufkin enrollees from CCC Camp P-57 began construction at the site, they were free to "enclose the foregoing described tract of land, or any part of same, with a good and substantial fence, to remove therefrom and to use any standing or fallen timber or other growth thereon, to place thereon watch towers and lookout towers and other property improvements, and, generally, so to possess, occupy and use said land as they state may see fit, as tending to the prevention and extinction of forest fires, and not otherwise."[232]

Some leases were harder won than others, like the one paid to the Langstons for the pole cab on Walker Mountain. Others came more easily, especially those negotiated with larger timber companies requesting cooperative fire protectorates. Nonetheless, the leases were signed as landowners acknowledged the benefits of the detection system, even if there was a tinge of sarcasm or a sense of humor in the process. Such was the case in 1961, when Director Folweiler wrote to the Bryans thanking them for their "interest in forest conservation" and their "public spirited attitude" in making the Cleveland Lookout site available to the TFS through signature of an April 1934 lease. Because the tower had been dismantled, Folweiler explained, the agreement was "automatically cancelled" and the land reverted back to their ownership. In reply, Eunice Bryan recalled:

> When Judge Love, of Cleveland, now deceased, came to us with almost "tears in his eyes," and said: "Charlie, nobody will let us have or use a small location on their land for a Fire Lookout Tower, Can you help us out?" I told him we were "big hearted," and besides, considered the Tower would be an asset to us and the Community, and that we would let them have use of an acre. Since this one acre has served your purpose, and you no longer require its use for a Tower, and have abandoned it, we take this method of confirming our original judgement that the tower has served as asset not only for our land, but for the whole surrounding territory, especially Kirby Lumber Co., who could not, and would not permit you to use a small tract out of their enormous holdings for location of a Tower.[233]

Extending the Core Network: 1936

At 33 degrees north latitude, and as the only tower in Red River County, the Negley Lookout feels geographically isolated from the rest of the state's once expansive fire detection network. In fact, the tower is so close to Texas' northern Red River border that a 1937 *Clarksville Times* article commented that from the top of the attraction, lights in Idabel, Oklahoma, were visible at night, and "when the atmosphere is clear," the Ouachita Mountains were discernible on the horizon.[234]

Finding time to visit the tower took years and necessitated a five-hour drive along a route that led me through Smith, Wood, and Franklin Counties. Passing through Bogata, the site of a Soil Conservation Camp, I circled around Clarksville and turned north before finally pulling the car off along the side of Highway 37 in Negley. Examining the tower from the roadside, I could only wonder what Bill Leatherwood, the patrolman stationed at the lookout for fourteen years between 1936 and 1950, would say during this inspection trip. The window lights were all missing from his office cubicle, and the floorboards below his vantage point had rotted away to reveal the faded clapboard interior of the walls and ceiling. On the ground, in the space once maintained below the tower, fifty years of vegetative growth concealed the stairway risers.

Initially, the TFS focused protective efforts in the heart of the pinelands, erecting forty-four lookouts in the Beaumont and Palestine quadrangles during 1934. "The need for fire protection," in twelve Northeast Texas counties,[235] which encompassed 3.5 million additional acres of hardwood, shortleaf, and loblolly pine, had "long been apparent to the Texas Forest Service," the *News* reported. "But," the notice continued, "just recently have funds been available to carry on this work."[236] With recent state appropriations and aid from the federal government, Negley became one of sixteen improvements constructed in 1936 during a campaign to expand fire protection into the Tyler and Texarkana quadrangles.

Notes from the surveyors planning the extension appear in a Division of Forest Protection field notebook, preceded by an introductory page on which is written, "The following pages contain location notes on tower locations made for northern extension, 1935, by William O. Durham, C. F. Ballard & Winston Kirkland—Locations numbered as made."[237] The Negley site appears thereafter, surveyed on March 25, 1936, by W. O. K (Kirkland). The property was owned by the Paris Grocer Company and was under the management of R. C. Lane, who wrote to White on March 27, saying, "The lease could not be completed yesterday as our president, who handles such matters, was not here. We are very glad to cooperate with you in this work and are enclosing three copies of the lease, which is in accordance with instructions from Mr. Kirkland."[238]

White replied three days later, thanking Lane and the president, Mr. Milling, for their consideration, trusting that the Forest Service's work would "be of benefit to you and your community." White then signed the letter as the "General Supervisor ECW."[239] Curiously, two other tower sites were also investigated in Red River County, one at White Rock and a second at Bagwell.[240] The White Rock site was situated southeast of Negley and would have offered a strong tie-in with the DeKalb Lookout in Bowie County. Apparently neither was built, and the Negley Tower was left standing by itself in the central portion of Red River County.

By April, the *News* mentioned that construction had already begun on lookouts in the northern district and was "progressing as rapidly as possible" under the supervision of division patrolman Murray E. Brashears, who oversaw progress from his headquarters in Marshall. To support the growing network, Brashears supervised eighteen forest patrolmen, one of whom was Bill Leatherwood, and sixteen other emergency patrolmen.[241] Brashears worked in a number of districts with the TFS at Maydelle, Rusk, and Marshall but left after five years to serve as Louisiana's state forester from 1940 to 1942. Many of the changes he implemented in Louisiana, such as mobile firefighting crews and the inauguration of short-wave radios, paralleled TFS programs.

When fires occurred in Red River County at the start of the ECW program, crews were dispatched from the CCC camp in Bonham, approximately 75 miles away.[242] News that the TFS had created a protection unit overseen by W. J. Leatherwood near Clarksville must have been met with satisfaction.[243] Following the construction of the tower, African American enrollees from the recently established Soil Conservation Service camp at Bogata erected a private telephone line between the lookouts at Negley and DeKalb, improving area capabilities to report and triangulate fires. When subsequent fires occurred, Company 2889 (C) enrollees were also available to assist Leatherwood, as they did a year later in suppressing a 135-acre fire.[244]

Other towers built as part of the campaign, such as those at Mims Chapel, Mt. Enterprise, Love's Lookout, and Deadwood, are still standing at their historic locations. Reconnaissance notes written for Deadwood, "Tower Location No. 13," identified a favorable location on land owned by the Werner Sawmill Company. The lease, signed during October 1935, left open the possibility that the structure would be of wood or steel construction. By November 1936, however, a steel Aermotor MC-39 had been built and connected by telephone to another lookout at Beckville with help from laborers at Soil Conservation camps in Northeast Texas.[245] In 1981 the Forest Service decided to maintain the lookout for communication purposes. The station is still owned by the state and stands in good condition, though it is leased by the American Tower Corporation. As a Depression-era fire control facility that maintains integrity of location and design, it qualifies for inclusion on the National Register.

Survey descriptions for the Mims Chapel station, meanwhile, made special note that "Tower Location 15" was about 9 miles west of Lassater in Marion County "near what is shown on map as Ero." The description continued, however, that "natives don't know Ero." Nevertheless, Ballard and Kirkland still decided to call the tower "Ero," and the name persisted on inventories until 1942, when Burnside wrote to White advising him that the name Ero, tower Number 55, should be changed to "Mimms Chapel" based on recommendations from Don Young.[246]

In his note, Burnside acknowledged that there was "no community near the tower by the name of Ero" and continued that "everyone knows the community in which the tower is located as Mimms Chapel, as there is a church by that name nearby." Director White approved the change on December 27.[247] Afterward, Don Young sent a letter of clarification to Burnside on January 6, 1943. "Mims," he said, is "used to refer to the church and school in that locality," and he advised that the spelling be revised to respect the place-name.[248]

Budgetary letters and purchase orders are exceptionally useful for inventorying the initial surge of towers constructed during the Depression (table 5). Tallying the ECW-era purchases in table 5 with a summary of towers assembled between 1926 and 1933 results in a cumulative list of eighty-eight structures that were erected up to 1941 (table 6). When reviewed alongside early TFS property inventories, it becomes clear that fifteen lookouts are missing from the state roster. The comparison is beneficial in determining the number and model of towers that may have

Table 5. An inventory of Depression-era lookout towers purchased between 1933 and 1941 has been compiled from budgetary letters and purchase orders. The difference between the number purchased and the number in TFS ownership provides an estimate of the quantity and model of towers transferred to the USFS.

Purchased	Manufacturer and model	No. in TFS ownership	No. missing
17	Aermotor MC-40 Platform ladder	13	4
27	Aermotor MC-39 Platform stairway	22	5
31	EMSCO TW External ladderway	27	4
3	Creosoted wood	3	0
Total			
78		65	13

Table 6. Steel lookout towers erected prior to the beginning of the ECW program using state and cooperative fire protection funds. The grand totals are a cumulative measure of towers from tables 5 and 6.

Number	Manufacturer and model	No. in TFS ownership	No. missing
1	Aermotor LS-40	1	0
9	IDECO OL	7	2
Grand total			
88		73	15

Table 7. Inventory of lookout towers that may have been relocated or reallocated during the early growth of the fixed point detection system: (A) Aermotor, (E) EMSCO, (IDECO) International Derrick and Equipment Company, and (U) Unknown. Towers with internal stairways are indicated by (IS), (IL) indicates models with inside ladders, and (OL) indicates outside ladders. US Forest Service drawing numbers, such as L-1000, are used when the bid winner and tower manufacturer are unknown.

	Tower name	Model	Notes
1	Chambers Hill	U	
2	Cyclone Hill	E-OL	TFS; ownership transferred to USFS
3	Devils	IDECO-IS	
4	Dreka (1)	L-1000-OL	TFS; ownership transferred to USFS
5	Forse Mt.	U	No surface expression of tower
6	Four Notch	L-1000-IS	
7	Geneva	U	Remained with TFS, relocated to Forse Mt. (?)
8	Gum Springs	U	
9	Hi Point	E-OL	Moved from White City (?)
10	Jackson Hill	L-1000-IS	
11	Moss Hill (Zavalla)	A-IL	TFS; ownership transferred to USFS
12	Mt. Zion	L-1000-IL	TFS Apple Springs; ownership transferred to USFS and relocated
13	Neblett	U-IS	
14	Pinckney (1)	U	Remains with TFS, possibly relocated
15	Pool	E-OL	TFS Farris; ownership transferred to USFS and relocated
16	Shepherd (1)	E-OL	TFS; ownership transferred to USFS and relocated to Liberty Hill
17	White City	U	TFS; ownership transferred to USFS

been allocated to the federal government and also offers an indication of those that may have been surplus for transfers within the state network.

Of the fifteen missing lookouts, one 99.75-foot lookout was taken down by 1936, and eight others were transferred to the USFS. Possibly, these included the four missing EMSCO towers at Hi Point, Farris (Pool), Shepherd (Location 1), and Cyclone Hill. Between 1937 and 1938, an additional two 99.75-foot towers were torn down. These were most likely re-erected, as was the case with the Apple Springs Lookout, which was transferred to Mt. Zion. By compiling information from fieldwork, early National Forest Service maps, and the details within the Willow Springs Collection, an inventory of uncounted lookouts can be compiled (table 7).

Firefighting

The curriculum at the University of the Woods included courses in firefighting. Sometimes the instruction was given by representatives of the TFS at the various camps, while at other times, it was provided by hands-on experience. From available records, it seems that classroom demonstrations were less impactful than time in the field, and enrollees participating in four hours' worth of drills at the Livingston Camp displayed poor "conduct and inattention" during one demonstration.[249]

The excitement, and fear, of being called to an actual fire left more lasting impressions. Fortunately, accounts of the experience are preserved in the diaries and oral histories of some enrollees, like those written by Connie Ford McCann:

> Tues. July 11, 1933: . . . Went to fight our first forest fire—22 in whole bunch. The truck—Model A—had poor brakes—the road was one of the most terrible I ever encountered—rough—crooked—etc. We started up a steep incline—the engine died. We started rolling backwards toward the creek—the brakes wouldn't hold so the driver threw it into compound low gear to stop—which it did. And so suddenly that about 10 of us in the back of the truck pitched out on the ground—along with two barrels of ice-water which poured all over us—spilled every drop—then the tools—axes—rakes + shovels also flopped out on the ground—everyone fell down—even those who stayed in the truck—we walked up the hill.
>
> We couldn't get to the fire in the trucks—had to walk about 2 mi + carry the tools.
>
> Our gang including two other fellows from another gang went north around the fire the rest—about 12 or 14—went south. We went along the edge—which—along here—was bounded by a branch—in several places we had to remove—or chop in two—logs + branches which crossed the creek (and could burning making it easy for the fire to jump to the other side)—into big timber. It had already burnt over about 2 or 3 square miles—second growth trees—acre after acre literally destroyed. When the wind blew hard, the fire going through these young oaks sounded like some great monster crashing through the forest—In fact it really is a monster—one that I care not to cross.
>
> We eventually came to a turn in the creek—here the fire cut across a little to the east—we whipped it out as best we could—our throats were parched—one could drink water—(we had our quart canteens)—and still his throat was as dry as toasted peanuts. Now the wind had changed + was blowing the fire towards us—the smoke was stifling we could hardly breath[e]—or see—I would open my eyes to where to step—then step with them closed—they

were watering so that I didn't know whether my face was perspiring or was it really water from my eyes!

Somewhere along the line we met a forest ranger—whom we stayed with until we left for camp. For my part I was mighty glad to see him—I figured that he would really know his so + so about this (fire is crossed out and replaced with) smoke eating business—which he did.

We had whipped it out and for a distance of about 1 ½ or 2 mi when it broke out behind us—before we discovered it—the fire had spread all along the edge we had just whipped out. This time we cut a fire line + back-fired the whole thing—and whipped it out along the edge. It was getting dark and we were all so tired + smoked up—it was decided that we get hold of the other fellows—whom we heard yelling across the fire—about a half mile at this place—and go back to where we had left the trucks to see about getting something to eat—were we hungry? O boy! The trucks—so far as we knew was supposed to have gone back to camp for eats + blankets—we were expecting to stay out all night. We all got together finally—and walked back to where we had left the trucks—the fire being well under control now—The other gang having cut it off on their side. It was about 4 or 5 mi to our destination—when we got there—a whole truck load of relief men from camp had arrived—and were we glad? I'm telling you. We all piled on the truck and beat it back to camp—it was 28 miles—we came back in the Dodge—army trucks—came through Yellowpine + Hemphill—this latter was a nice little town—county seat. We arrived in camp at 10 o'clock—there was awaiting us a hot meal—cake—water-melon—fruit and I ask you—did we eat?[250]

The *Eighteenth and Nineteenth Annual Reports* published in 1934 indicate that early CCC firefighters, perhaps enrollees like Connie Ford McCann, were often a hindrance to the regular TFS field force. The men had "no concept of a forest fire," the report lamented, nor did the enrollees know how to extinguish one with the tools at hand.[251] Despite the early pessimism, CCC fire details were working efficiently by the end of 1933. This was certainly the case when Lufkin Camp enrollees "had the opportunity to participate in fighting a forest fire." The call came in to headquarters, the *Lufkin Daily News* published, "and in less than five minutes after the report was made the fire-fighting detail was on the way to Davisville."[252] The 5-acre fire was on Cameron Lumber Company property, and within three hours it had been successfully extinguished.

Planting

It's an interesting experiment to harvest seeds from below the scales of a pine cone, dry and plant them, then watch as they germinate. Initially, the light green seedlings are thin, but the delicate branches look decidedly like pine. It's a trick for someone without a green thumb, but with patience and a bit of practice it is possible to nurture a handful of woody-stemmed, branching yearlings that are strong enough to transplant into the ground. If the loblollies planted in my side yard are representative—and I think they are—within ten years' time the juvenile trees can reach double-digit heights. They are ideal for reforesting barren landscapes, being thinned for pulpwood fifteen years hence, or harvested for saw logs forty years in the future.[253]

The awakening conservation consciousness brought about by visionaries such as Roosevelt and Jones created tangible, albeit small, changes in the Piney Woods. Some companies, like the Angelina County Lumber Company and the Delta Land and Timber Company, began cooperating with the TFS in reforestation experiments. Angelina County Lumber Company tried first in 1929 and 1930.[254] That same year, the Delta Land and Timber Company planted 4 acres with alternate rows of longleaf and slash pines, as well as another acre in loblolly, across the millpond from the whining saws of the company's Conroe operations.[255] The seedlings were produced by the state's tree nurseries, which began operating first at State Forest Number 1 (Siecke Forest) in 1926 and State Forest Number 2 (Jones Forest) in 1928. By 1929, the nursery at State Forest Number 1 was capable of producing three hundred thousand seedlings annually.[256] While a seemingly large number, the output fell short of the supply necessary for reforesting the Piney Woods.

Fortunately, Roosevelt's "Tree Army" offered assistance in just a few years' time with overwhelming production and planting programs. CCC boys were in the woods, gathering pine cones and shipping them to the Stuart Nursery on the Kisatchie National Forest near Alexandria, Louisiana. There, the seeds were extracted and planted, and the bumper crop resulted in the production of 48 million seedlings between December 1, 1935, and May 1, 1936.[257] In Texas alone that year, 3.65 million young trees were planted in the four national forests. The following season, the Southern Region of the USFS produced 57 million seedlings from December 1936 to March 1937. Again, TNFs benefited by receiving over 10 million seedlings,[258] keeping crews of "tree troopers" at the Turpentine, Nancy, Milam, Pineland, Bannister, Apple Springs, Jasper, and Ratcliff Camps busy planting.[259] Alone, the Turpentine Camp planted over 3.2 million seedlings. The 1938–39 program saw additional gains, and the Turpentine and Nancy Camps were out again with "swinging trays and planting bars" to start an additional 6 million trees.[260]

Through most of the years of the Depression, trees came from federal nurseries in other parts of the Southern Region at the Stuart Nursery in Louisiana, the Ashe Nursery in Mississippi, or the Ozark Nursery in Arkansas. These harvests were supplemented by comparatively small yields produced at other nurseries administered by the TFS. But beginning in 1940, CCC contributions accelerated East Texas' local reforestation efforts when two hundred enrollees and twenty-one staff constructed seventeen buildings at the Indian Mounds Nursery in Alto. During the camp's first season, a "tiny but dense forest" of 1.2 million 6-inch-tall pine seedlings was produced locally, and plans called for incremental increases in production from 6 million to 10 million pines in several years' time.[261] Many of the seedlings were sold to farmers for reforestation and soil conservation on private holdings, on a cost basis of $2.50 per thousand.

The breaking wave had finally washed ashore. "The pines now growing in the Alto nursery," announced the *Alto Herald*, "will not be marketable for at least ten years but the government's forest program is long range, designed to protect future generations against soil and timber wastage."[262] Sterile lands denuded by clearcuts and fire were now being given a second chance and were finally being brought back from their die-off point. A pivot from the cut-and-get-out attitude of the bonanza era had occurred, and Texans were finally reforesting with a vision of the future. To be successful, though, the young growth had to be protected from foragers and fire. Now, more than ever, lookout structures and staff would be instrumental as the young plantations grew. Jointly, the two became the symbols and public face of the reforestation effort, the guardians of the replanted Piney Woods.

Mapping: Visibility Surveys, Geography, and Forest Types

Arguably, some of the most important contributions performed by CCC forestry workers were scientific in nature. Accurate maps, of which there were few, were required to make the fixed point detection system successful. As late as 1932, Bulletin 23 reported that "the Division [of Forest Protection] has been severely handicapped in securing adequate maps."[263] Without knowing the distances and elevations between points, it was impossible to triangulate or judge the extent of blind spots. Without reliable maps, it was also impossible to determine the azimuth of an incipient blaze or plot the most direct route to a fire. Initially, mapping tasks were performed by USC&GS personnel or forest engineers like Winston Kirkland. He was quartered at the Conroe Camp during October 1934, "mapping roads, telephone lines, and other improvements finished by enrollees of this camp."[264] A year later, Kirkland was north in Timpson "to make a survey of roads in this section, particularly those affording access and connections between a series of steel towers to be erected in this section,"[265] while Calvin F. Ballard worked near Woodville in Tyler County. But it was difficult to keep pace with the construction activity at the various camps, the *Montgomery County News* acknowledged while Kirkland was working in Conroe, and "work in this line is far behind the construction efforts of the CCC."[266] Perhaps most significantly, accurate maps were also necessary for recording data, such as the size of a wildfire, the type and condition of soils and outcropping rock exposures, land-use attributes, or "type mapping" the age and variety of trees.

To accelerate the work, engineering classes began to appear at CCC camps throughout the Piney Woods as the educational program evolved. One such tutorial began in February 1935 at the Conroe Camp when mechanical drawing and forestry courses were offered.[267] Another appeared at the Jacksonville Soil Conservation Camp during 1936 and included instruction to fifteen men who assisted the camp engineer. The curriculum began with a "month's work of elementary mathematics in order to give the students a sound background for other subjects" and met twice weekly. It also included homework meant to fill two additional evenings each week. Beyond teaching surveyor's math, "other subjects touched in this class are drawing, mapping, care and use of survey instruments."[268] Bill Hartman would recall in 1998 that he taught night classes in math and drafting twice a week at the Lufkin, Alto, and Houston Camps after 1939.[269]

Foster would have praised the effort, having struggled to accomplish the same vision without adequate financial and human resources. "Underlying the whole survey," the *News* explained in 1940,

> has been the preparation of an accurate base map locating as key points the seventy lookout towers maintained by the Texas Forest Service in East Texas. From these towers main highways have been located and mapped in. So accurate has this work needed to be done that an error greater than one foot in two thousand was not permitted. Secondary roads were next mapped. This completed the

work on the key base sheet. The network of roads thus served for locating starting points for the type mapping work.[270]

Triangulation was key in this effort, either from USC&GS towers like the one at Bannister or from fire lookouts. Johnny Columbus and Columbus Lacey remembered the Bannister Tower during an interview with the USFS in 1996: "They had a tower here, they called it the Geological Survey, the Triangle Survey, they had this tower here which they got the readings from all the other towers at night, by night."[271] Photographs in the TFS archives capture these efforts, albeit from slightly later during the 1940s. Characteristically, the Hooks Lookout appears in one such image taken in October 1943, with a Bilby surveying tower surrounded by a wood-framed work platform adjacent to it (fig. 29).

Outside the Chireno Lookout tower, it was clear and windy. Calvin Ballard and an assistant had spent the day shuttling along dusty roads in the district, rushing to stage equipment and carry electric lights up the narrow ladders to the tops of several lookouts. Now it was almost eight in the evening, and the pair sat in tired silence listening to wind shrill past the guy wires attached to the tower. After a hot and intensive summer field season, the passage of this early October front teased at the colder months ahead. Silently on the southern horizon, lights started to flash at increments as the automatic timers they had positioned earlier in the day began to illuminate neighboring towers.

Sighting the distant lights—first at Jackson Hill and Etoile, then at Etoile and Mitchell—they worked their transit and recorded an initial reading on the gridded pages of a stiff fabric field book. Meticulously, they repeated the process eleven more times, working without a break until nearly ten o'clock when the lights finally blinked out. Tonight, however, the job was doubly difficult because of the wind, compelling them to record in their notes that the initial readings and check shots were more accurate. "The other readings," they reported, "were taken with bad wind."[272] Later, in more comfortable surroundings, Ballard would sum and average the observations to compute the azimuths and distances between lookouts with the utmost precision to obtain a result with a variation of less than five minutes.

Figure 29. The Hooks Lookout stands adjacent to a Bilby surveying tower with inner and outer structures in 1943. Image courtesy of Texas A&M Forest Service.

As Ballard triangulated tower positions that October, he also unknowingly left the earliest written record revealing that Jackson Hill had been erected, most likely in 1934 by laborers working on state-managed CCC projects. Later, with the establishment of Texas National Forests, Jackson Hill was probably transferred to the USFS. It became a primary lookout and substantial guard station on the Angelina National Forest and was staffed until the 1970s, whenever the danger rating rose above Class III. Today, the site is overgrown and the tower Ballard once sighted has been removed. The government's hasty abandonment inadvertently preserved the cinder-block shell of two generations of relay buildings and the tower piers, however.

For Siecke, White, and other members of the technical service, such as D. A. Anderson, D. Young, and A. T. Chalk, the progress in mapping forest types must have been met with satisfaction as the comprehensive survey progressed and white space disappeared. The effort was initiated at the Maydelle and Trinity Camps, where ten trained crews consisting of a "compassman, rear chainman, and tallyman" cruised a three-mile-long strip of land daily.[273] These strips began from base lines triangulated from USC&GS triangulation stations and were supplemented by secondary lines created at one-quarter-mile intervals that increasingly covered more East Texas counties as the Woodville, Livingston, Jasper, Huntsville, and Humble Camps joined the effort.[274]

Observations made by the "tree troopers" contributed materially to comprehensive surveys published by James Cruikshank and James Cruikshank and I. F. Eldredge.[275] By 1940, the work had reached its halfway point and 5.25 million acres had been mapped.[276] During 1941, at nearly the end of CCC era, mapping two-thirds of the Piney Woods had been completed.[277]

The Lookout Yard: Forest Infrastructure and Forest Improvements

Tangible pieces of ECW projects have become part of our national heritage. Scenic roadways, rustic structures, imaginative design features, and deliberate choices made to enhance the relationship of buildings with the landscape are all commemorated today.[278] Some of the most recognizable local examples include the cabins at Bastrop State Park and the refectories at Caddo Lake and Palmetto State Parks. These structures conform beautifully with the landscape through the selection of organic building materials and the use of low horizontal lines that make the structure appear to grow from the ground.[279] Scenic roadways that wind around or connect parklands are also celebrated, like those uniting Bastrop and Buescher or encircling Tyler State Park. Even the masonry associated with culverts, dry stone walls, and dams are honored for their craftsmanship.

But there are other, less appreciated examples of ECW architecture. The day-use pavilions at Ratcliff and Double Lakes offer local examples, as does the strategic placement of rocks in the spillway at Boykin Springs that act to create rich audiovisual elements for the viewer. Other structures emerged on Texas state forests through CCC labor, including recreation and meeting halls, outbuildings, and cottages for the "use of the lookoutman who is also the caretaker for this [No. 4] forest."[280] Improvements here, and in the national forests, seem to have received the least amount of attention. Completely overlooked are the cultural attributes of the lookout and the lookout yard.

Even the first *Cultural Resource Overview of the National Forests in Texas* (1983), which was meant to be a summary of "the current level of knowledge of the archeological and *historical* resources unique to this part of Texas," makes no reference to the importance of CCC or ECW architecture, despite the inclusion of other historic features like cemeteries and Aldridge Mill.[281]

Mark DeLeon's contemporary report, *Cultural Resources Overview of National Forests in Mississippi*, recognized the significance of such construction but noted that until recently, "historical-era resources have been overlooked." DeLeon provides a credible definition of cultural resources as any "sites, artifacts, structures, or features which remain from human activity in the past. In Mississippi, a site older than 50 years or a standing structure built prior to World War II is considered a cultural resource." He suggested that these features are significant because they are associated with events that made a contribution to the broad patterns of the nation's history and allow for the interpretation and reconstruction of past lifeways, which provide society with a sense of perspective. Importantly, DeLeon wrote that "these cultural resources are both fragile and non-renewable. They are easily destroyed by the activities of vandals and amateur collectors, by legitimate construction work, and by scientific investigation."[282] Ironically, they can also go unrecognized or be destroyed by the organizations meant to preserve them. As historian Jeffrey Owens marveled, "Someday we will wonder where all the . . . CCC projects . . . went."[283]

These failures appear to be a consequence of the bias for identifying and working with prehistoric sites, coupled with the unrecognized significance of American or Depression-era cultural elements. To be fair, some New Deal cabins and warehouses were modeled after approved, standardized designs and were not viewed as having distinctive characteristics.

Specifically discussing the historical significance of fire towers, Mark Spence pointed out that South Dakota's Rankin Ridge Lookout is not "stylistically or materially distinct in any meaningful way" from others in the region and concluded that "industrial prefabrication of interchangeable steel parts" meant that the lookout is no different from any number of structures emphasizing fire suppression.[284] Likewise, Richard Malouf recognized that the towers "remaining on the Mark Twain National Forest are essentially all the same. All are steel, 100-feet tall, with 7' x 7' cabs."[285] Yet Spence recommended Rankin Ridge's inclusion on the National Historic Lookout Register, while Malouf's opinion was that Mark Twain National Forest lookouts were potentially eligible for inclusion in the National Register.

Unfortunately for Texans, progressive foresters and archaeologists elsewhere in the United States were already taking steps to identify, inventory, and preserve similar cultural resources while those in the Lone Star State did not evaluate these resources. Eastern regional forester Larry Henson appointed a Working Group on Cultural Resource Management in Region 9 during 1984 and recommended a "thematic evaluation of Forest Service administrative sites, fire lookout towers and CCC camps."[286] By 1989, the Southwestern Region had also completed a Cultural Resources Management study of lookouts in Arizona and New Mexico.[287]

Without proper acknowledgment, many CCC-era structures fell because of time, vandalism, or disregard. Texas' guard stations, the lookout towers, and the buildings meant to support them have nearly disappeared.[288] Period photographs and site investigations indicate that these compounds were elaborate, built with style, function, and purpose. In the Southern Region, this was the result of Regional Forester Joseph Kircher's effort to retain qualified observers by combining salaried fire protection duties with the opportunity for subsistence farming at the lookout site.[289] In this scheme, tower staff demonstrating a familiarity with livestock and agriculture, and who had a family, could utilize small houses, outbuildings, and fields close to the fire tower when not involved in presuppression or suppression activities. Many tower sites, especially those on the national and state forests, were built with dwellings, barns, chicken coops, warehouses, or garages for the use of the observers.

John Huffman's experience with the TFS provides insights into the expectations for an agent at the state forests. Huffman began working for the state as an emergency patrolman in 1930 but advanced to become the lookoutman at State Forest Number 1 by the autumn of 1937. He received an annual salary of $700 to protect state forest and private holdings

Figure 30. The Kirby State Forest (State Forest Number 4) dwelling and Aermotor MC-40 in 1936 two years after construction by Woodville Camp enrollees (A). An original Aermotor placard on the corner post is visible today (B). Looking west at the tower in 2016, the structure overlooks the forest donated to the state by John Kirby under the stipulation that it be maintained for research and as a source of revenue for college loans (C). Image A courtesy of Texas A&M Forest Service. Photographs B and C by the author.

within his patrol area but was allowed to live with his family in a residence on the forest. The home was provided with electricity and running water, and he was permitted to tend the garden near his residence at times when it did not interfere with his official duties.

Similarly, enrollees from the Woodville Camp working within State Forest Number 4 (Kirby) constructed a cottage for the forest's caretaker and lookout, erected an Aermotor MC-40, built a toolbox, and completed a warehouse (fig. 30). They also built the fire lanes that dissect the property. Photographs demonstrate that the dwelling stood in the lookout yard through at least the mid-1960s, but maintenance costs and liability must have increasingly made the state-owned accommodations a burden. It is unclear from available records when it was removed.

The Kirby Forest Tower still stands today and fulfills several criteria of age, significance, and integrity of location, even though the cottage for the observer has been removed. The original tower, complete with an Aermotor placard, sits in a small yard surrounded by the forest donated to the state by one of the most "flamboyant, empire-building" lumbermen in Texas.[290] Kirby's gift signifies a utilitarian conservation philosophy, whereby proper management could result in profit and regenerative forest growth. The structure is also significant because of its association with Depression-era conservation projects and is one of the few existing examples of an Aermotor MC-40 platform-ladder design known to the author.

Likewise, Alfred Chatham's career provides a good example of how USFS lookouts lived and worked. Chatham was employed with the TFS as a horse-mounted patrolman during the mid-1920s, before he transitioned to seasons as an observer at the Bath and Farris Towers. With the expansion of national forests in Texas, Chatham resigned from the TFS in 1937 to begin a lifelong career as the lookout at the Four Notch Guard Station. His wife, Gladys Lee, was also a long-term Forest Service employee, and the pair worked together as the lookout and dispatcher at the complex. The station, now a hunter's camp and trailhead for the Lone Star Trail, was once an extensive compound with the lookout tower, a dwelling, large warehouse, garage, chicken coop, and pump house.

Chatham helped build the residence. To the north lay the tower (fig. 31), and beyond, along a commonly surveyed line, the warehouse sheltered patrol trucks, a fire plow, and the firefighting equipment for dozens of men. Farther north, a fenced field protected neatly plowed, east-to-west-oriented rows planted and cultivated for the family's produce. Two hundred feet northeast of the tower was a much smaller combination garage and woodshed. West of the lookout was the pump

Figure 31. The Four Notch Guard Station was photographed by Bluford Muir shortly after it was constructed in August 1938 (A). The design of the four-room dwelling, plan B-2300, was duplicated across the US Forest Service's Region 8. Other examples are recognized in the Angelina National Forest and Mt. Magazine in the Ouachita National Forest, Arkansas (B). Image A courtesy of the National Archives. Photograph B by the author.

house, weather station, and chicken coop. Kircher's symbiotic model provided a place for Chatham's family to live and grow, while the USFS retained a dependable lookout and dispatcher with lifelong knowledge of the district. Chatham and his crew knew the territory so well that one memorandum noted, "Most of our men have been around long enough so Chatham can tell them exactly where the fire is and they can go there without looking at the map."[291]

There were other similar sites across the forest, the designs of which were prepared to blend the stations into the surroundings with cost in mind. Chatham's home, a standard Forest Service design, was duplicated elsewhere in Texas and across Region 8 (see fig. 31). In fact, a standing example appears intact within the lookout yard at Mt. Magazine in the Ouachita National Forest in Arkansas.

Texas National Forests were being developed at nearly the same time that the US Department of Agriculture released a collection of *Acceptable Plans, Forest Service Administrative Buildings* in 1938. While use of the plans was not compulsory, consulting architect W. Ellis Groben highlighted useful design characteristics that possessed a Forest Service identity and could be harmoniously adapted to "local characteristics and natural environments of the various Regional subdivisions."[292] The collection included DeFord Smith's drawings for the Mt. Magazine Lodge, a design that he had considered to be one of his most notable projects. Smith, a graduate of the University of Pennsylvania, was hired as the regional architect for Region 8 in 1934 and spent the Depression designing picnic shelters, residences, offices, bunkhouses, lodges, and offices for the USFS.[293] In 1935, he completed forty-three designs. During the next four years, he would design no fewer than sixty structures per year.

The publication of the Forest Service's *Acceptable Plans* drew from successful styles across the organization and was meant to provide comprehensive planning that discouraged "promiscuously located" groups of buildings that were constructed without adequate coordination.[294] Purposeful design planning provided many advantages, Groben pointed out, including access to utilities and easy ingress and egress. It also allowed for surveillance of the compound and the separation of public and service structures to avoid interference or interruption. Organized designs also considered landscape studies and afforded better harmony with the building and the natural landscape.

"Once the comprehensive scheme is established," Groben advised, "then the Project may be considered from the standpoint of planning the respective buildings to fulfill their particular requirements."[295] Progressive, scientific planning was preferred for each building to eliminate dark hallways and avoid structures that created rooms that would be used as passageways. This was a common mistake, Groben pointed out, when a living or dining room became the hallway between the entrance and the kitchen.

The site's orientation was also important to the comprehensive plan and was "not necessarily confined solely to the building itself" but also to "the arrangement of the various rooms." The amount of sunshine, daytime temperatures, prevailing winds, and seasonality all had to be considered. In southern climates, Groben recommended that "where the heat during the day is excessive and where there is little or no cold

Figure 32. Groben's clever use of an azimuth circle combines environmental information with recommendations for various living and working spaces. Image from *Acceptable Plans, Forest Service Administrative Buildings.*

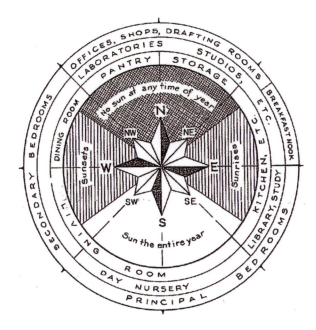

weather, shade is depended upon as a surcease from the rays of the sun. Here the windows are frequently made smaller to exclude the heat, and bedrooms are located to receive the cool, night breezes to facilitate restful sleep."[296]

Groben ingeniously summarizes his descriptions with a single illustration (fig. 32). He drew a compass rose broken into quadrangles through the use of cardinal directions. A series of concentric rings radiate outward from a central star. Into the innermost ring, he variously shaded the orientations that received full sun, no sun, or partial sun during the morning and evening. Outside this ring he then drew a series of progressively larger circles that suggest the various living and working spaces best suited for each of the orientations. It is clear from mapping several lookout compounds that these utilitarian recommendations influenced Texas' guard stations.

Notable among them is the Pool Lookout site. Initially, the tower was located several miles north at Farris, but it was moved by enrollees from the Richards Camp. The *Navasota Daily Examiner* published an update from the camp during March 1937, bringing context to the enrollee experience and announcing completion of the tower. The camp was "in full sway," author and TFS forester Knox Ivie wrote, after a period of wet weather and "other disappointments which naturally arise from time to time."[297] When the camp was established in July 1935, it was described as one "of the most dense thickets in the surrounding territory."[298] Even today, the location is secluded, and it is easy to imagine how a cadre of young men would feel being surrounded by impassable, rain-swollen roads that prevented "traffic except by foot or caterpillar."[299]

Once moved, the 100-foot-tall tower near the Pool church guarded 71,542 acres of government land and was equipped with a telephone and maps, Ivie continued, so "fire finding is made easier and more perfect." The story explained that "at times when there is great danger of fire a man is kept stationed in this tower on lookout." When smoke was spotted, the observer could locate the source with the help of cross-shots from other surrounding towers before telephoning the fire crew at camp. The firefighters were "ready to go on a minutes notice," and while they were away a second crew was organized and held ready for action. With this system in place, the article boasted, there were only thirteen fires during 1936 that burned 87 acres, this "less than 1–10 of 1 percent of the total" protection area.[300]

The transfer of the Farris Tower had been facilitated after the government acquired land situated on a prominent rise that sloped downward toward the southeasterly flowing Pole Creek. The hillside has considerable relief: a dissected plateau bound by the creek and two ephemeral tributaries draining nearly perpendicularly toward it. In this configuration, the axis of the hill on which the Pool Tower was erected runs northeast-to-southwest, north 40 degrees east.

The compound that was built around the tower conforms to Groben's plan and has been extensively studied as part of this project. A dirt track provides ingress and egress from FM 149 and passes by the lookout before contouring around the hillside to the observer's house and the garage. These structures are tucked away downhill and out of the way of the publicly accessible tower. Though now destroyed, the foundations indicate that the buildings were not aligned. The simple, L-shaped dwelling was oriented toward the road, which in turn paralleled the contours of the hill. The front and side of the structure, with a large screened porch, faced the southeast and would have been sunlit the entire year. In contrast, the square, two-car garage that sat slightly farther downhill and faced the northwest and northeast, is largely protected from the sun during most of the day (fig. 33).

The cottage was sparse and small. Downhill and away from the lookout, it hardly distracted from the surroundings. Conformable with the comprehensive development plan, it was utilitarian and architecturally pleasing. Within the appropriate margins of cost and function, it was also duplicated on the Davy Crockett National Forest and at Yellowpine, Tenaha, and Cyclone Hill.

The relocated EMSCO tower was completed by Richards enrollees

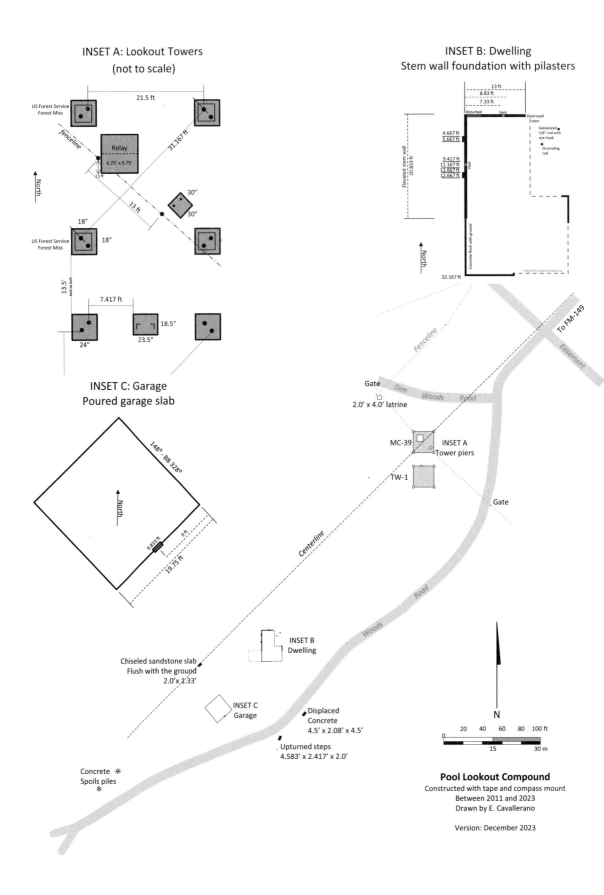

who may have been supervised by David Anderson, based on the neatly inscribed initials "D. A." that are still visible on the southeastern pier for the tower. During his career, Anderson worked for both the US and Texas Forest Services. In addition to the piers, the platform landing for the ladderway is also preserved on the northern face of the tower's base, partially concealed by wax myrtle.

In 1963, the original lookout was replaced by a surplus Aermotor MC-39 decommissioned from Polkville, Mississippi.[301] Thus, two sets of tower piers with discrete mappable attributes can be inspected in the lookout yard. Remains of the second tower lie a few paces to the north. The baseplates that secured the corner posts were left in the piers when the structure was abandoned, and they preserve black stenciled lettering, reading "U.S. Forest Service, Forest Miss." and "U509." This marking is especially telling, as component "U509" conforms to the bearing plates used for a 99.75-foot Aermotor MC-39 lookout. The position of the landing platform for the stairway riser is consistent with this design. Also within the tower's footprint is a concrete slab for a relay house. The observations are meaningful and reinforce the central role of lookout towers in the decades after the CCC as well as the necessity of modernizing the stations. Both factors underscore continued occupancy at the complex and the protocols used to fulfill a midcentury USFS fire management strategy.

Pool station was a primary lookout throughout the 1940s and 1950s, appearing regularly on memoranda of understanding between the USFS and TFS. Under these arrangements, the post was staffed anytime the fire danger rose above Class III. Before the transition to radios for communication, the tower also served as an interagency exchange with a direct telephone connection to the state's Bath Lookout tower.

The need for dispatching from the Pool compound may have diminished in time, especially after the Raven District office building was constructed in 1966. Two years earlier, the district ranger communicated with the forest supervisor requesting permission to dispose of the well house, chicken coop, barn, and fence at the site,[302] and later cooperative fire control agreements make no reference of the Pool Tower being used for communication services. At that time, all interagency communication was to occur through the Four Notch Guard Station. It seems likely that the USFS left the tower unoccupied during the late 1960s, and there are indications that the TFS began stationing observers there as early as 1962, though records become more frequent between 1967 and 1969. This, along with construction of Lake Conroe, may have made detection from the tower less important in later years.

Even without the tower or associated structures at Pool and Four Notch, each was designed using Groben's principles to maximize comfort

Figure 33. Map of the Pool Guard Station. Public structures are closer to FM 149. Note that the garage and dwelling are conformable to Groben's line (thick, dashed centerline), drawn with respect to a flat sandstone monument at the southwestern end of the compound. Map drawn by the author.

Figure 34. The first-generation EMSCO lookout tower and weather station at Cyclone Hill ca. 1950. Note the radio antenna on top of the tower. Radios were installed on the Texas National Forests during 1949. When the USFS upgraded its tower network during the mid-1950s, this tower was sold to the TFS and re-erected at the Vidor Duty Station. Image courtesy of the Texas A&M Forest Service.

and utility in the hot, humid forest conditions. Likewise, the Cyclone Hill Station, which was originally supervised by the TFS but ceded to the USFS as Angelina National Forest took shape, demonstrates the orderly approach preferred by the USFS. The hillside received its name when a cyclone destroyed a 200-acre corridor through the forest around 1897.[303] Because of the storm, the timber in the area was too immature to log during the early twentieth century and must have been an attractive, maturing longleaf forest by the time CCC crews constructed the compound below the tower. The station boasted a dwelling, a weather station, and several outbuildings meant to support fire control operations and supplement the observer through agriculture and husbandry (fig. 34). Eventually, the tower was replaced, but the hillside remained an important link in the federal detection system until the early 1970s. The tower was removed in 1978, but the second-generation tower cab is on display at the Forestry Museum in Lufkin.

Chestley Dickens and Betty Huffman
The 1940s

Alsey's Fire—as Told by Cromeens

It was on a windy day in March that I looked up to see
A curling black smoke go up that seemed to laugh at me.
I put on my old gold-rimmed specks and looked at it again
I gazed up at the cloudless sky and wished that it would rain.
I took my metal alidade and pointed it to the smoke
The reading was just ninety-eight and I thought something must be broke.
But ninety-eight was correct and correct I should have known
So I sighed a bit and turned around to my old black telephone.
I called up Mr. Cotton and asked if he could see
The fire over to his left that was meant just for me.
Yes, he could see it curling up his reading was one-o-two
So I put three and three together and knew what I must do.
So I crawled down from my perch and down the ladder I went
In my old car I shoved a rake but I noticed that it was bent.
Well, it would do for this one fire so I took it off to go
To the hottest thing in this wide world and the one I hated so,
When I got there not to my surprise the fire was burning bright
The grass was burnt to an inky black and the poor pines, they were a sight
I quickly got my rake at hand and vainly began to rake
It was for me and my fellowman but mostly for his sake.
I soon got it raked around and sat down to rest
I sighed a sigh of relief and knew I'd done my best.
When the fire had gone almost out I slowly arose and said,
"I am so tired I cannot walk, I wish I were home in bed."
I got into my car again and homeward I swiftly rode
I felt that some old fire bug had exacted a heavy toll.
When I had eaten a bite to eat and had lain down to sleep
I knew that some old fire bug would make another leap.
But not tonight, I really hoped for I knew that I would go
If a fire broke out anywhere and I'd call the fire bug a so-and-so.[1]

SIX CCC CAMPS were under TFS supervision at the beginning of 1939, and lookout construction projects continued into the new decade (table 8).[2] Enrollees had completed the Burkeville station by March 1940, bringing the total number of towers to seventy-one.[3] Experiments continued with

Table 8. Lookout towers constructed between 1940 and 1949.

1940	1941	1942	1943	1944	1945	1946	1947	1948	1949
Bon Wier	Emilee	Karnack 1 *from Newton*		San Augustine *from Denning*				Barnum *from Peach Tree Village*	Buna
Burkeville								Camden *from Hortense*	East River *from Humble*
		Based on relative dating: Yellowpine 2 (1948)						Wolf Hill 1	Evadale *from Kirbyville 1*
									Mercy 2 from Haleyville, Alaba

wooden structures too, and the *News* reported that the next lookout to be constructed for the state would be a monstrous 120-foot wooden tower with a catwalk surrounding the cabin in Protection Unit 9. Another creosoted tower would be erected the following year, this time at Emilee.[4] Meanwhile, the Alto Nursery was providing Texas-grown seedlings to farmers and foresters while the forest mapping efforts progressed satisfactorily. Camp operations were routine.

The eleven-year service record of the wooden lookout at Emilee, itself a temporary logging community, highlights the excessive funds and labor available during Depression-era "make-work" programs. Tyler and Polk Counties were saturated with lookouts by 1941 when the *News* announced the new protection unit in the county. The tower was built by Woodville Camp enrollees according to specifications duplicated at other wooden tower sites at Weches and Lufkin. Most likely, redundancies in the fire control network and increased maintenance costs led to the salvage of the Emilee Lookout. By 1952 the lease for the site was canceled.

Four forestry camps remained at the start of 1941 in the period before the bombing at Pearl Harbor, Hawaii, on December 7. Immediately, the country entered an expanding global conflict, and the nation's "reservoir of manpower" was diverted to defense and preparations for war. By June 1942, the CCC had been terminated, and the program's accomplishments were celebrated in *Texas Forest Progress, 1941–1942*.[5] All told, forestry camps had constructed 2,147 miles of telephone lines connecting seventy-eight fire towers. Enrollees had built 2,810 miles of road, completed timber mapping across the Piney Woods, and devoted 110,791 work-days to fighting forest fires.[6] Looking back in 1945, Director White reflected on the role of the CCC in a letter to William H. Parker. In it, he said that "all of us have been pioneering forestry in Texas, however, the only real money that was ever put into Texas on forestry was put in by the federal government during the CCC days, when many millions of dollars were spent on forestry in East Texas."[7] Certainly, much had been achieved.

Immediately, the TFS was concerned with meeting its firefighting obligations. The *News* reported in July 1942 that studies were being conducted to determine the feasibility of "using lumber and mill employees on forest fire fighting during critical emergencies." The situation was urgent, the article summarized, noting that "abandonment of the CCC camps and shortage of other sources of manpower caused by Army and war industrial demands have created a serious situation."[8]

During the conflict, the government feared sabotage and envisioned enemy combatants igniting forest fires along dry, rural roadways. Approximately 90 percent of the timber being harvested was consumed by the defense industry, a 1942 *Timpson Daily Times* article reported.[9] Fire must be kept out of the woods, and propaganda campaigns with racially charged cartoons quickly developed. They exclaimed that the country was "at the mercy of Nazis saboteurs,"[10] and they proclaimed that "Tokyo Loves an American Forest Fire."[11] Each urged citizens to be vigilant for suspicious activities and help by volunteering for the Forest Fire Fighters Service (FFFS), a citizens' group to help combat forest fires. TFS foresters and patrolmen participated in these efforts by organizing volunteers and loaning approximately forty-five hundred hoes, fire flappers, rakes, and axes to East Texans for the duration of the war.[12]

The fire control program evolved out of necessity to overcome wartime shortages. Tractors were purchased to mechanically create firebreak perimeters where people had once done the work, and the agency experimented with aircraft surveillance during 1941 and 1942 to supplement lookout towers and forest patrols.[13] During March 1943, a single Civil Air Patrol airplane conducted a ten-day trial to determine the effectiveness of such reconnaissance, spotting eighty-five fires and relaying the information to ground crews.[14] These experiments were encouraging enough that a Piper Coupe airplane was purchased and stationed in Lufkin during October 1943.[15] By December, White had signed agreements formally activating the Texas Forest Patrol Service branch of the Civil Air Patrol.[16] The following year, 46,000 miles were flown for fire patrols, and sixty personnel with eleven aircraft stood ready to supplement "the lookout towers on days of high fire hazards and poor visibility." The *Clarksville Times* reported:

> A system of forest fire control is being worked out for the East Texas region, which extends from Red River to the Gulf, that is expected to greatly reduce losses. Airplane patrols will be in operation extensively, planes belonging to the Texas Forest Service and the Army covering the vast area daily. The roof of the house

occupied by each volunteer fire fighter will be marked with a large "V." Ground patrolmen's car tops will be marked with the number of the fire tower around which they operate. The tower north of Clarksville is No. 62. W. J. Leatherwood, local patrolman, has that number on the top of his truck. The top of the fire tower will display these numerals in large size.[17]

Albeit slowly, the TFS began nudging toward alternative methods for fire detection during and after World War II. Comparative studies were made in 1945 and 1946 to test between a system of fully staffed lookout towers and plane patrols for several districts. The results demonstrated that if a plane consistently observed the same area during fair weather, the tower system maintained a "small but barely significant edge in effectiveness."[18] Not surprisingly, tower effectiveness decreased as the visibility worsened and lookouts peered through the haze. At the same time, though, the capabilities of an aircraft observer decreased only slightly, and observations from the patrolling plane were superior to those from staffed towers.

Another 1947 study compared aircraft observers against the lookout-smokechaser system during normal operations when airplanes did not begin to patrol until fire occurrence demanded. On days when the fire danger increased above a Class III rating, there was an average fire occurrence of 19.3 fires per day. Of this total, only 7.3 fires were reported by towers, while 14.5 fires were reported by plane. Just a few months earlier, the *News* also reported that "four years of study and experimentation have proven the value of light aircraft in detection and reconnaissance, particularly under high fire danger conditions and in periods of reduced visibility." Economically, aerial patrols were cheaper than the lookout network, but the same *News* article summarized, "When danger of fire is slight, however, towers suffice."[19]

In many ways, the aerial surveillance programs benefited from the capabilities of veterans returning from World War II. In 1945, the *Silsbee Bee* printed an article announcing that Theodore Busselle, a lifelong Warren resident who spent his thirty-year TFS career as a smokechaser, patrolman, and foreman at Kirby State Forest, had participated in planning meetings for "concentrated aerial forest patrols" in Tyler County along with J. O. Burnside, B. D. Hawkins, and District Forester John Foster. The aerial program was only two years old, but it had already flown 79,000 miles, and "the past experiences" of the aerial foresters like Foster were helping "develop new techniques."[20] Certainly, innovations suggested by the district forester were bringing exciting changes to fire detection, Foster having just been honorably discharged from the US Army Air Force where he flew thirty-five missions over Germany as a B-24 Liberator pilot.

Wartime conditions also led to changes on the ground. Don Young wrote to Chief White in September 1942. Along with his letter, he enclosed a map of Caddo Lake State Park created after Knox Ivie surveyed a lookout tower location. The site, he assured the Division of Fire Protection chief, "was determined to be the best location for the tower. It affords an excellent view of the Longhorn Ordanance [sic] Plant as well as the rest of that section." Though another site was available at the park with slightly higher elevation, Young continued, "much of the publicity value would be lost."[21]

A lookout tower had been planned for Caddo Lake State Park during the Fourth ECW Period, but it was never built. The need for protecting a new military installation near Karnack must have spurred decade-old memories among the ECW planners who surveyed the recreational area again. With a map in hand, White corresponded with the executive secretary of the State Parks Board and referenced the two agencies' "cooperative efforts . . . during the days of CCC Camps." With the nation at war, White continued, "it becomes necessary for the Texas Forest Service to afford better and more intensive forest fire protection in connection with the powder plant, adjacent to the Caddo Lake Park and Karnack."[22]

The Texas State Parks Board ratified the proposal to construct a 100-foot steel lookout on a "high point just south of the main lodge within the Caddo Lake Park boundary" on November 7, 1942.[23] The EMSCO lookout that was erected at Caddo Lake State Park shared its identification number, No. 32, with the tower at Newton, from where the structure was moved. Later Fire Control inventories also remarked that the tower was moved from "Newton to Karnack State Park."[24] The lookout stood at Caddo Lake until 1952. Then, over safety concerns, park renovations, or the changing vision of the State Parks Board, the tower was dismantled and moved north into Marion County by O. C. Braly to become the Smithland Lookout.

Simultaneously, the end of emergency conservation work programs and wartime demands changed the future of the Hooks Tower. The station was erected in 1936 to provide fire protection to approximately 277,000 acres in eastern Bowie County. In this capacity, it worked to triangulate fires "with similar improvements to the south and west in Texas and to the east and north in Arkansas."[25]

As early as June 1941, a Forest Service letter issued by Burnside was sent to the liaison officer for the Eight Corps Area concerning a large shell-loading plant that was being planned for the district, and he inquired about the possibility that the Hooks Tower would be within the new reservation's boundaries. Speculation, the letter explained, had increased rents in the area to the extent that R. V. Smith, the forest patrolman assigned to the lookout, could no longer afford to rent a house in the community. His monthly payment, the letter continued, had increased from $10 to $30—about half of the $66.66 salary Smith received. Because of the increased cost of housing, "Mr. Smith desires to build himself a house on the one-acre tower site in order to save rent." The lessor and the Forest Service had no objections to the plan given the circumstances, but the service wished to clarify "what action the Federal Government will take in connection with our tower and tower site."[26]

A bureaucratic back and forth followed with roundabout referrals and non-commitments. District Forester Don Young was unyielding during a meeting with the commanding officer of the facility in August 1941, even suggesting that relocating the tower from the Hooks site might necessitate the use of "more than one tower to give us the same coverage."[27] The colonel asked for a written estimate for the cost of providing equal protection to the district in the event that the tower was relocated and later informed the TFS that "the tower at Hooks would definitely have to be moved." The colonel compromised, however, indicating that the "Government would pay for the moving of the tower and necessary telephone line connections to the new site."[28]

By September 1941, though, the army rescinded its offer to finance the cost of relocating the lookout and restringing the telephone lines to a new site "about 3.5 miles northeast" of the reservation.[29] There was frequent communication between District Forester Young, Chief White, and Director Siecke as the month continued, and Young seemed anxious to finalize arrangements, either for the tower to remain in place through a cooperative agreement or for it to be moved so that it could continue operations in the district.

White and Siecke forcefully petitioned local leaders and the army. Siecke, who always had the reputation as a man who "knew politics and how to get things done," wrote to White, coaching a response from his chief that September. In his letter, Siecke indicated that Young "seemed to feel that Senator Beck and Representative McCann would object if we withdrew our protection activities from eastern Bowie County, and that such procedure might also result in the loss of interest on the part of the Texarkana Chamber of Commerce and civic organizations."[30]

The Division of Forest Protection applied pressure too, and White composed a letter to the army:

> The Texas Forest Service has always cooperated with the Army in connection with their CCC Camps, Army Maneuvers and manning of our lookout towers for Aircraft Observation Posts and will continue to do so to the best of our ability. Since we have no funds available for making these changes in our fire protection improvements, we figure that it is no more than right that the Army should pay for the changes made necessary by their occupation of the Shell Loading area.[31]

Siecke also wrote to the special assistant to the attorney general that, "offhand, it strikes me that the Texas Forest Service tower and telephone lines, if kept in place, and used on a cooperative basis, would be a distinct asset in promptly detecting forest fires. I am hopeful that some arrangement can be worked out along this line." Tactically, as always, Siecke concluded his letter by reminding W. D. McFarlane that "in the Twenties when you were in the Texas Legislature, I talked to you occasionally relative to the State Forestry activities, and consequently, I feel more or less free in writing you concerning the above mentioned proposition."[32]

Privately, Siecke had doubts that the tower would remain on the reservation and expressed his concern to White at the end of September: "The War Department certainly would not want the tower, 10 miles of telephone line and State controlled personnel inside the fence. I believe we are wasting a lot of time by miscellaneous correspondence without first finding out what the War Department plans are."[33]

The lobbying worked though, and on October 10, 1941, Don Young received a letter indicating that employees needing a military pass to the tower should report to the Texarkana National Bank Building. The army had finally "decided that the tower would be a help to them and that it could be left in place,"[34] but it stipulated that the phone would need to be redirected through the military switchboard so that they could "listen to all conversations going over the phone lines from the tower to our other men. In other words, they want to check up on our towerman to see that he is not a fifth columnist."[35]

With the resignation of R. V. Smith, who went to work with the Defense Corporation, and the tower location secured, Emmett Turner moved from Douglassville to become the patrolman at Hooks.[36] A year later, Chester Johnson applied as a crewman and listed the status of his eyesight on his application as "extra good." But he did not last, and a February letter terminating him concluded, "You have received payment which you have not rightly earned."[37]

When the TFS announced reorganizational changes to the lookout network after the war in 1948, the Hooks Tower was recommended for a new location "outside the Army ordnance tract."[38] It appears that the change occurred slowly, however, and USC&GS descriptions place the lookout at the depot through 1952. Internal memoranda written with regard to the District 1 budget in March 1952 also indicated that the "removal of the Karnack and Hooks towers" could not be afforded without a supplement. Capital must have been secured shortly afterward though, and a lease for the Redwater site was signed in May 1952. Independently, a 1954 USC&GS description for the Hooks location noted that the "lookout tower is completely removed."[39]

Aircraft Warning Service

The expansive network of fire lookout towers constructed during the Depression proved effective in fire detection and forest protection. Observers had sweeping views of vast territories and the ability to quickly locate and communicate the position of any menacing burns. After Roosevelt's declaration of war in December 1941, however, the nationwide fire tower system assumed a new defensive role as military planners realized that the posts could also be used to spot aircraft, determine flight paths, and estimate flight altitudes. The Aircraft Warning Service developed around these objectives and called on citizens and forestry organizations on the home front for help.

Bill Hartman remembered being sent to Mobile, Alabama, for what he described as secret training on how to utilize lookout towers as aircraft warning posts,[40] and throughout the war the TFS made all telephone systems and thirty lookout towers available to the military for detecting airplanes as well as fires.[41] After signing a citizen-loyalty affidavit on May 7, 1942,[42] Lookoutman John Huffman at Kirbyville received his appointment as AWS key observer directly by letter from Director Siecke in June 1942.[43] He and other lookouts were required to work eight hours a day, six days a week in the watch towers unless telephone line repairs forced them to descend. Each was instructed to show preference for spotting and reporting aircraft, even at the detriment of their fire detection and suppression responsibilities. Most important, these employees were bound to keep aspects of the program confidential unless they were in discussions with other TFS or AWS personnel. It was forbidden to disclose information on the location of lookouts being used for the service, how many towers were activated, or the types of communication systems involved. Finally, visitors were no longer allowed to enter lookout cabs or be within earshot where they might overhear conversations.[44]

If aircraft were spotted, "flash messages" were immediately transmitted to the military describing details of the sighting.[45] The Scottsville Tower was one of three lookouts activated for aircraft warning purposes in September 1941, and full-time patrolman G. W. Rose was responsible for selecting observers to staff the tower through the night. During one military training maneuver called the "Battle of Shreveport" during the winter of 1941, "fast reporting" from the nearby Jefferson Lookout provided information to Houston on squadrons of aircraft flying overhead less than a minute after being spotted. One sighting, reportedly, was timed to take fifteen seconds, and the *Pineywoods Pickups* commented that perhaps during planned exercises the following spring, "all our towers [will] have the exciting experience of reporting those planes."[46]

The program offered small, temporary salary increases for lookouts, but it operated only during the early portion of the war. By January 1943, the TFS sent a General Circular letter to employees such as Francis Ott at the Cleveland Tower informing him and others that the program would expire at the end of the month. Only two towers at Kountze and Smart School were going to be maintained in the warning network after that date.[47]

"Emergency" Roles for Women

Chestley Dickens's heart fluttered on October 24, 1943, as she met fellow TFS employee and photographer Sherman "Jack" Frost at State Forest Number 4. In front of her stood the still gleaming 100-foot Aermotor MC-40 that had been erected only nine years before. Frost clicked the first image of Chestley, with a wide grin on her face, from the ground as she ascended a narrow ladder inside the frame of the lookout. After developing the negative, Frost would write a brief description, "Mrs. Claude Dickens," referring to Chestley by her husband's name, "emergency lookout woman." Then, ascending the tower, the pair stopped again. Frost, ahead of Chestley now on the second landing platform, focused his camera lens downward to take another photograph as she confidently ascended the ladder to "man" the station (fig. 35), despite the objections of old-time "fire dogs" convinced that women would never be fully qualified to carry out such essential forest protection duties.[48]

Unusually, Chestley Dickens was not the first woman to work at the Kirby Forest Tower. Earlier that year, thirty-five-year-old Mabel Pitts, a mother of two and the wife of Patrolman Enoch Pitts, accepted temporary employment as an emergency lookout with an annual salary of $225.[49] Enoch had been employed with the Forest Service since 1937.[50]

Figure 35. Chestley Dickens ascends the ladder on the Aermotor MC-40 tower at Kirby State Forest in Tyler County during 1943. Staffing shortages during World War II created opportunities for women to enter the male-dominated profession. Image courtesy of Texas A&M Forest Service.

and the family lived in the cottage constructed by the CCC below the tower on the grounds of the forest. When Pitts's assistant, Smokechaser Thomas McGowan, was drafted into the army during 1942,[51] Mabel was offered a job at the tower in February 1943 until such time that a "suitable man" could be employed on a full-time basis.[52]

Mabel's "emergency" role involved staffing the lookout when her husband was called away for fire suppression work so that adjacent *towermen* at the Woodville and Kountze Towers could still secure cross-shots when smoke was spotted. She was advised to contact her male colleagues in the event that a fire occurred while she was on guard and ask Smokechasers Busselle, Gordon, or Woods for assistance. Her appointment was short-lived, though, as she and Enoch resigned in May, "having never enjoyed any more the association of a group of men from the ranks of the highest office of the Texas Forest Service."[53]

Domestic labor shortages meant that there continued to be opportunities for Mabel, Chestley, and a handful of other women such as Betty

Huffman and Lettie Hughes, who were becoming some of the state's first female lookout observers. Lettie, just twenty-eight years old at the time,[54] accepted employment with the TFS in November 1943.[55] She was paid $30 per month to serve at the Tabernacle Tower (fig. 36), when she was called on by Patrolman James Kimbro.[56] She was expected to work a minimum of thirty-five to forty days per year, and her contract also dictated that she was responsible for suppressing any small fires that might be discovered.

Lufkin's "demon tower building crew" erected one lookout in Shelby County by 1934, and it is probable that the Tabernacle Tower was completed through their efforts. The station was an important vantage point in the shortleaf pine territory encompassing Protection Unit 10 but also communicated directly with the USFS tower at Dreka, making it an essential relay between the state and federal fire control systems. Hughes, and Kimbro, would have been obligated to operate the phones and supervise the tower during high-danger periods to fulfill the memorandum of understanding between the two agencies.

The ascent by ladder to the top of the EMSCO tower must have been unnerving for lookouts. Attempts were made to improve safety at towers with outside ladders in the decade after Lettie's employment when Director Folweiler approved the use of "woven or chicken wire [cages] around ladders on ladder-type towers." The addition was acceptable, he said, "when the user of the tower believes that the wire makes climbing safer." For normal tower use though, he thought, "the installation of the

Figure 36. A 1952 photograph of Tabernacle Lookout demonstrates how wire safety cages were installed around lookouts with outside ladders to protect an observer while ascending (A). The Tabernacle Tower in 2022, showing the remnants of the metal ribs supporting the cage (B). Image A courtesy of Texas A&M Forest Service. Photograph B by the author.

wire is to be avoided."[57] The remains of one, the design of which is typical, still cling to the rungs of the Tabernacle Tower. There, the struts supporting the wire are built from two bent pieces of steel. One member, approaching the shape of a static recurved bow with elbowed ends, is fixed to every sixth or seventh rung on the ladder. A second piece of steel has been bolted to the elbowed edges of the first member to create an arc that surrounds the ladder, onto which the wire could be fixed.

Kimbro was employed with the TFS in February 1935 and served his twenty-nine-year career in in Shelby County until he retired in 1964. The tower he staffed with Lettie, albeit briefly, is still owned and maintained by the TFS as a radio relay. Despite the 1960-era relay house and remnants of the safety cage, the structure is in excellent condition and worthy of preservation. The lookout surveys a sparsely populated portion of San Augustine County dominated by commercial pine forests where land use has hardly changed since James Kimbro or Lettie Hughes reported for work. It is inevitable to think of these "guardians of the Piney Woods" driving along the single-track dirt road through pine plantations, and the tower should be nominated for inclusion on the National Register in recognition of revolutionary women observers who succeeded in traditionally male roles alongside career TFS crewmen.

Farther south in Newton County, forty-eight-year-old Betty Huffman had been hired nine months earlier to be a lookout at the Kirbyville Tower.[58] Her husband, John, at that time a lookoutman and AWS observer, had been with the TFS since 1930,[59] and both lived in the dwelling at the state forest. When questioned about experience in forestry or lumber during her application, Betty responded, "My knowledge of the forestry work and my experience helping my husband who is an employee for State Forest #1 will qualify me to do [what the] job application requires. Have had quite a bit of fire fighting experience." When the questionnaire asked both women if they were able to climb the lookout at their own risk, both wrote "yes." Betty, for her part, added, "I have had this experience."[60]

Regardless of male convictions to the contrary, other women were proving themselves in essential duties across the Piney Woods. For example, a severe ice storm in 1944 menaced the detection work and destroyed 855 miles of telephone lines, disrupting communication for weeks. With men overseas and approximately 40 percent of the phone system destroyed, women went to work repairing severed wires and installing new insulators and brackets. These crews also provided the backbreaking labor needed to take trees off the lines and clear the rights-of-way.[61]

Despite encouraging results from studies focused on the effectiveness of aerial patrols, the *Sweetwater Reporter* announced a reorganization of lookout towers in 1948 aimed at improving ground-based detection.[62] As mentioned, the Hooks Tower would be moved outside the ordnance plant to Redwater, and two new towers would be built in Jasper and Newton Counties. Meanwhile, the tower at Hortense would be moved to Camden, and the Peach Tree Village Lookout would be re-erected at Barnum. Other changes would take place too, and the Humble Tower, once with views of the San Jacinto Monument, would be moved northward to the East River site. Finally, Betty and John Huffman's lookout at Kirbyville, the first tower erected in the state, would be moved and replaced by the relocated Sheffield's Ferry Tower to provide additional height over the quickly growing forest.

Administrative Changes

Administrative changes were also occurring during the decade. E. O. Siecke retired during October 1942 to care for his ailing wife. White, at fifty-six, was selected as the third state forester for Texas and was the chief executive through the most critical periods of the war, when "employees were difficult to come by and resources were limited."[63] While the TFS survived the war years, Chapman points out that "degenerative factors" began to "tear at the very fabric of the Texas Forest Service." Notably, he assessed White's lack of strong leadership ability and suggested that the director allowed the "Service to drift slowly away from the cohesive structure and spirit that Siecke had worked hard to establish."[64] Though White was often described as a "kind-hearted man," by some estimates it might have been challenging for him to hold people accountable, and he outwardly appeared aloof and apathetic about the role. Morale suffered.

A letter written by White to William Parker in 1945 demonstrates his temperament. Parker had been employed with the TFS in Polk County since 1929 at both the Buck Tree Cab and Livingston Lookout. The tower was built during 1934 by Company 840 enrollees stationed on a 15-acre campsite that had previously been used as a Presbyterian encampment ground just inside the city limits.

With the onset of World War II, the Livingston Lookout was referred to as Palm 88 and used as an AWS post staffed by Parker. Surprisingly, the regional signal officer and federal officials abandoned the station comparatively early, during November 1942, citing that Livingston-area volunteers had organized their own AWS network. The change resulted in a reduction of supplemental funds to the TFS, and the salaries of several patrolmen, including Parker, decreased to prewar levels when they were stripped of their "Key Observer" titles.

Evidently, Parker's troubles continued with the passing of his wife in 1945. During Parker's bereavement, J. O. Burnside wrote to him to extend his sympathy:

> Please permit us to share in your compassion. Only the passing of time and having our friends around us can tend to deface our memories on these occasions. I will be unable to attend the funeral this afternoon, however, Messrs. Ivie, O'Quinn, and Lawrence will be there. I am asking them to hand you this little note personally which conveys my heartfelt sympathy.[65]

Parker wrote to White the following month, and while Parker's letter to the director did not survive, it can be assumed that the mourning employee solicited his supervisor for a salary increase. White's ethos is captured by his response:

> Received your letter of the first and certainly was glad to hear from you. I have been hoping to take a trip down that way and see you, but so far have not had the opportunity. I agree with you that all your hard luck came at the same time, and know that you must be very lonely living by yourself, however, I admire your courage to keep going and am glad to know that you are not sick. With reference to your pay, now, my dear friend, I know that your pay does not amount to so much as an emergency patrolman, only $220 per year, however, that is all that we can allocate for this position, as you know the last legislature did not agree to increase our appropriations to any great extent. . . . You know we all appreciate the fine work you are doing and have done in preventing fire damage and keeping the forests growing so that the communities could prosper, and trust that someday we can reward you in a better manner. So long. Your friend, W. E. White, Director.[66]

With White's promotion, his assistant J. O. Burnside assumed responsibility of the Fire Protection Division.[67] Burnside was a graduate of Colorado State University and arrived in Texas during the Depression after having worked as a junior forester on the Arapaho and Ozark National Forests.[68] In Texas, he initially supervised national forest CCC crews at the Coldspring, Richards, Nancy, and Ratcliff Camps fighting fires, planting trees, collecting seeds, building lookouts, and mapping forest cover.[69] While at Camp Nancy, the story goes, Burnside remembered a "young forester from one of the northern states" who could not bear the ticks, chiggers, and heat of a summertime day and "temporarily lost control and took a naked stroll through the streets" of town. At the height of the CCC period, Burnside left the USFS in 1936 to work for the state.[70]

After his 1942 appointment as the head of fire control, Burnside oversaw fire response efforts and supervised field personnel until his retirement on May 31, 1971.[71] A letter he wrote during the early years of fire protection at Alto is telling of his character. Evidently, one of the station's lookoutmen was relieved of duty for drinking on the job. When potential employers contacted Burnside seeking recommendations for the individual four years later, the chief was compelled to write to the former observer. In his letter, Burnside explained, "We could not possibly make a statement to any person who had intentions of employing any member of our personnel which statement was incorrect. Neither could we afford to cover up anything . . . which might come out at a later date." With that in mind, Burnside recommended squarely, "Will you therefore write me a letter outlining your activities in connection with that habit since your appointment was terminated. . . . With that exception I can freely say that you were a good employee."[72]

Burnside later advocated for the extension of the lookout network into the Lost Pines and post oak "fringe areas" of the state. During his tenure, he "steered Texas' firefighting team out of the hand tool era, into the age of mechanization—a difficult transition for many elements of our society during the postwar period. Fire flaps, axes and rakes were replaced with crawler tractors for plowing firebreaks. Telephone communication was phased out to make room for two-way radios." In 1971, Burnside remembered, "Back in the early thirties, the principal means of detecting wildfires was a tall pine tree fitted with climbing spikes and a 'crow's nest' on top. Our work force was limited, but they were stouthearted and dedicated; they had to be."[73]

Responsibilities for a forest lookout were better articulated during the decade after the towers were erected and the CCC labor pool disappeared. Employment contracts for lookoutmen like Neal W. Kincel, Barney Parrish, and Cleo Horn portrayed their obligations with urgency. "As lookoutman," Kincel's 1944 employment offer noted, "the whole efficiency of the fire fighting crew depends on your discovering the forest fires and accurately locating them as they occur and communicating with your Patrolman, Smokechaser, or crew."[74] His supervisor, Patrolman William Horn, meanwhile, would advise Kincel when he was required to staff the tower, where he was expected to stay for a minimum of eight hours per day. Parrish's orders were identical, but his contract added: "No set hours of work can be given you as weather and fire conditions will control the time when you go to work in the morning and the time you leave your official post at night. When going on duty in the morning always take your lunch and eat same on the job as this time of the day is generally when fires begin to show up and burn fast."[75]

Aside from tower duty in 1943, Horn was obligated to patrol for fires when visibility was poor, maintain the roads and telephone lines within his district, visit schools, and investigate fire violation cases. Whenever the work was satisfactorily completed and there was no imminent fire danger, he was then at liberty to spend his time on personal business.[76]

Unusually, Groveton resident and TFS employee William Powell's service contract identified his official post as the "USFS Trinity County tower," suggesting collaboration and integration between the two organizations. Powell started as an emergencyman in 1940 and worked as a lookout, firefighter, and crewman within the "ice damage area" after severe winter weather severed the state's telephone communication system.

Though there were two federal towers in Trinity County by then, it appears that Powell was posted to the recently built creosoted tower at the Piney Duty Station based on communication with J. O. Burnside at the time of his resignation in 1947. The Piney Tower was constructed at an isolated spot on the southern edge of the Davy Crockett National Forest while the federal fire control network was still evolving. With land acquisitions slowing, the Apple Springs Lookout was relocated to a new site at Mt. Zion, and the Nogalus Tower was built high above the treetops to accommodate the rapid growth anticipated by studying the area's site index. The remains of the original wooden Piney Tower can be found along the southeastern side of Forest Service Road 528 B, identifiable by a series of piers with a three-point anchor rod design. They lie adjacent to set of piers from a second, surplus steel tower moved from South Carolina and in service until the 1970s.

Best and Worst Fire Years

Some of the worst fires in Texas history occurred around Conroe in Montgomery, San Jacinto, Liberty, and Walker Counties over ten days from September 24 until October 3, 1947.[77] The state was excessively dry, and the heavily forested region was ignited by a suspected, but never apprehended, arsonist. On September 26 alone, aerial patrols reported sixty-seven fires. This, as the TFS later reported, would have made each of the district's eight patrolmen, including Bob Williams, responsible for suppressing eight fires each. With an inadequate firefighting force working solely with hand tools, Burnside asked White to appeal to Governor

Beaufort Jester for emergency assistance in suppressing the inferno. Quickly, helicopters, a C-47 cargo plane, and P-51 Mustangs were dispatched from military installations at San Marcos and Ellington Field to assist. The fires brought unwelcome publicity to the TFS and demonstrated the vulnerability of the protection scheme. The governor visited the emergency fire headquarters in the control tower at the Conroe airport on September 29, conferred with county officials and woodland owners, and made an aerial inspection of the counties. In all, 459 fires burned over 55,000 acres.[78] White later reported that more than 10,000 column inches of newspaper space had been devoted to the flare-up in Texas newspapers alone.[79]

After the dust had settled, Burnside and Nort Baser compared the effort to David's fight against Goliath—again.[80] The TFS "began battling its enemy with weapons little better than a slingshot," the *News* explained, "and the threat of destruction by fire is just as husky a Goliath as ever." To justify the TFS response, Burnside noted the department received $173,000 from the state for fire protection. This, he pointed out, was subsidized by $141,000 in federal aid and $39,000 from contributing landowners. Even though these were the largest appropriations ever granted, Burnside conceded, "These sums limit us to a basic field force of one patrolman to each 100,000 acres over the greater portion of our extensive protection area!" In a move to reassure the reader, the chief announced that new aids were being developed to improve protection work. Notable among them were modulated radios and "war-born Jeeps and light but high performance aircraft." In penance, Burnside pointed out that more than a million acres of timber were burned annually before the TFS offered protection. "A particularly dry 1947 caused us to suffer a setback," he reasoned, and the losses during the Conroe fires inflated the statistics. But, he concluded, fire prevention in Texas "compares favorably with other southern states," and the public "must recognize its responsibility" in preventing forest fires.[81]

The fallout from the fire season was severe. White resigned at the end of March 1948, only six years after becoming director. He had been with the TFS twenty-one years and was leaving, he said, to be closer to his ailing mother in Florida. It was never given as an official reason, but the recent fire situation and the director's perceived shortcomings had taken their toll. Perhaps as a matter of state convenience, he was also being accused of the unauthorized use of a state-owned vehicle.[82] Adding insult to injury, the announcement of White's retirement and news of the fires appeared together in the March 1948 issue of the *News*. S. L. Frost was to become the acting director of the TFS while a search committee looked to fill the vacancy. One of Frost's priorities, Texas A&M University president Gibb Gilchrist said through the *News*, was to thoroughly study "the organization and future plans for the Forest Service."[83]

Burnside's optimism that decreasing fire activity through 1948 and 1949 "could maintain the downward spiral on the loss chart" would be difficult to achieve because of unusually dry conditions, and fires continued to devastate the western Piney Woods the following fall.[84] In October 1948, the fire weather was so severe that Livingston's fire chief drove through the city's streets using a loudspeaker to call for volunteer firefighters.[85] Smoke billowed thousands of feet into the air, and fires roared at treetop level through 15,000 acres of Polk and Tyler Counties while over one hundred men answered the fire chief's call and rushed to the blazes in two Forest Service trucks.

Fortunately, the 1949 fire year was the least damaging in TFS history. Burnside acknowledged that favorable weather contributed to a successful year but added that "the weatherman cannot be given all the credit."[86] He pointed to the improved visibility the fire tower reorganization afforded and the increased use of aircraft as factors in the early detection of fires. Other technological adaptations, like the use of Jeeps, were also successful.

Burnside's surplus "war-born Jeeps" were quickly being converted to woodland firefighting vehicles. The state-of-the-art units were heavily grilled and equipped with a hydraulic hoist, winch, and 200 feet of flexible steel cable complete with tripout hooks. Most important, a middle-buster plow was attached to the rear of the vehicle and could be dropped to plow a 3-foot-wide, 6-inch-deep furrow that quickly created a firebreak.[87] Beautifully restored, one is on display at the Texas Forestry Museum in Lufkin.

The written narratives that followed the terrible fire losses during 1947 and 1948 had the necessary impact. By August 1948, increasing funding from private timber owners and the state and federal governments ballooned the Forest Service's budget, and it was possible to double the service's mechanized equipment. The *Rusk Cherokeean* reported the impact of the additional support. Prior to the increase, the TFS maintained twenty-two Jeeps, fifty-eight radios, and three planes. With the newly available funds, twenty-five additional Jeeps, thirty-six radios, and two new aircraft were ordered.[88]

The changing times were skillfully captured in a 1949 photograph of the Central School Station (fig. 37). The lookout, sometimes known as the O'Quinn Tower, was built by enrollees at the Lufkin CCC camp to replace the Pollok Lookout Tree, which had been surveyed on August 17,

1931. Close to the airfield and Fire Control headquarters in Lufkin, the station was frequently publicized by the TFS. In an often duplicated image, the tower is surrounded by a young pine tract and looms above a surplus Jeep. Overhead, a low-flying, recently purchased airplane buzzes the tower. Meant to demonstrate the increasing role of aviation, mechanization, and communication, the TFS was a leader among southern forestry agencies and was eyed enviously by the TNFs.

Despite the new technology, fire protection was inadequate, Burnside said in the wake of the active fire seasons around Conroe. "We have a long way to go even yet. With the addition of the new equipment, the Texas Forest Service will still have an average of only one mechanized unit for every 200,000 acres of timber."[89] Still, the increased use of machinery caused redundancies, as was the case for Jefferson patrolman George C. Reed. The thirty-five-year-old Reed started with the TFS as a lookout in 1937 and became a patrolman in 1939.[90] His July 1948 termination letter clearly demonstrated the changing times:

Figure 37. Fire control equipment in 1949. The Central School Lookout tower and fire Jeep form the setting for Pilot D. A. Terry's flyby. Image courtesy of Texas A&M Forest Service.

> Reference is made to recent conversation and explanation to you by Mr. M. V. Dunmire and Floyd A. Bramlett when visiting with you on July 23. During their conference with you the plans for affecting some changes in District No. 1 were explained. I feel certain that after the explanation you can understand the situation and are in entire agreement with the plans. During the conference of Mr. Dunmire and Mr. Bramlett with you the plans explained included the placement of several additional mechanized fire fighting units in District No. 1. Although it will not be possible to place 2-way radio with all equipment this year possibly in the near future radio equipment can be added. Other plans include the placement of other equipment in that area as rapidly as possible. While these changes for the use of mechanized equipment necessitates the cancellation of your appointment and the appointment of some others as patrolmen, effective in your case the evening of July 31, 1948, we will attempt if you desire to use you in the future as much as possible on short term work as the need develops. In making these cancellations we want you to definitely understand there is nothing personal and we certainly appreciate all of your long term of service in the capacity of which you have performed. As rapidly as possible equipment

can be placed in the area and made available to the area at which time I believe you can then visualize the increased fire protection it will render to the forest protection work. We certainly solicit you [sic] continued cooperation and we feel that you will furnish it. As was explained to you your Texas Forest Service equipment will be picked up in the near future including the telephone instrument.[91]

A similar letter went out to Barney Parrish, the lookout at Hortense. Motorized units continued to be purchased into the early 1950s, but preferences for even more powerful crawler-type tractors soon made even the Jeeps obsolete. The unit at the Smart School Duty Station was typical. A two-ton Chevrolet truck with a tilting bed was purchased in 1956. The five-roller, John Deere, crawler-type tractor that was outfitted with the truck could be loaded onto the bed of the Chevrolet and provided rapid transportation within the patrol areas.[92]

While the transition from Jeeps to tractors may seem insignificant, the fire control tactics employed by the TFS were truly innovative and adaptable. For example, TNF communications from 1958 cite the TFS's use of the grid system as the reason it implemented the methodology for locating fires from federal towers. When it came to plow units, one 1959 memorandum stands out in demonstrating the differences between organizations: "The U.S. Forest Service has been a leader in forestry, of which protection is an integral part, but we have fallen behind state and private agencies in the field of mechanized fire fighting equipment in this region of the country. Many advancements have been made in this field and it is embarrassing to come in contact with other agencies in regard to this aspect of forestry."[93] Another memorandum, composed a day earlier, by a second district forester simply wrote, "As has been agreed by most everyone, the State rigs are much better."[94]

Other technologies also modernized the agency, and though the Forest Service often boasted about the extent of the private telephone system and the ability to "holler over" nearly 2,000 miles of circuits,[95] the truth was that the system had severe limitations and was costly to maintain. Forest Engineer Millard Lawrence, who had traveled the state securing leases a decade earlier, was now burning up the road restringing crossed wires or repairing broken insulators and brackets.[96] The insulators especially were targets for vandalism, and in 1938 alone, fifty-seven minors admitted to shooting at them. "No court action was taken in any of these cases," the state forester explained in the annual report, "but the boys were severely lectured by the law enforcement officer and reprimanded by their parents."[97]

Weather was also a nuisance. A hurricane downed communication lines for more than a week in the area north of Houston during September 1941.[98] Fortunately, the event occurred during the ECW period, and the Humble Camp was given emergency authorization to spend five hundred workdays helping repair the lines within the limits of the camp's work area. A few years later, the damaging winter ice storms also exposed the fragility of the system, this time in the midst of a wartime labor shortage. Without help from the CCC, observers were receiving small increases in salary to restore the phone network within the "ice damage area." William P. James's supplemental appointment at Alto is typical, and he was responsible for overseeing a crew:

> Equipped with a portable telephone. See that the telephone is tied on the line so that a crew can be reached by the lookout at all times. The crewleader or other employee will call in to the lookoutman concerning the fire situation at intervals of not longer than 15 minutes on days when fires will occur. When the crew is directed to work away from the telephone line on days when the woods will burn, instruct one man to remain with a telephone so as to keep in touch with the lookoutman.[99]

The fires in 1947 and 1948 gave Burnside an opportunity to highlight the success of radios during severe conditions. While the state maintained great telephone networks, he noted the "men at the scene of a fire are often far from these phone lines. Mobile radio units make possible time-saving transmission and reception of necessary information between crews and headquarters or aircraft."[100]

The TFS began experimenting with radio communications in 1936 when D. A. Anderson and an officer at the CCC camp in Trinity purchased two transceivers. As Anderson recounted in 1968, the ice storm in 1944 occurred during the "critical days of World War II," when "the supply of both telephone and radio equipment was meager."[101] Because of the defensive need for lookout communications, the War Production Board authorized the purchase of limited radio equipment, which proved much more economical and improved efficiency. In 1949, when surplus military equipment was more readily available, the TFS began transitioning to radios across the system using funds from the sale of the telephone lines. Alone, the disposal of the first 483 miles of telephone lines provided enough revenue to equip twenty-six of the seventy-five towers with radios.[102] Furnishing the remaining towers and newly acquired surplus military Jeeps would take two years.

The TNFs were also transitioning to radios at the same time.[103] Radio equipment was first installed on the Angelina and Sabine National Forests,[104] but *The Palacios Beacon* reported in March 1949 that ranger vehicles in the Sam Houston National Forest were all outfitted with two-way radio equipment that would allow them to keep in contact with the forest's lookout towers.[105] Evidently, the $10,000 upgrade was successful, and three months later, the *Rusk Cherokeean* covered a story concerning the apprehension of a "fire bug" that occurred "when spotters in watch towers radioed location of a series of suspicious outbreaks of flames. Foresters on the ground followed the arsonist and finally caught him."[106]

The 1949 interagency memorandum of understanding between the USFS and TFS outlined the tower staffing schedule and identified the primary lookouts. Critically, it was written during migration to radio communications and pointed out that "both Services are in the process of working out the most efficient communication system." For redundancy during the transitional period, the two parties agreed that "the radio system will be supplemented by certain commercial exchange connection to dispatcher tower and certain other 'key' towers" in the event of fire.[107]

Included on the Sam Houston National Forest were connections between Pool and Bath Stations and a line between Huntsville and the Willis and Four Notch Towers. Within the Davy Crockett National Forest, a connection would be maintained between Ratcliff and Weches Towers, as well as others between the Nogalus, Piney, and Shady Grove Lookouts. Communication between the Jackson Hill, Etoile, Moss Hill, and Huntington Towers would be maintained on the Angelina National Forest, while circuits between Dreka and Tabernacle would persist on the Sabine National Forest along with lines between Chambers Hill and Bronson. Across the four TNFs, "all other inter-service lines will be abandoned."[108]

Albert D. Folweiler

The memorandum of understanding with the USFS was signed by Albert D. Folweiler only one year after he had accepted employment as director of the TFS in January 1949. Folweiler held degrees from Pennsylvania State, Yale, and the University of Wisconsin and had worked for the USFS, Louisiana State University, and International Paper and worked under George Patton during World War II.[109] He had already coauthored a textbook, *Fire in the Forests of the United States*,[110] and brought with him a firm, demanding style. As one former department head remembers today, the "entire agency turned on Folweiler's iron will and his impersonal, unforgiving, hard-nosed management style."[111]

Immediately, Folweiler began to initiate changes. In the February 8, 1949, staff meeting the director outlined a new organizational plan for the TFS that reduced the six divisions to four departments and dropped the title of chief.[112] The reorganization also changed the name of the Division of Forest Protection to the Department of Forest Fire Control, a telling move about the new director's militaristic philosophy toward fire management.

Burnside remained as head of the Department of Forest Fire Control and oversaw a section charged with presuppression, suppression, and law enforcement activities. Under his umbrella was also an engineering unit, meant to support work across all of the Forest Service. Folweiler kept the headquarters for the FCD in Lufkin but decentralized the field force by creating six administrative districts at Marshall, Maydelle, Lufkin, Woodville, Kirbyville, and Conroe. Stationed at each were a district forester, fire supervisor, clerk-dispatcher, radio technician, mechanic, pilot, and the lookout crews.[113]

Folweiler led the agency until 1967.[114] Burnside, meanwhile, remained head of fire control. Memoranda and meeting minutes suggest an efficient professional association between the pair, though some directorates hint at fractures in the men's personal relationships. "Burnside submitted in typed form three documents arranged in excessive detail accomplishments by his department and plans for the future," the director wrote just after joining the agency in the July 1950 meeting minutes.[115] In another instance, Folweiler addressed Burnside: "It's my impression that you have been supplied with a copy of the manner in which we propose to spend the funds. . . . Had you referred to this material, you would have found that the provision has been made for a tower."[116]

The director "took a few minutes to drive into the tower" while passing Bon Wier, he told Burnside on a later occasion during 1963. His inspection revealed a short, rutted, and improperly drained roadway, and he complained to Burnside that any FCD inspections should "include not only the condition of the tower itself, but also of the roadway leading to the tower."[117] The first reply from Burnside in an increasingly heated rebuttal acknowledged that a nearly year-old inspection indicated that the "site conditions, including signs, roads, gates, cattle guards, fences," were satisfactory. Any decay since then, Burnside suggested, had resulted because of rains. "Nevertheless," he continued, "the district will be instructed to remedy the situation."[118] To this, Folweiler replied that he had "some experience over many years on road maintenance" and advised that "the roadway to the Bon Wier Tower can never be maintained satisfactorily until such time as there is some drainage arranged for. This is not now present and has never been present. This would be the first step in improving the roadway."[119]

Or, finally, in the tiresome summary of an agenda item in November 1965, "Burnside presented some data that the district foresters were already acquainted with."[120] That Burnside left such an immense trail of memoranda and correspondence documenting the evolution of the lookout system likely stems from his teetering position, and Patrick Ebarb notes that the "fire program was bogged down in myriad directives, memoranda, and bureaucratic ennui" at the time he took over as head in 1971.[121] But it is an enormous gift in chronicling the history of the lookout network.

7 Guardians of the Piney Woods
The 1950s

A Man Looks over His Kingdom

I am climbing up the ladder
Very earnestly in prayer,
Step by step; I'm here at last
One hundred feet up in the air!
Now the door I quickly open;
Carefully, I climb inside—
And still I stand; I gaze
With heart o'erflowed in pride.
Stretching vast and wide before me
With a glory that's Devine [sic],
Stands the Kingdom I'm to care for
Three thousand acres—all pine.
And as I look into the distance
Gazing as far as eye can see,
I breathe a prayer: May God
Take care of me!
It could happen, you know,
At most any noon hour.
But I am now secure
In a Forest Lookout Tower!
Far below in its beauty,
Spreads the timber and field.
Through the air comes the hum
Of an automobile.
There's a poor horse below
That has never been shod.
I am filled, now, with the presence
Of a most powerful God![1]

THE COUNTDOWN LEADER ticks the seconds away, queuing the music and introducing the viewer to our protagonist, Bob Jackson, "Guardian of the Pineywoods."[2] The motherhood and apple pie film begins with a walk through Bob's farmyard to milk cows and feed turkeys, stopping, of course, with bowls of fresh milk for the family pets, Spot and Matilda. Unquestionably, Jackson's homestead and his subsistence lifestyle are relatable to the audience, and it is easy to imagine a classroom of contemporary East Texas schoolchildren tensely drawn into the storyline. As it unfolds, we are led away in a surplus Jeep to better understand the duties of a TFS crewleader.

Jackson is part farmer, educator, watchman, and firefighter, and we begin our patrol with him at the local fishing hole where he reminds picnickers to be extra careful with their campfire. After picking up "crewman and helper John," we head straight to an International lookout tower and watch our "guardian" climb hand over hand to the top. Foreshadowing signs of trouble, the crewman evaluates the fire weather at the danger rating station and flashes three extended fingers to the observer in the tower, revealing that it is a Class III day. Late in the afternoon, Bob finds the first white whiffs of smoke, and intense music begins to play. After an alidade reading and a call to the dispatcher, who cross-checks the location using compass bearings from two adjacent towers, we race to the scene with Jackson. The dispatcher, meanwhile, calls a local tree farmer and his son, who are happy to help and leave home with a flap and a rake they "always keep right handy."

Crouching, and drawing battlefield lines in the sand, the TFS crew instructs the volunteers as they plan their attack. Jackson then extends the fire plow from the back of his Jeep to create a perimeter around the blaze. The suspense only grows as an aerial patrol flies overhead to monitor the fire and direct the ground crew while the volunteer group uses flaps, rakes, and pumps to suppress the flames. The Piney Woods are saved.

Remains of a campfire indicate the source, and Bob asks how "anyone could be so careless with so dangerous a thing as fire." The fourteen-minute film is propaganda paradise, mesmerizing children and reinforcing Smokey Bear's call for care. It still works today and even compelled two modern-day youngsters to watch it twice. The *Guardians of the Pineywoods* message was duplicated in several regional newspapers during 1957 and early 1958, suggesting a coordinated educational effort. In each article, the district's personnel were introduced, often with accompanying pictures, and the descriptions focused on demonstrating how the crews were more than just rural firemen.[3] While the film created the perception of success, the truth was that by 1955 approximately two-thirds of TFS funding was being spent on the suppression of anthropogenic wildfires.[4]

Crewleaders and crewmen took the place of the lookout, patrolman, and emergency patrolman after Folweiler's 1949 shakeup.[5] Together, the two-person crews were assigned to a "duty station," encompassing approximately 180,000–200,000 acres around a lookout tower in close proximity to the employee's homestead. Proximity to the lookout was, actually, the first requirement in a typical help-wanted advertisement for an entry-level crewman. One, written for the Timpson Tower, asked that the applicant live within 12 miles of the station, be between the ages of twenty-one and fifty-five, and have had a minimum of eight years of education. The remaining qualifications were that they pass a physical examination and have experience operating trucks and tractors.[6]

In 1955 there were eighty-one lookout towers and eighty-seven crews. Forty-nine of these fire crews worked with a two-and-a-half-ton truck and crawler tractor, while the remaining thirty-eight pairs used a Jeep fitted with a middle-buster plow.[7] A stable rural population across most of East Texas meant that the labor environment was interested in supplemental work.[8] Farmers, retired military personnel, former oil-field workers, or the self-employed could perform their primary duties and engaged with the TFS during their downtime. They facilitated school lectures or posted fire prevention signs during periods of low fire danger and shifted their focus to tower staffing or fire suppression during periods of anticipated fire activity.[9] Folweiler insisted, however, that the employees drop their other responsibilities "when the Texas Forest Service demands his time."[10]

For their trouble, in 1950 crewmen received between $1,200 and $1,680 each year, and TFS crewleaders earned an annual salary between $1,750 and $2,400,[11] with the expectation that they work a minimum of sixteen hundred hours each year.[12] Compensation for a temporary forest guard on the Texas National Forests in 1948 was similar, carrying a $900 wage.[13] For comparison, an experienced, college-educated department head at the TFS earned between $5,500 and $7,000 annually,[14] while a forest supervisor working for the USFS earned between $5,000 and $6,000.[15] Things had changed very little by 1962, when a crewman received an annual salary between $1,284 and $1,752 after a six-month probation period. Crewleaders also saw only modest wage increases, receiving around $2,568.[16]

By midcentury, a person's career might progress from the responsibilities of a crewman to a crewleader through experience and the availability of vacancies created through retirements or resignations. Additional incentives were constructed to motivate employees through the "Outstanding Crewleader Awards," which were presented to only two high-performing crewleaders each year. The awards were earned through excellent performance in all phases of TFS work, including fire detection and control, equipment maintenance and operation, and influential educational activities.[17] Winners were presented with a raise and received a circular lapel pin marked with the words "Texas Forest

Figure 38. The Outstanding Crewleader Award was given annually to two crewleaders, each of whom received a lapel pin in recognition of their service. The appreciation was earned by demonstrating excellent performance in all phases of work, including fire detection and control, equipment maintenance and operation, fire prevention and educational work, and the encouragement of small landowners to practice proper forest management techniques. Image courtesy of Texas A&M Forest Service.

Service Outstanding Crewleader" (fig. 38). The commendations were so coveted that Burnside once mentioned during a staff meeting that he had seen the letters of recognition, which accompanied the award, framed and hanging in some of the crewleader's residences.[18] Folweiler decided during the same meeting that, in the future, letters of recognition would be mailed unfolded so that they could be framed if the recipient chose.

Crewleader Floy Creel was one such "Guardian of the Pineywoods" scanning the horizon above the shortleaf forests around Alto during the heyday of fire protection during the 1950s. His educational efforts are captured in numerous *Alto Herald* articles, where he reported on county fire conditions. He answered nine fire calls last week, a September 1954 summary reported, five of which were on highways. "A careless flip of a cigarette or cigar butt are the main causes of these fires," Creel pointed out, and he reminded everyone to use their ash tray.[19] Frequent reporting must have worked, and four years later Creel thanked "his friends for [a] good fire record," proudly announcing that his precinct had not responded to a fire in thirteen months. When brush or field burning, he reminded everyone, "catch a still day and be sure that you have a good safe fire lane. Burn early in the morning or late in the evening. Never gamble with fire. It is very tricky."[20]

Creel's tower at Alto was one of seventeen platform-ladder MC-40s delivered to the TFS during 1933 under contract with the USFS. In addition to newspaper campaigns targeting area citizens, the tower itself served as an educational piece and was frequented by visitors. In fact, a ledger preserves the names of all tower guests registering in the cab between 1946 and 1958. The list includes reverends, lumbermen, farmers, county agents, postmasters, and even W. T. Hartman along with USFS employees from Atlanta. The original 1934 lookout still stands in private ownership, and the structure meets the criteria of age and significance for consideration on the National Register.

Noah Platt also kept a record of all known visitors to the Smith Ferry Lookout. Platt, who appeared as an actor in *Guardian of the Pineywoods*, reported in 1963 that the list of visitors he had kept in the station since 1934 had swollen to some four thousand names.[21] Like Alto, the Smith Ferry Tower is privately owned but eligible for inclusion on the National Register.

Other crewleaders emphasized education, and frequent public engagements recognized employees who were nominated for Outstanding Crewleader Awards. Bird Mountain crewleader Lawrence Johnston, for instance, was recommended for the award because of his work with youth groups and elementary and secondary schools. What stood out to the review board, however, were the instrumental roles he played in an FFA Club and a girls 4-H Club in forestry projects. As the *Rusk Cherokeean* noted, "To our knowledge this is the only girls organization actively working an on-the-ground forestry project."[22]

Many of Texas' first crewleaders had spent time in the CCC or the military. Harold Peck, for instance, was a Canadian World War I veteran who survived a gas attack in 1917 and served at Kountze from 1952 until 1957. He was one of two crewleaders chosen for the Outstanding Crewleader Award in 1956, in recognition of his exceptional work in fire prevention, suppression, and forest management. Shortly afterward, he was promoted to the district crewleader position and served with the Forest Service until 1963, four years before his death. Director Folweiler expressed his gratitude to Peck upon retirement, writing, "It is our opinion that the public in your area will long remember your association with the Service because of your devotion in carrying out your duties. We are sure that the image of the Texas Forest Service, in the minds of the people in the Kountze crew area, is at the current high level because of your constructive efforts."[23]

The *Guardian of the Pineywoods* film and accompanying newspaper stories were being published when the first wave of career lookouts and fire patrol employees were retiring. One wave in particular, at the end of December 1956, resulted in the combined loss of 193 years of experience as Alsey Cromeens, Arthur Williams, George Hightower, and John Thigpen departed.[24] Alone, each had worked thirty or more years with the Forest Service. Cromeens had been there so long, in fact, that one of the tree cabs in Walker County was known simply as the "Cromeens Tree Cab" in survey notebooks.[25] When a nearby site became available for a steel tower in September 1933, the lease was signed by W. J. Fails, and Cromeens went to work at the Fails Duty Station.

Ider Kelley, Herbert Hare, Allen Travis, Walter Phelps, and Joe Rich were also among those leaving.[26] Kelley's career at Zion Hill was featured in the *News*. He began as a patrolman on February 22, 1927, and rode his horse between 15 and 25 miles each day watching for forest fires. As related in the article, "It was more or less a one-man and one-horse proposition." Like Jeff O'Quinn, Kelley remembered the days of being "classified as a meddler" and "found it difficult to locate a farmhouse where he was welcomed to spend the night." Dedicated to his work, at the age of sixty-seven, he was still climbing "the hundred-foot lookout tower with his helper, Herbert Hall" to look for smoke.[27]

Clearly, these men had a lasting impact on East Texas forests and the agency. Burnside, writing personally to Claud Purvis, commended his

TFS career, which began indirectly through the CCC in 1933:

> For many years you have participated actively in the affairs of the Texas Forest Service. You have put in long hours and have exerted strenuous efforts to fulfill the duties and obligations you assumed for the Service. You saw the Service when we were working with limited resources, and you have seen it grow. Doubtless, on many occasions you have wondered if the efforts exerted were "worth the price." It has been through your interest in the work of the Service, and others like you, that the Service has been enabled to attain many of its work objectives. Soon you will retire from active Service duty, all of us will miss you, but none of us will forget you.[28]

Purvis, as an LEM in the Lufkin Camp and as a TFS employee at Livingston, was busily constructing lookout towers during 1935 and 1936.

Lookout! Deaths and Near Misses

In many ways, the Piney Woods remained an insular neutral ground of small hamlets surrounded by large, recovering forests when the first generation of lookouts retired. The isolated conditions contributed to injury and near misses at several towers because of wildlife, access to health care, or accidents. The earliest recorded event occurred at the end of March 1950, when Mrs. Gene Anderson was out with three companions from Longview.[29] The four had climbed to the top of the Hallsville Tower, bouncing back and forth up the stairs of the MC-39 when the incident occurred. At thirty, she was out for a day of adventure and was just beginning to absorb the sweeping views when she fell. The tower was not staffed at the time, and County Attorney William Taylor ruled her death accidental.[30]

A second medical event occurred in 1958 when forty-year-old lookout Al Wicker was climbing his tower at Evadale. Wicker had just been employed as a crewleader,[31] and he had worked only four months before he "became ill" one afternoon. He went home to rest but passed away two hours later from an apparent heart attack, leaving four children and his wife.[32]

Encounters with wildlife caused other dangers, as was the case for Union Hill lookoutman Earl Strickland in 1955. The tower was located on a site "about halfway between Hughes Springs and Linden" on property owned by J. F. Mitchell, and when the lease was signed in September 1935, the tower bore the family's name, though it later become known as the Union Hill Lookout. Strickland worked at the duty station between 1942 and 1957, and his most eventful shift was detailed in a September 1955 *News* piece. After a day in the tower, everything was peaceful and Strickland was ready to go home. With the tower cab locked for the night, Strickland descended the steps but heard a "peculiar barking, snarling and scratching noise. Looking down he was shocked to see a fox trying to start up the tower steps towards him." He decided the fox had contracted rabies and retreated to the protection of the cab. Time passed, but the fox did not go away, finally prompting him to call the district office by radio to request help. "Educational Officer H. K. 'Dude' Byron dashed to the rescue, stopping only long enough to pick up his shotgun on the way," the article related. "In the tradition of Davy Crockett," Byron managed to shoot his prey with only one shot from 'Old Betsey.'"[33]

Cleo Perkins was startled by wildlife nine years later at his Neches post. Cleo, who retired after thirty-one years of service with the TFS, had "the shock of his life," as reported by *Have You Heard*. "Cleo was nearing the top rung of his tower," when he looked up at the cab and "was greeted by a large ring-tailed wildcat." Because "80 feet in the air is a poor place to argue about right-of-way," the article continued, "Cleo evacuated in the 'fastest 10-foot descent' he'd ever made."[34]

Employment Opportunities

The decade saw continued limitations on employment and advances. For instance, the director offered opinions on the preferred age of fire crews during the November 1956 staff meeting. Folweiler stressed that he seriously questioned whether a crewleader over the age of forty-five should be employed and added that a "man too young" who served as a crewleader might also "distract from the dignity of the position. In the Director's opinion, the ideal age for a crewleader [and crewmen] would be from 35 to 40 years of age."[35]

Later minutes from the September 1961 meeting highlighted the poor retention statistics of crewmen between the ages of eighteen and twenty, who stayed only one or two years with the TFS before accepting other employment.[36] Because of training costs, it was likewise decided that "a great deal of caution should be exercised in employing anyone under 20–21 years of age." In reviewing the employment form for a nineteen-year-old Evadale crewman in 1966, Burnside even advised the district forester that "it is doubtful whether the training you would give him is worth your time. You are well aware of the sad experience of retaining the services of very young men."[37] The district forester's reply, however, summarized the difficulty in securing any reliable personnel, given the nature of the work and salary. "We are aware of the disadvantages of employing young men," he refuted, saying "all the people in this area that meet the qualifications we require, are either employed or are not interested." Regarding the candidate's age, the forester continued, "Age does not necessarily indicate stability—Harris' predecessor was 47 years old and worked about one month."[38]

District Forester Spinney also raised the question during the October 1950 staff meeting of whether or not women observers could be employed on a full-time basis. The conversation was coming on the heels of the "emergency" hire of women during World War II and was likely initiated to find dependable workers interested in accepting low-wage, part-time work as people began migrating from rural districts. Director Folweiler replied that there was "no objection whatsoever to using women as substitutes for men to do lookout work." The most important objective, he went on, "was to have those towers manned by lookouts where such action is mandatory because of responsibility to either the United States Forest Service or participating landowners."[39]

These discussions paved the way for women like Juanita Westbrook to work on a seasonal basis. Westbrook, who lived near the Negley Tower, began her employment for the TFS in February 1953 and described her job as the loneliest in the world. She was officially titled a radio operator, relaying messages between district office, Jeep mobile units, and a "key" tower near Daingerfield. Three times a day, the *Timpson Weekly* news explained, she would climb the dizzying steps of the tower to watch over 58,000 acres of Red River forest during shifts ranging between four and eight hours. Most people, the story continued, "who start give up about half way to the top."[40] Her son Jerry Westbrook later worked at the tower, first as a crewman, then as a crewleader, between 1961 and 1973.

Meanwhile, Jeff Hunter in 1952 and C. L. Springfield in 1956 were breaking ground as pioneering African American crewmen around the time that *Brown v. Board of Education* was being fought in the Supreme Court (fig. 39). Though integration steps were modest at first, they reflected the Texas Forest Service's awareness of the benefits created through equal opportunity in fulfilling their mission. Pictures survive of both men working with colleagues and posing at their duty stations.

The Daingerfield Duty Station, where Springfield was assigned, was activated in 1951 after the Daingerfield Chamber of Commerce, the City of Daingerfield, and Bowie County purchased property on top of a prominent hillside west of the city.[41] The TFS erected a secondhand oilfield derrick at the location the following year, which sported a cabin and

catwalk designed by O. C. Braly.[42] The tower was originally staffed by Crewleader Connor, but by 1956 Carl Moursund and his crewman, C. L. Springfield, assumed responsibility for the station.

Hunter once staffed the Jefferson Tower, which was constructed 3 miles west of the city to protect shortleaf forests in central Marion County during 1936 when the fire protection network expanded into northeastern Texas. Not far from the community, the tower was frequently used for tours and, in time, became an important relay station for the Marion County ambulatory services. With the watchful eyes from several nearby residences, the tower did not receive much vandalism, and the TFS balanced the risks and benefits of maintaining the structure into the mid-1990s.

The tower was finally sold in 1998 and was supposed to be removed within 120 days of the purchase. Nevertheless, the lookout was not salvaged until 2003, when it found a new home as a cell tower at the intersection of Lafayette and Washington Streets in downtown Jefferson. Though the station is closed to the public, the Depression-era tower can be appreciated from the city and should serve as a monument to observers such as Jeff Hunter and C. L. Springfield, who made significant, but heretofore unrecognized, contributions to the protection of the Piney Woods and the diversification of the American workforce.

Figure 39. Crewmen such as Jeff Hunter at the Jefferson Duty Station (A) and C. L. Springfield (B) were pioneering African American lookout observers for the TFS during the early 1950s. Images courtesy of Texas A&M Forest Service.

Reliance on Air Patrols

By 1951, the TFS owned and operated five aircraft and contracted others during periods of high fire danger.[43] While the agency sold all of its planes by 1959, contract flying emerged as a successful model for inexpensive fire detection in the Piney Woods. The *Wood County Record* confirmed in 1953 that an increased efficiency in smoke spotting was being realized through "contract air craft patrol," which "enabled more rapid dispatching of suppression equipment to fires." Curiously, the article reported that "air detection has proved so satisfactory during the past year, particularly over Gregg, Northern Rusk, and Eastern Smith Counties, that the installation of fire lookout towers has been postponed indefinitely."[44] Still, the Church Hill, Arp, Meadows, and Starrville Towers were built in these counties nine years later.

Aerial studies during the previous decade confirmed the effectiveness of plane patrol on high hazard days, and the procedure gained acceptance quickly. In fact, a discussion at the December 1953 staff meeting focused on "ways to reduce the time spent by crewleaders on towers." District 4 forester Arthur Green advised the group that they used plane patrols almost entirely for fire detection, noting that "the towers are used only for checks when the planes are not in operation."[45] Moreover, he felt that there were times when the crews were staffing the towers when they might be better employed on maintenance or educational work. He concluded by suggesting that the fire hazard and burn index be used as the best criteria to determine when a crewleader should do educational work, staff the lookout, or rely on aircraft detection. When the discussion continued a year later at the December 1954 staff meeting about when to use aircraft or lookout towers for fire detection, District Forester Green again stated that his district staffed the towers primarily on low danger days or when an aircraft was grounded for maintenance. As the danger rating increased and visibility worsened, Green noted, two planes were assigned to aerial patrols. The most vulnerable fire period, he reported, was between 11:30 a.m. and 1:30 p.m., when residents were burning and left their fires unattended for lunch.[46]

At the same time, the USFS contracted the TFS to provide aerial fire surveillance over the TNFs and report any smokes to the appropriate USFS dispatcher. The December 1953 memorandum of understanding between the two agencies adopted a hybrid model in which ten federal lookout towers would be staffed for primary observation during fire season. In step with Green's recommendations, the agreement noted that "plane patrol may be substituted in lieu of tower manning . . . particularly during periods of low visibility or low risk."[47] In speaking of an option to replace the Tenaha Tower in 1959, Albert Mandeville even suggested the secondary tower was of limited use because of the "service we received from the State Forest Service from the airplane detection."[48]

One of the main disadvantages of the tower system was that it could be incapacitated when smoke or haze reduced visibility. Ironically this was realized as early as 1932, when Bulletin 23 noted "the heavy lying smoke caused by the numerous fires practically paralyzed the lookout system."[49] As the Gulf Coast industrialized and the FCD discussed the most suitable location to erect the Vidor Tower in 1958, the condition was recognized again. In discussing the proposed sites, it was noted that it would "give the lookout a good chance to differentiate between grass or marsh smokes, which at certain times of the year are very prevalent, and wood smokes which are much rarer. Also, being farther south, it may help in seeing through or into the industrial haze which is so prevalent along the coast."[50]

Other problems were also recognized, and the network was often understaffed during peak fire periods. As crews left a tower to respond to a blaze, their absence would create holes in the organization precisely when accurate lines of sight were needed to pinpoint a fire. Adding to these inefficiencies, the News also acknowledged that firefighters were deployed to towers without regard for the fire load or historical incidence patterns. This resulted in a work imbalance, where about 50 percent of the crews suppressed nearly 80 percent of the fires.[51]

Despite the inefficiency of tower detection and the success of aerial surveillance, the lookout network continued to develop during the 1951 to 1953 biennium because of an expansive state appropriation that increased the Forest Service's budget by 52 percent. Eighty percent of the increase, the News indicated, was earmarked for annexing an additional 800,000 acres into the forest protection scheme.[52] As part of these changes, the state acquired a surplus, three-legged, Bilby tower from the USC&GS and erected it west of Saratoga close to where W. E. White and B. D. Hawkins had positioned a tree cab in 1932.[53]

An early photograph of the tower is a curiosity (fig. 40). The image depicts a three-legged structure supported only by light, triangular braces. The cabin at the top of the tower is hexagonal, with six lights on each window. Below, a narrow board extends outside the frame of the tower and offers entry to the lookout from the base of the cab. Where most USC&GS descriptions summarize four-legged structures, records for the Saratoga Lookout noted that it is "a 3-legged steel structure supporting a metal cabin on top."[54]

Figure 40. The three-legged Bilby tower at Saratoga in 1950. The structure was purchased from the USC&GS and modified for use as a fire lookout. The corner-post ladderway and light construction made it unpopular with fire crews, and the tower was replaced by the Aermotor MC-40 salvaged from Livingston. Image courtesy of Texas A&M Forest Service.

The shaky structure was immediately unpopular, and the district forester suggested replacing the tower using the surplus components from Livingston (Location 1) that were stored at the district headquarters. Among the complaints, modernization was recommended because of the danger in climbing up the structure's corner post and the difficulty for the crew in accessing the cabin without rest platforms.

Before any changes were authorized, Burnside called upon Braly, requesting "the benefit of your experience, knowledge, and [any] recommendations" that may improve safety at the structure.[55] Braly visited the site later that month. Based on his examination, he replied that the "inspection shows that it can be modified at a lot of expense, and you still wouldn't have much of a tower. It is made of light material and is weak.... I wouldn't climb the tower if there was over a ten (10) mile per hour wind."[56]

Leases were also signed for the Wolf Hill (Location 1) and Ariola Stations during the spring of 1950. Paxton and Redwater followed and were all built by cannibalizing watch towers from other portions of the system. New towers sprung up too, and in 1952, six were erected using repurposed oil derricks (table 9).[57] These, the *News* reported, were favored instead of purchasing steel lookouts from major manufacturers because of "the critical steel shortage,"[58] which made allocations to the state uncertain. Perhaps more important, the cost of the derricks was far less than the branded alternatives. In an effort to further drive down expenses, each of the secondhand towers was surmounted by a waterproof plywood cab designed by none other than O. C. Braly.

The Magnolia Tower serves as an example. The 96-foot EMSCO derrick was completed at the margin of the Piney Woods in western Montgomery County, in a region the TFS had desired to protect since, at least, the CCC's Thirteenth Period work plan in 1939. Braly erected the tower and constructed a custom-built cabin on top of the vantage point, which he surrounded by a catwalk. Once completed, Bannon Damuth and Henry Martin became long-term observers at the tower. Damuth

Table 9. Lookout towers constructed between 1950 and 1959.

1950	1951	1952	1953	1954	1955	1956	1957	1958	1959
Ariola from Fred	Paxton from Rocky Hill	Cairo Springs	Elysian Fields from Yellowpine 1		Dreka 2 from Arkansas NF	Cyclone Hill 2 from Vilas, Florida (?)	Livingston 2 from Soda	Candy Hill to Grapeland	Elkhart from Bath
Saratoga 1	Salem	Daingerfield	Yellowpine 3 from Mississippi		Liberty Hill USFS 2 from Ozark NF, Arkansas (?)	Nogalus 2 from Ozark NF, Arkansas		Vidor from Cyclone Hill 1	Piney Woods (Piney) 2 from South Carolina
		Gilmont						Weches 3	
		Magnolia						Wolf Hill 2	
		Mauriceville							
		Redwater from Hooks							
		Smithland from Karnack 1							

received his firefighting instruction as a member of the CCC and worked at Magnolia for nineteen years (fig. 41). Martin, once a farmer and USFS employee, worked at the tower for sixteen years.

Recycling surplus oil-field equipment became so economical, in fact, that later in the decade wooden lookouts at Weches, Wolf Hill (Location 1), and Bon Wier were replaced with derricks from Talco Oil Field. The correspondence for these construction projects captured conversations on the potential relocation of the Weches Lookout to the site of the little-documented Pine Springs Tree Cab and benchmark.[59]

The Candy Hill Tower was built during the 1958 campaign to replace the aging wooden lookouts, further expanding fire services into the western portions of Houston County. The TFS explored two potential sites around Latexo and Grapeland for the tower. The first was midway between the two communities. The second was approximately 1.5 miles north of Latexo and had a slightly lower elevation.[60] In discussing details, Director Folweiler compared the cost of an Aermotor stairway tower (LS-40) with one with an inside ladder (MC-40), noting that "an inside ladder type of tower costs about $1100 less, but I personally would prefer the sort of ladder with which an oil derrick is equipped rather than the inside ladder type. I consider them more dangerous to climb than an outside ladder."[61]

No decision on the type or location of the tower had been reached three months later, and the director planned a day of fieldwork on

Figure 41. Bannon Damuth on the Magnolia Tower catwalk in 1963. The lookout was a repurposed oil-field derrick. Image courtesy of Texas A&M Forest Service.

September 3, 1958, regardless of weather, to sort out the details. A few days later, the 122-foot Lena Wims #6 derrick and three others were purchased for $200 apiece, and a lease for the lookout was signed for the site 1.5 miles north of Latexo.

By 1961, the original lessor had sold the property, and the new landowner contacted the TFS requesting that the structure be moved. Problems festered into the following year, and the TFS regraded the road to the tower and added drainage in an attempt to appease the landowner. By 1964, the new owner, "by error or design,"[62] locked the Forest Service out of the premises by changing the gate lock. The following month O. C. Braly removed the tower and rebuilt it at "a site originally selected as an alternative."[63] At the new Grapeland location, the tower piers were arranged "on a north-south, east-west axis, so that one side will face due north."[64] After seven days of cement curing, erection of the steel parts began. Once the tower was complete, Braly finished the woodwork and insulation within the cab.

The tower is still owned by the TFS for use as a communications relay and is the only Nashville Bridge Company structure left in the detection network. Field observations indicate that the corner posts at Grapeland are fully one inch wider than those of a comparable 120-foot Aermotor tower, like the one at Kirbyville. The wider posts, more rigid members, and asymmetric design, with an A-framed opening on one side meant to accommodate drill pipe, all hint at the structure's role as an oil-field derrick. Likewise, the outside ladder and safety platform about halfway up the structure are also characteristic of drilling derricks.

Attached to the ladder is a Saf-T-Climb device, a notched carrier rail that extends the length of the ladder. The TFS tested the device in 1967 and quickly retrofitted other outside ladder towers with the climbing aid. The kit contained ring clamps to fix the rail to the ladder rungs, a Saf-T-Lock sleeve, and a belt, into which observers could harness themselves before ascending the tower. The device was installed at Grapeland during June 1969.

Despite eligibility because of age, the former derrick does not appear to meet the standards set forth by the National Register, at least not for a fire lookout structure. The tower is significant, however, in that it demonstrates the FCD's ability to expand protection by maximizing its budget and utilizing alternative fit-for-purpose structures. Because the Grapeland Duty Station represents the only remaining example of a Nashville Bridge Company lookout, it should be photographed, described, and assessed by a qualified engineer before being declared surplus or removed. Perhaps the intertwined histories of O. C. Braly, the TFS, and the legacy of former observers can offer some protection at the community or state level.

Uncertainty and Renovations on Texas National Forests

The future of the TNFs seemed uncertain during the early 1950s. Folweiler announced during a December 1953 staff meeting that Hubert Harrison, manager of the East Texas Chamber of Commerce, advocated sale of federal lands within the state.[65] The Chamber of Commerce was acting on the belief that public scrutiny advanced healthy democracy, and the group raised concerns over the value, need, and administration of the forests. During Folweiler's staff meeting, he explained that the news coincided with President Eisenhower's plan to minimize the federal government's role in business but advised that the TFS had "no direct or indirect concern" in the matter and should avoid "any discussion or arguments on this question."[66]

That same month, Harrison contacted Sherman Frost to make a study on the "problems of the national forest land ownership in East Texas." Frost was well acquainted with Texas forestry, having worked for the USFS acquisition surveys during 1934 and 1935, as well as with the TFS between 1936 and 1948. Frost accepted the project and framed his argument against federal ownership in the opening paragraphs of his report: "The 658,000 acres of national forests in East Texas, purchased from private owners during the depression in 1934, were not established primarily to rebuild forest lands but to aid unemployment. Congress did not vote on their purchase directly but the funds were made available by President Roosevelt from emergency conservation appropriations."[67]

To understand the history of federal land acquisitions in Texas, Frost explained how appropriations were made under the Emergency Relief Act of 1933. At the time, Roosevelt was authorized to spend about $4 billion on relief measures, around $350 million of which was earmarked for emergency conservation work. Through other executive orders signed on May 20 and July 1, 1933, and December 1, 1934, an additional $30 million became available to purchase lands for forest conservation. As the 1934 report of the National Forest Reservation Commission pointed out, the June allotment of $20 million for the acquisition of forestlands "was motivated primarily by the desire to create the best obtainable conditions for effective work in the Civilian Conservation Corps and by local populations in acute need of relief employment."[68]

Peter Stark provides a more detailed summary of Roosevelt's executive orders.[69] The first meeting of the National Forest Reservation Commission in May 1933 focused on acquainting the cabinet with the work of the committee. During the meeting, the committee chairman and secretary of war, George H. Dern, admitted he had just learned about the National Forest Reservation Commission and that he was its chairman. Nevertheless, he recognized that his difficulty in finding enough projects for CCC work programs might be solved if more federal land could be acquired on which to place additional camps. With support from Forester Robert Y. Stuart and the committee, the National Forest Reservation Commission passed a resolution endorsing the recommendation contained in Public Law 73–5: that $25 million be allocated to the Department of Agriculture for use in carrying out land purchases already approved by the National Reservation Commission.[70]

Roosevelt approved $20 million with Executive Order Number 6135, "Purchase of National-Forest Lands," on May 20, 1933. This allocation was reinforced in his June 7 Executive Order Number 6160, "Administration of the Emergency Conservation Work." Then, with the new fiscal year on July 1, 1933, Roosevelt repealed both orders and issued a new one, Number 6208, reauthorizing the $20 million for 1934. In the meantime, Executive Order Number 6181, dated June 24, 1933, lifted the restriction that the $20 million be expended only with the preexisting national forest purchase units. This simple revision allowed the money to be spent on any purchase units thereafter established by the secretary of agriculture and approved by the National Forest Reservation Commission. Finally, another $10 million allocation was made for the 1935 fiscal year on December 1, 1934, under Executive Order 6910-A.[71] Importantly, Stark noted:

> These allocations did not carry the requirements of an ordinary congressional appropriation for purchase of forest lands under the Weeks Law, namely, that the lands be purchased for the protection of watersheds and for timber production. They only carried the

requirement that the procedures for purchasing land for conservation work on the national forests as established by the National Forest Reservation Commission be used in expending the amount.[72]

Regardless, Frost also took aim at the watershed protection provisions of the Weeks Law in developing his argument against federal landownership in East Texas without considering the implications of Executive Order Number 6181 or that the lands were acquired after the introduction of the Clarke-McNary Act. Because the TNF tracts were "situated mostly in flatwoods sections, far down the course of main rivers," he reasoned, "it requires considerable imagination to associate them" with watershed protection in the way that the law was designed for forests in the East.[73] Frost's interpretation of watershed protection also weakens his case and creates a noticeable flaw in his conclusions. Quite simply, a watershed or drainage basin is the catchment region from which water drains into a stream system. A simple map check of the Palestine or Beaumont quadrangles proves that the Neches, Sabine, San Jacinto, and Angelina Rivers are integral parts of the forest or forest boundaries. That many acres of TNF land were drowned during reservoir construction projects reinforces the fact that the forests were in proximity to the principal drainage basins in East Texas. More significantly, the area's association with navigable waterways was instrumental to Foster's effort in securing matching federal funding two decades earlier.

Frost also analyzed the language of S.C.R. No. 73 and pointed to the special forestry committee created in Texas during 1923 to study forest tax and land purchasing problems. When the committee submitted its report to the Thirty-Ninth Legislature in 1925, the panel concluded that "the acquisition of large areas of land in Texas by the federal government would not be in accord with factors of state policy subscribed to by the majority of our citizenship."[74] Specifically, when S.C.R. No. 73 was authorized by the Texas legislature a decade later, the federal government was permitted to purchase only the various forested tracts in an "effort to relieve unemployment."[75] Piggybacking on this idea, Frost suggested that there were no provisions in the Weeks Law that permitted the government to purchase forests for such a purpose.

The *Texas National Forest Study* makes some compelling arguments in favor of diluting Texas' federal timberland holdings. For a long time, the Texas National Forests—like all the national forests—were in the "business of growing and selling timber." Gary Snyder once described it as a "shopkeeper's view of nature,"[76] while Jack Kerouac thought the USFS "boasts so proudly of the number of board feet in the whole forest."[77]

Pragmatically, Frost recognized that an argument could be made that the TNFs encroached upon private enterprise. "To what extent," he contended, has this "business activity, operating on public tax money, become restrictive, competitive, or injurious to private tax-paying enterprise?"[78]

The study also compared the revenue counties would have received from land taxes with the unpredictable cash flow generated by government timber sales. Here, Frost argued that the variable income supplied to the counties, 25 percent of the total timber sales in that county, made it difficult for planners to forecast budgets. Moreover, a portion of the capital meant for education arrived in counties with higher forest cover, lower population, and fewer schools, while other counties within the national forest boundaries that were more heavily populated received less capital from timber receipts.

Frost had helped appraise the original timber lots before they were purchased by the government, and he contended that not all of the parcels were "devastated," as the federal government claimed. While some areas of the Piney Woods were barren at the time of purchase, there were other commercial properties that had regenerated into healthy secondary forests. Frost pointed out that the cutover dates for many Houston and Trinity County tracts were as early as 1908 and 1922. By at least one private estimate, these properties had a minimum of 3,000 board feet per acre by 1934. Timber management plans developed during the late 1930s for both the Davy Crockett and Sam Houston National Forests also suggest that the tracts supported healthy growth. Contrary to the perception of a ravished landscape, it was the "depressed market demand for lumber . . . high taxes, timber supply plentiful enough to keep prices down, and general business depression [that] were important reasons which led many Texas mills to liquidate their operations."[79]

The study's recommendations were clear. It would be in the public's best interest for the lands to be sold "back into private ownership in units such as to promote the best economy." Without precedent, the recommendation also advised that each sale should "carry a covenant in the deed providing for continuing adequate forestry management" and that "the lands be kept open to hunting and fishing."[80] Recreational areas administered by the USFS, the report advised, should be donated to the counties or State Parks Board. It seems likely decades later, however, that these provisions could cause conflict with landowner rights and individual freedom.

Certainly, the TNFs were on the defense. A pamphlet, titled *Facts about the Forest*, was printed to articulate the mission of the USFS. The purpose of the forest, it related, was to "protect, grow, and harvest the timber crop; to promote watersheds so as to prevent erosion to preserve and develop public recreation areas; and to encourage fish and wildlife restoration." Not mincing words, the booklet continued that the "invitation and consent" for the government to purchase land was approved by an act of the Texas legislature, though no mention is made on the scope of the original legislation.[81]

It appears that the relationship between the USFS and TFS did not suffer between the mid-1940s and mid-1950s during the push to reclaim the territory, probably because of Folweiler's preemptive moves to establish a neutral agency position. If anything, the duties of both organizations became better articulated through detailed memoranda of understanding and the division of responsibilities. At Bronson Lookout, for example, Johnnie Page's contract made clear that he was to cooperate with personnel in the adjacent federal lands:

> Information concerning the present boundary which separates your patrol district from the U.S. Forest Service lands has been furnished. That boundary between your district and the U.S. Forest Service holdings is known as the "fire action boundary." As a representative of the Texas Forest Service you will be responsible for the protection from fire of all lands within the "fire action boundary" in your district while the U.S. Forest Service is responsible for the protection of National Forest lands on their side of the "fire action boundary." In cases where cooperative protection unit acreage lies inside the U.S. Forest Service "fire action boundary" in or adjacent to your district, such protection unit lands are to be protected by the Texas Forest Service. You will cooperate with the U.S. Forest Service representative to the fullest extent possible both in suppression and detection of forest fires. Should either organization find itself in need of some cross shots from towers which may not be occupied, such towers will be occupied upon request within the limits of the ability of each Service at the time. Should you perform any suppression work on fires burning National Forest lands, you will promptly furnish to the U.S. Forest Service dispatcher the necessary fire report data. In cases where a fire burns both National Forest lands and State or private lands, the U.S. Forest Service will furnish this office with a report covering the private acreage burned.[82]

The memorandum of understanding also required the TFS to staff the Bronson Tower on all days when the fire danger rose above Class III. Thus, the TFS installed a danger rating station there in 1950.

Though not in direct response to the 1954 *Texas National Forest Study*, decreased federal funding, positive changes in land-use patterns, and advancing forestry methods led to a reduction of TNF acreage during 1956. While the proclamation boundary for the Angelina Forest retained its 391,300-acre size, for example, the gross acreage of the forest was reduced to 284,831 acres. Likewise, the size of the Davy Crockett National Forest was reduced to 281,104 acres, the Sabine to 380,957 acres, and the Sam Houston to 235,484 acres.[83]

Perhaps the balance of USFS objectives has shifted, if only slightly. Frost's criticism that "data is not kept at the Texas National Forest headquarters at Lufkin" and is accessible only by traveling to the "regional office in Atlanta, Georgia, 700 miles away, or Washington, D.C." still holds true.[84] Fortunately, the TNF survived, and black-and-white photographs in the Stephen F. Austin Digital Collection document renovations and technological improvements in the federal lookout system that occurred during the same decade.

One intriguing picture taken during September 1950 carries the description "dwelling at the Apple Springs Work Center—constructed from two smaller cabins—Trinity District—Davy Crockett National Forest." The image shows two guard cottages conforming to Plan B-700, comparable in design to those at Cyclone Hill or Yellowpine but salvaged from Mt. Zion and Piney Stations and joined by a flat-roofed, screened breezeway. The implication, of course, is that secondary lookout stations were being left idle or that observers were moving out of the woods away from compounds at some primary guard stations by the 1950s. Other images in the Center for Regional Heritage Research at Stephen F. Austin State University's collection also capture Ranger Robert F. Irwin as he instructs crews on how to properly operate the new, mechanized fireplow tractors.

Other federal lookout towers were receiving a makeover at the beginning of the decade. The IDECO tower at Yellowpine, initially part of the state's detection system, was back in TFS hands and repositioned at Elysian Fields following a USFS decision sometime around 1948 or 1950 to redesign the duty station several miles to the northeast, where they erected a live-in tower (fig. 42). A few years later, the wooden lookout at Nogalus was also updated using a surplus conventional steel tower. Meanwhile, EMSCO towers at Cyclone Hill, Dreka, and USFS Liberty Hill were removed and replaced over safety concerns for the exposed ladderways on the outside of the structures. Some of this surplus was purchased by the TFS, such as the Cyclone Hill Lookout, which was re-erected to protect additional "fringe" acreage in Orange County.

Figure 42. Fire Dispatcher Albert Jones checks a smoke from the spacious cabin atop the 100-foot CT-4 lookout at Yellowpine on the Sabine National Forest during September 1950. Three years later, Jones identified several large vertical cracks in the tower legs, and a thorough inspection was concerning enough that a secondhand steel structure manufactured by McClintic-Marshall was transferred to the duty station from Mississippi. Photograph by Daniel O. Todd. Image courtesy of the National Archives.

The Aermotor MC-39 that superseded the first USFS Liberty Hill Tower persists at the historic duty station near the original tower piers on the west side of FM 2025, thanks largely to a radio relay and weather station at the site. One of only two standing federal towers in Texas, the Liberty Hill Lookout is regionally significant and worthy of local preservation. As a second-generation structure, however, it does not meet the strict criteria for inclusion on the National Register. Even so, the tower should be refurbished and opened to the public for demonstration and education, as it is within reach of large urban centers like Houston and is adjacent to the Double Lake Recreation Area. The tower's proximity to the Big Thicket Work Center and its location alongside a major thoroughfare mean that the structure can be easily patrolled and protected.

In part, the continued use and infrastructure replacement at USFS Liberty Hill and other stations are documented by two generations of piers at many federal lookout sites and the configuration of the anchor rods, landings, and relay slabs, which are all mappable attributes. Most of the succeeding towers were transferred from other Region 8 forests, such as Dreka and Nogalus (Tower 2), which arrived from Arkansas, and Yellowpine (Tower 3), which arrived from Mississippi.

8. Unit 308
The 1960s

ALFRED FOLWEILER'S INSISTENCY on standardization and documentation are preserved in unusual ways. One example is his requirement that crewleaders keep summaries of how, and where, they were spending their time.[1] With the continuous threat of being audited, crewleaders like E. O. Lowery of Unit 308 in Etoile diligently wrote daily entries in their red, stiff-covered, moiré clothbound diaries.[2]

The Etoile Lookout, which predated ECW programs, was erected on the summit of a high hillside about 10 miles south of Chireno as the state's sixth tower in 1931. E. O. Lowery and his son, Mervis, both worked as crewleaders at the station. E. O. Lowery began his service career as a seasonal employee in 1944 and became a crewman in 1956.[3] He was promoted to crewleader a year and a half later and held that position until his retirement on December 31, 1965. He passed away shortly afterward in June 1966.

Mervis assumed his role as crewleader that year. He was interviewed by the TFS about his experiences in 2015 and recalled working with his father. The climb to the top of the IDECO tower "was straight up," he told the interviewer, and "he hated it from the word go." Without a harness, "you'd climb up so far, about twenty feet or whatever, and they had a step out. And you could rest a minute and then you'd climb up, step back over to the next one." At the top, Mervis awkwardly had to "hang on that thing [the ladder] with your arm and unlock that door."[4]

The lookout was in Mervis's life since childhood, and he remembered putting his feet in the pockets of his father's overalls and holding on to reach the top. When Mervis was crewleader, he recalled climbing the tower four times in one day and using a bucket as a latrine to avoid the trip up and down the tower during rest breaks.

By 1957 the IDECO showed signs of age, and between sixty and eighty members required repair. At the same time, crews reworked the ground wires to prevent rusting. But the structure was becoming unsafe, and during 1961 Director Folweiler approved replacing the tower with a more modern, taller Aermotor. The change occurred in 1963, and benchmark descriptions captured the renovations. A 1968 entry noted that the directions to the station were adequate, "except the lookout tower has been torn down. A new lookout tower has been erected about 135 feet southeast of the station."[5]

This tower served until it was listed as surplus in 1981.[6] But without bids on the structure it continued to stand into 1983, at which point a negotiated sale was authorized. Later that year, the structure was sold for $200. The tower was not removed immediately, though, and the corporate landowner permitted the TFS to release the lease with the tower standing on the premises in 1984.[7] Later, it was lifted by crane and hauled southward in one piece by truck, where it still stands today.

The Aermotor LS-40 preserves an original "Chicago, Ill." placard, a novelty because it may have been one of the last towers manufactured in Illinois before the company moved its headquarters to Broken Arrow, Oklahoma, in 1964. Observers can also find the TFS property inventory number for the Etoile Tower, "11,125," stamped on one of the tower's corner posts. While the tower has been relocated, is comparatively new, and stands in private ownership, it is a tangible portion of the state's lookout legacy and the Lowery family's commitment to the Etoile crew area.

The journals E. O. Lowery kept were state property, Folweiler proclaimed in 1960,[8] and were to be retained by the crewleader for a minimum of one year before they were submitted to the district office, where they would be kept for no longer than three years. Throughout the year, the diary was never supposed to be far from the briefcase, expanding envelope, and binder issued to each crewleader. Inside was a copy of the "Manuals of Instruction for Crewleaders," extra FCD-6 fire reporting forms, educational materials, and equipment such as a compass, an extra grid for pinpointing fire locations, and a damage meter.[9]

The short diary entries, written by Lowery in flowing, sometimes difficult-to-discern cursive, capture his days.

- Sunday, February 14, 1960. Plane patrol
- Thursday, June 2, 1960. Went to Chireno tower. Was OK. Tower duty. Was a smoke south of tower. Went to Braum Ferry looking for it. C-39 said it was at Moore farm. Was contained.[10]

Four years later, Lowery wrote:

- Sunday, October 25, 1964. Tower duty. Got word there was a fire in San Augustine Co. 633.50–1 at 1900. Got hold of Hill. 312. Put it out.
- Thursday, December 24, 1964. Left pick up helper. 1200—around tower. 1300 left tower. 1500 around hdq [headquarters]. Cont. [contact] hunters on way to tower warning fires.
- Friday, December 25, 1964. Left pick up helper. 1200. Plane patrol until 1500. Around Hdq 1600.[11]

What becomes clear from these entries is that Lowery was increasingly splitting his time between tower duty and plane patrols during the early portion of the decade.

Maintenance and Network Expansion

Figure 43. The IDECO lookout at Jones State Forest (State Forest Number 2) in 1945 (A). F. R. Balthis stands at the base of the tower. Ivie and White recorded their names in the curing concrete piers while constructing the original tower on August 30, 1931 (B). The artifact was in situ as late as 1985 but has since been destroyed. Corrosion and swelling of the folded angle braces compromised the integrity of the original structure despite efforts to stiffen the struts using surplus angle iron (C). Images courtesy of Texas A&M Forest Service.

As many of the original lookouts showed signs of deterioration and crewleaders like Lowery were augmenting their time in the air, an agenda item titled "The Place of Lookout Towers in Forest Fire Detection" was scheduled for the February 1962 staff meeting. The discussion focused on the necessity of replacing or repairing some structures if they were to remain in service. Burnside and Folweiler reviewed an opinion written by O. C. Braly, again noting that he had "much experience in the maintenance of oil derricks." Braly's conclusion was that it was "almost as cheap to replace the towers as to make the major repairs needed."[12]

But the question was bigger than the cost of replacement. The director had asked Burnside to initiate a study and report on requirements for each district if air patrols were used as the sole means of detection. Was it really worth the expense of reconditioning the towers? Burnside reported it was, based on the analysis of several years' worth of data. According to his estimate, an additional 9,693 hours of flying time would be required over the 3,586 hours budgeted in the 1962 fiscal plan. The cost, he said, would approach $100,000, well beyond the $37,100 allocated for this purpose. In addition, he determined the annual depreciation of the towers and tower radios occurred at a rate of $23,120 per year, assuming a forty-year life for the towers and a ten-year life for the radios. The conclusion reached by the director after studying the opinion was that the most economical method of fire detection continued to be a combination of lookout towers and airplane patrols. "In view of this," the meeting minutes noted, "there will start this year a replacement of at least those ten towers that have deteriorated most."[13]

Two contemporary documents, as well as correspondence between Burnside and Braly, capture the development of these results. The first, a report, titled "A Visit to Texas Forest Service Watchtower at District No. 6 Headquarters South of Conroe," surmises that "ten towers are now showing evidence of deterioration." The majority of these towers were "predominantly but not exclusively" within the southern part of the forest area. Eight were some of the state's first International lookouts, one was an Aermotor, and one was an EMSCO.[14]

The report focused on the maintenance history of the International lookout at Jones State Forest and was written exactly thirty years after the tower was erected.¹⁵ The piers for the structure had been poured in the presence of Knox Ivie and W. E. White on August 30, 1931 (fig. 43), and by November the *Rusk Cherokeean* announced the completion of the station. In its reporting, the newspaper noted that "more intensive fire protection for State Forest Number 2, near Conroe, and for the surrounding woodlands within a radius of 12 miles became effective the first week in October with the completion of a new steel 90-foot tower just inside

Unit 308   **185**

the northern boundary of the state forest."[16] The following year, the TFS annual report boasted that the tower was "thoroughly tested" when "several fires on or threatening the State Forest were 'picked up' by the lookout, reported to the Superintendent, reached, and extinguished before 1/4-acre in area had been burned."[17] During these early years, cross-shots were often obtained through the cooperation of another lookout stationed at the cupola of the courthouse in Conroe.

By 1953 tower members began to deteriorate, and approximately sixty underwent some degree of strengthening or replacement.[18] Like it was for most lookouts, the construction of the tower was relatively simple, and each of the four corner posts was kept in position by horizontally and diagonally positioned angle braces. These braces, the inspectors observed, had been pressed together at both ends, folding the angle iron to form a flat surface. During construction, the folded side of the flattened braces were positioned upward and bolted to the corner columns or the relevant members.

Not unlike other towers with this construction, deterioration appeared to be chiefly in the folds at the end of the angles, "apparently as the result of the corroding away of the zinc coating and subsequent rusting of the steel." Where this happened, the iron oxide accumulations forced apart and swelled the steel to the point that the structural integrity of the tower was at risk. Interestingly, the report recommended using "tubes rather than angles for braces" or angle braces without folded corners.[19] In fact, EMSCO, L. C. Moore, some Aermotor, and some International designs did just that.

The second collection of inspections related to the repair or replacement of the system's aging towers is a series of brown field notebooks.[20] Inside are a tally of part numbers and the measured size of the swells observed during Draftsman Obie Wayne Havard's visits to each of the lookouts. After the data were compiled, Burnside wrote a memorandum to the director suggesting that repairs be made on the DeKalb, Woodville, Kirby Forest, and Ariola Towers. He also suggested that eight towers be replaced, those at Elysian Fields, Bird Mountain, Cushing, Jones Forest, Elkhart, Etoile, Liberty Hill and Mauriceville. The tower at Evadale, which had been moved from State Forest Number 1 and perched on a 20-foot extension in 1949, would be dismantled and reconstructed using the 1949 base with a new 80-foot tower.[21] The director approved these recommendations in May 1962. Two of the replacements, Elysian Fields (Winston Ranch) and Etoile, have been moved from their original positions but stand in private ownership.

Twisting back and forth up fifteen short risers to the top of the 100-foot Winston Ranch Lookout makes it clear why observers preferred the longer stairways and triangular landing platforms of the MC-39. Stretching the risers from corner to corner up the frame eliminated six landings for a tower of equal height and offered more direct access to the base of the cab—not that the extra effort to reach the top seemed to bother me, or my children, during our visit.

Everyone was smiling as the windows opened and a breeze rushed through the clapboard room, drawing our eyes from the interior to the rich green landscape that merged into the purple horizon outside. If the unusual opportunity to become a card-carrying member of the Texas chapter of the Ancient and Honorable Order of Squirrels was not enough, then the uninterrupted view of the Winston's 3,400-acre tree farm emphasizing the economic role of forestry in Texas could only make us reminisce about what the scenery must have looked like for midcentury observers.

Small wonder our host received the 2014 Leopold Conservation Award, presented to private landowners who demonstrate voluntary conservation through the ethical and scientific management of working lands. Part of the habitat improvement program for the Winston Ranch involves prescribed burning, which has encouraged a mixed southern pine forest, native bluestem, and wildflowers to reemerge on what was, at the time of purchase in the 1970s, a sandy watermelon patch. Along with creekside hardwoods, the tract provides nesting spaces for wildlife, habitat for pollinators, and a refuge for migratory birds. From the top of the tower, one can only imagine what Aldo Leopold, who spent a field season as a Yale student in the forests near Woodville, would have written about a "tramp" through Winston's tract or a survey over the pines from the tower.

The LS-40 lookout is in remarkable condition, another testament to the landowner's pride in the land and the history of East Texas. Constructed at Elysian Fields in 1963 and moved to the Winston property in three segments during 1999, the tower still boasts its original Aermotor placard and the state's property number, stamped subtly on the same corner post. The only renovation to the structure appears to be the cabin's windows, which were probably replaced because of vandalism before the tower was acquired by the Winston family.

This alteration, together with the fact that the tower has moved from its original location and postdates Depression-era construction projects, excludes it from inclusion on the National Register. The educational value of the tower, its setting, and its sense of place make the Winston Ranch Lookout regionally significant, and the variation in design hardly distracts children—or adults—who want to more fully appreciate the social and economic role of forests in Texas.

Table 10. Lookout towers constructed between 1960 and 1969.

1960	1961	1962	1963	1964	1965	1966 1967 1968 1969
Arp	Barnum (Chester) *from Barnum*	Cushing 2	Pool 2 *from Polkville, Mississippi*	Grapeland *from Candy Hill*	Adams Hill	
Church Hill	Karnack 2 *from Scottsville*	East Mt.				
Kenefick *from Cleveland*	Moscow 2 *from Moscow 1*	Grayburg				
Latch	Saratoga 2 *from Livingston 1*	Leesburg				
Maydelle 2 *from Maydelle 1*		Tenaha 2 *from Fullerton, Louisiana*				
Meadows						
Pine Mills						
Shepherd 2 *from Chireno*						
Starrville						

Lookout renovations and construction continued during the 1960s (table 10) despite the recognition that air patrols were often more effective at fire spotting. Funding provided by an appropriation bill passed during the first special session of the Fifty-Seventh Legislature assured completion of four steel lookouts to protect 500,000 acres of post oak forest in Wood, Camp, Upshur, Franklin, and Titus Counties.[22] In turn, towers manufactured by L. C. Moore went up at Leesburg, East Mountain, Latch, and Pine Mills. The L. C. Moore derrick towers were purchased secondhand and are unusual because of their tubular leg construction and baseplates.

Towers were also cannibalized during the decade to provide expanding coverage and protect additional fringe acreage. Folweiler made this philosophy clear in a 1957 letter to Burnside: "It is my opinion that duty stations on the periphery of a district need a tower worse than duty stations on the interior."[23] For this reason, the Cleveland Tower was moved to Kenefick, and Chireno was moved to Shepherd (Location 2). The latter tower was part of Lowery's crew area, and his 1960 diary frequently noted security trips to the Chireno Lookout. The frequency with which he visited the tower was the result of the inconsistent assignment of a permanent crew at the location. Beginning in the mid-1950s, the Chireno

Station had been out of active service, and there had been "considerable vandalism." A 1956 memo from Ken Burton even suggested that there was "no use in keeping [the] tower in usable condition under such continuing maintenance costs when it is not being used."[24] The tower was pressed back into active service when Folweiler received the news, but it was soon moved to extend coverage to the margins of the network.

The Chireno Tower, an Aermotor MC-40, came down on December 3, 1960. Lowery's diary captures the days leading up to the transfer:

- Thursday, December 1: Went to Chireno tower to cont [contact] Mr. Singleton about moving tower out over his land. Plane patrol in p.m.
- Friday, December 2: Went to Chireno tower for tower take down. Plane patrol.

Then the district forester informed Burnside on December 6 that "the Chireno tower was dismantled and moved from site on December 3, 1960."[25] That Saturday though, all Lowery wrote in his diary was "mother passed away. Port Neches." She was buried the following day.

There were other optimizing moves throughout the early 1960s. The Moscow and Barnum Towers were relocated to be closer to all-weather roads and power supplies, while the Scottsville Tower was moved south of Karnack to replace the lookout at Caddo Lake State Park. Also, the Maydelle Lookout was moved off Fairchild State Forest to a new location north of town, and the three-legged Bilby tower in Saratoga was replaced with the 1934 lookout salvaged from Livingston (Location 1).

Because of the decision to operate both the tower and aerial patrol systems, other routine maintenance continued during the 1960s. Included within the program were reapplying putty to the window lights, repainting the interior of lookout cabs, and repainting the exterior members of the tower to prevent rust. District 5 wrote to Burnside in August 1962, seeking advice on the "best practices" for painting the roof and exterior walls of the cab. Burnside's short memo is whimsical. "The other districts have been doing their own painting," he reassured District 5. "This has been accomplished by placing 2 x 10s through the windows and tieing [sic] them down so they will not tip up. A long handle is then attached to a brush which enables the man to paint the roof without getting on it." He concluded the letter by suggesting, "If your personnel do not want to do this work, you will have to advertise for bids according to the procedures outlined in the Fiscal Manual."[26]

Memoranda and agenda items increasingly discussed tower isolation during the late 1950s and early 1960s, indirectly indicating that the towers were being left unstaffed. Folweiler had written to Burnside in March 1957 to remind him that there had been several "depredations" in District 1 during the past several months in addition to the radio theft at Mt. Enterprise.[27] Folweiler recommended preparing a letter to all district foresters requesting that each place a chain across the road leading to the tower site. The following week Burnside complied, writing to districts that "many of our tower sites can be isolated by placing a chain across the road so that the tower cannot be reached without the removal of this barrier." But he continued, "Naturally the barrier will have to be stout. The very fact that a barrier is there will invite some people to attempt to beat it down."[28] District Forester Green at the Woodville office replied, "We are of the opinion that unless the access road to a tower can be blocked, some distance from the tower, there is very little use in fencing it. If people can drive to a tower they wouldn't mind climbing over a barricade." In the same letter, Green noted that several towers in his district were already isolated with fences and that the Livingston Tower just south of town had been fixed so that the bottom section of ladder on the Aermotor MC-40 could be pulled up and suspended, "thereby isolating the tower."[29]

Changing times were increasing unauthorized visitation and driving up maintenance costs. At least one person had perished from a fall from such a structure. Part of the problem was that the TFS continued to erect and maintain highway identification signs at each lookout as a form of advertisement for the existence and mission of the Texas Forest Service. In 1961, Ken Burton commented during the November staff meeting that perhaps it was best to remove the highway signs as a means of reducing "honorable or dishonorable" visitations to the lookout towers. The director disagreed, however, and felt that signage was useful advertisement and offered that "just because a fence is constructed around the base of the tower merely means that visitors who have depredation in mind, or would be tempted without a fence, are discouraged." Furthermore, he thought, if anyone wanted to visit the tower while in use, the personnel should certainly welcome the visitor. In another amusing moment, and with "a general lack of enthusiasm for the proposal," Folweiler even authorized Burton to experiment with the use of "Warning, High Voltage" signs on tower fences to deter would-be vandals.[30]

"A Crewleader's Wife"

Retirements continued throughout the 1960s. Albert Holder was one of the "Guardians" featured in a 1957 *Polk County Enterprise* article describing how crewleaders were essential to protecting woodland areas.[31]

Holder and his family had been farmers in Polk County for over one hundred years.[32] He had portions of his holdings cultivated as a tree farm, and his father, William Price, had been a forest guard.[33] When he retired, his wife composed a poem for him and Roy Gay, the long-serving crew-leader at Liberty Hill.

Retirement for Two

Two "Ole Pros" have left the Forestry Force.
They have fought many fires and have finished their course.
Side by side they worked when tools were crude,
And trudged through the forest without water or food.
When the fire was found, they never thought about
Quitting or leaving until it was dead out.
Many times they climbed towers higher than trees,
No matter how hot or cold the weather may be.
They looked over the forest for miles around,
And the fires they missed, nobody found.
Their own cars they drove to fires day or night,
Over rough, winding roads that were a terrible sight.
There was no tractor, truck, or radio,
Only a flap, a rake, ax, and a hoe.
The salary was small, smaller than you would guess,
And very few people have ever worked for less,
When called, they were always ready to go
And have assisted crews from Alto, Silsbee, and Conroe.
They received no awards, asked for no special favors
But their records tell the story of all their labors.
So don't be looking in the Great Hall of Fame
Because that isn't the place to find their names,
But they both deserve a rousing ovation
For their contributions to this great nation.
"Hats off," they said, "to those who fill our places,
May they, too, total fifty-nine T.F.S. years and take the smutty faces,
Keep all their equipment in first class shape,
Call in on time and never be late.
Please keep the Smokey Bear Poster tacked on a tree,
You'll find he is a real friend to you and me."
You may ask why I know these facts are true,
Well, I am the wife of one of these two.
They both have lived a very clean life

And I count it an honor to be a crewleader's wife.
When the Forest Supervisor checked them out today
He said, "We'll miss you, Albert Holder and Roy Gay."[34]

Holder's duty station at Pinckney was erected during the early years of the tower building campaign in 1934 but received a high identification number in the statewide inventory (No. 63).[35] The discrepancy arose because the first "County Line" tower, built at the site of a tree cab, was quickly disassembled and replaced by a taller structure in 1937. The change is recorded in early USC&GS descriptions and in a letter written by White to the president of the Carter & Brother Lumber Company. In the February 1, 1937, communication, White stated that the TFS planned on "erecting a 120 ft. tower in place of the 100 ft. tower on Highway #45 near the Polk-Tyler County boundary line. If this is done, it will be necessary for us to ask you for a tower site lease on this tower, said improvements being necessary to top the virgin timber which will make for better location of fires. Above construction will be done with CCC labor at no cost to your company."[36]

Plans to increase the height of the lookout were met with satisfaction, and Carter noted in his reply that "we are very glad to know that you are erecting this tower because that particular locality between Woods Creek and Horsepen Creek has always given us a great deal of trouble."[37] White then assured him that the TFS would make "every effort to get the tower up by the end of the month."[38] A 1943 photograph of the tower shows the 120-foot Aermotor MC-39 and the piers of the short-lived 100-foot structure (fig. 44).

In 1979 the tower was proposed for elimination. Later communication in May 1980 suggests that certain towers, like Pinckney, should be prioritized for removal as they had become "eyesores." The lookout was purchased by former observer W. A. Holder and was removed by 1981.

Meanwhile, Roy Gay spent nearly thirty years working at the Liberty Hill Lookout and had lots of stories by the time he was featured in a 1978 *News* tribute. He was one of the last in a generation of subsistence farmers, the article related, who "took on the added duty" as a part-time firefighter. In one particular incident, Roy remembered being chastised by Burnside for plowing irregular fire lines. In his version of the story, Roy remained quiet until he began backfiring the sinuous line. As he and Burnside progressed past a large pine stump, Roy recalled telling the Fire Control head, "You see that? If I had plowed inside that and set it off, it would have gone up like a torch. You just learn to look for things like that when you've been out in the thicket for awhile." According to Gay,

Burnside then "allowed as to how" maybe Roy knew more than Burnside thought he did.³⁹

Gay's lookout was the fourth fire tower erected in Texas at a time when the economy began to collapse during 1929. The 114-foot structure was completed "about three miles southeast of Knight Post Office" once a cooperative fire protection agreement was consecrated between the state and the W. T. Carter & Brother Lumber Company, Kirby Lumber Company, Texas Long Leaf Lumber Company, and John Henry Kirby. Along with another lookout at Camden and four tree cabs, intensive protection was afforded to an area "sparsely populated and practically all timbered with a well stocked stand of second growth pine and considerable areas of old growth shortleaf and longleaf pine."⁴⁰

The TFS recognized the educational value of the tower and was eager to showcase the facility. Lookouts like Gay soon began hosting scouting trips, such as one instance reported by the *Corrigan Tribune* during 1931

Figure 44. The 120-foot Aermotor MC-39 at the Pinckney Duty Station in September 1943. Note the piers from an earlier, shorter tower in the foreground. After brief service between 1934 and 1937 the original lookout was dismantled and transferred to raise the elevation of the vantage point. Image courtesy of Texas A&M Forest Service.

when a patrol from Beaumont camped at Indian Village, visited a turpentine camp still operating in Tyler County, and stopped at "the fire tower above the Big Thicket."[41]

But his primary role was fighting fires, and Gay remembered fighting with "nothing more than a rake and a flap." In the years before mechanized fire plows, he said, crews would usually improve any natural barrier they could to create a firebreak. While creeks and roads were obvious choices throughout the agency, Gay recalled relying on the hog trails that crisscrossed the Big Thicket. For his effort during the early years, Gay earned a mere $54.16 per month.[42]

Fire protection remained a priority as war loomed during mid-November 1941. At the time, Polk County maintained eight lookout towers, one tree cab, four patrolmen, ten emergency patrolmen, twenty-six guards, five smokechasers, and two lookoutmen. The dedication of these "part-time" fire crews helped keep the acreage burned below the best year on record in a decade. Gay, for his part, reminisced in 1978 that he still awoke every morning and "wanted to go get in that truck. . . . I did it so long. . . . It seems to just be a part of me. . . . I always enjoyed it."[43]

The lookout required extensive repairs in 1957, when between sixty and eighty joints were replaced. Deterioration continued into 1961, and a report issued that September indicated that more than 50 percent of the structure was affected by joint swelling or previous splicing. The bulletin convinced the director to approve replacement of the tower in 1962, when a 120-foot Aermotor LS-40 was purchased to modernize the station. Presumably, O. C. Braly erected the replacement, as documentation indicates he delivered the aging International lookout to the vacant lot behind the FCD headquarters in Lufkin during April 1963. Gay would have enjoyed the last two years before retirement, using his "tower climbing legs" up the *stairs* to the summit of his new lookout.[44]

A number of crews cycled through the duty station after his retirement, but only Charles Cain is known to have stayed at the post for longer than a decade. The tower was proposed for elimination in October 1979 and was purchased the following year. The out-of-state buyer never removed the structure, however, and the property reverted back to the state. After another unsuccessful sale, the lookout was sold in 1983 and removed.

Other crewleaders were also retiring, such as Wakefield observer Simeon Williams after twenty-three years of service in 1963,[45] and Emory Covington, who served thirty-two years before resigning in 1968.[46] Covington was a smokechaser when the only available firefighting tools were flaps and rakes, and like Cromeens, his employment predated the tower.

In September 1935, when the TFS signed a lease with farmer W. E. Tyer to extend fire protection onto lands not yet reached by construction crews from the Lufkin Camp in 1934, Covington would become the first towerman at the Timpson site (fig. 45).[47]

A *Timpson Daily* article written during March 1936 recognized Covington when it ran the headline that "Texas Forest Service Erecting Large Steel Tower near Timpson" at the site of W. E. Tyer's farm, and by June, Assistant Forester S. L. Frost had arrived to "inspect the large steel towers erected at various points by the Service."[48] Records do not survive, but the author speculates that the structure was moved from elsewhere in the network, as it was an EMSCO TW-1 model erected during 1936 when most new lookouts conformed to Aermotor MC-39 specifications.

Figure 45. Emory E. Covington worked with the TFS for thirty-two years. He was photographed at the beginning of his ascent up the EMSCO lookout in Timpson during 1952. Note the EMSCO nameplate on the horizontal strut to his left. Image courtesy of Texas A&M Forest Service.

When Covington retired and climbed down the tower for the last time, the *Timpson Times* simply said, "Covington will long be remembered by the people in the area he has serviced."[49]

Looking Ahead in Fire Control

Burnside was nearing the end of his career in August 1966 when his essay appeared in the staff meeting minutes. Simply, it began, "Where are we going in the control of wildfires in Texas? The answer may require a crystal ball. There must be successful adjustments with the future." His ten-year vision is remarkably clear, however, and Burnside foresaw more complex fire control problems brought about by greater public pressure and urbanization. "The urban sprawl," he noted, "will gradually move northward from the Orange, Beaumont, and Houston areas. Slowly, the areas of Vidor, Silsbee, Sour Lake, Cleveland and Conroe will merge with adjoining areas" to become a "continuous population complex."[50] Only a firefighter would choose the word "complex" to describe the merging communities. Then, Burnside suggested that the TFS would be faced with eight potential crossroads and proposed policy adaptations that included changes to communication, detection, suppression, and personnel.

With an understanding of the past that developed over the years since the Depression, Burnside summarized the evolution of detection policy. Initially, fires were suppressed by a combination of smokechasers and detectors (lookouts) operating from a network of linked fire towers. By the 1940s, a pilot supplemented the lookout network or was used in lieu of the towers. More and more, the public, or patrolling service employees, were reporting fires. "Without doubt," he summarized, "these systems may be improved." But Burnside read between the lines of Lowery's diary. "I'm not ready to predict that plane use for fire detection will make the lookout tower obsolete. There will be a place for both means of detection *for several years yet*. Other means of fire detection are already in the works."[51]

Burnside's prediction that aircraft would dominate detection work within a few years was correct. In 1970, the area around Conroe converted primarily to aircraft for fire detection with excellent results. The total detection cost for two years, the cost per acre, and the average fire size were lower than those in other districts relying primarily on lookout towers. Philosophies had changed, and the *News* now advocated that the aerial detection program was "a dynamic aggressive system compared to the old tower system."[52]

9. A Lot of Smoke but No Fire
The 1970s

Gallant Men of 213

Rob and Charlie
And M. L., too
Was a Texas Forest
Fire fighting crew.
They were the best
But you know that.
They'd fight a fire
At the drop of a hat.
I went over one morning
Just to see
Them gas up
Old 2 . . . 1 . . . 3.
I was standing there
And they were talking to me,
The radio said
"Henderson—2 . . . 1 . . . 3."
He gave the block
And he gave the grid
Around the corner
That big truck slid.
Everyone around
Knew everything was OK
Because 2 . . . 1 . . . 3
Was on the way.
If you see smoke,
Go by and see
The gallant men
of 2 . . . 1 . . . 3.[1]

THOUGH BURNSIDE RECOGNIZED the changing times as he looked ahead in fire control, there were still discussions on how to best optimize the lookout network as the decade of the 1960s closed. Most urgently, the department wished to relocate the Fern Lake Tower. A lack of commercial power, the theft of the electrical generator, and depredations to the isolation fence and cab made the current site unattractive, one memorandum explained. A better-protected, more economical site was available

10 miles away where the lookout could be re-erected to provide expanded visibility into portions of the crew area that were prone to fires in the past.[2] Despite the cost of relocating the tower, the FCD recommended to the director that the project "be given first priority at this time." In some sense, the solicitation was a litmus test, and the newly appointed director's unsupportive response foreshadowed things to come. "There seems to be a lot of smoke generated by this tower," Paul Kramer concluded, "but not much fire. What is the possibility of dismantling and storing the tower, and using plane detection for a while to see how it works out?"[3] Burnside's hopes for a tower project evaporated. All Fire Control could do was reply that "if the Fern Lake tower is not to be moved to a new location within the district I would recommend that it be left at its present location. The effectiveness of aerial detection can be determined without taking the tower down."[4]

Despite the debate, the TFS continued to invest in the station, and "history was finally made" on February 26, 1971, when power was connected to the tower's electrical system. As *Have You Heard* reported, "Crewleader Raymond Barker, who has been starting the battery charger for the past 20 yrs., is particularly happy about the changes. So is his crewman of 12 years, Robert Haney."[5] Even with the improvements, maintenance problems persisted. Windows were broken, the door to the cabin was beaten off, and the cab had been set on fire. The recommendation by 1975 was that the tower should be condemned as soon as possible. That September, sale of the tower was approved "as is—where is," and by February 1976 the tower had been removed.

Folweiler retired in 1967, and Kramer, the fifth state forester, made noticeable policy shifts toward decentralization and aerial fire detection.[6] Kramer received his training at Washington State College and Yale University and had worked for the Tennessee Valley Authority, the Ohio Forest Survey, the Soil Conservation Service, and the USFS.[7] In 1948, he joined the TFS as a forest products technologist, becoming the head of the Forest Productions Department in 1955 and associate director in 1966.[8]

With Burnside still at the helm of Fire Control, though, the department continued to flounder "without a sharp vision or clear objectives,"[9] envisioning additional tower expansions and a largely symbolic project to bring fixed point detection to the Lost Pines region and cedar-oak forests west of the Piney Woods. The Lost Pines, a drought-hardy, geographically isolated strand of loblolly pines, had been studied in 1967 to determine the infrastructure needs of Bastrop, Lee, and Fayette Counties.[10] Meanwhile, fringe forests long eyed for protection in Titus, Smith, Henderson, and Anderson Counties were also surveyed for suitable tower sites. With support from Speaker of the House Gus Mutscher and Representative Charles Jungmichel, the 1969 legislature funded the expansion,[11] and Burnside took the occasion to write to Braly. "We are contemplating the erection of seven fire lookout towers within the next several months," he said in July; "advise if you will be interested in erecting these towers." Before signing with an uncharacteristically sharp tone, perhaps prompted by urgency or age, Burnside added, "We will be expecting a call from you within a few days."[12]

As fixed point detection systems became largely outdated, especially within the TNFs, some of the lookouts the FCD contemplated erecting could be cannibalized from other agencies. Fortunately for the TFS, the USFS offered three surplus towers to the state in 1967. Burnside, on behalf of Kramer, responded quickly to accept the towers they offered gratis.[13] First to fall were the Hi Point, Devils, and Moss Hill Lookouts, which Braly diligently delivered to Lufkin by August 1, 1968.[14] Ironically, when Braly backed his truck into the storage lot at the Cudlupp Forestry Center in Lufkin, he returned the Moss Hill Lookout to the state after a thirty-four-year absence. The tower had originally been a TFS lookout but was transferred to the federal government after the creation of the Angelina National Forest. Later in 1971, the USFS also offered the International towers at Mercy and Moores Grove to the TFS.

Six years earlier, the area around Moores Grove was developed into the New Waverly Job Corps Conservation Center, an attempt to relieve domestic poverty using a model popularized by the CCC thirty years before. The site was originally operated by the USFS in cooperation with the Office of Economic Opportunity,[15] and it served 224 corpsmen when it reached full capacity in 1967.[16] In 1971, the site was leased to the Gulf Coast Building and Construction Trades Council under a special use permit, which began operating the center to rehabilitate and motivate young men in a work-study environment.[17] At the same time that the site was being leased, the USFS transferred the lookout tower to the TFS, perhaps over liability concerns. A state assessment of the structure completed before the handover listed the tower as an IDECO design.

Before Braly arrived at the Angelina and Sabine Forests to decommission the towers, George Chase and Floyd Bradford inspected each one using form FCD-1, the "physical condition of lookout towers and sites." Their notes provide temporal maintenance records for the federal lookouts and seem to indicate that the structures had not been maintained for some time. The tower at Hi Point, for instance, does not appear on the spring or autumn staffing schedules in either the 1949 or 1953 memoranda of understanding between the United States and Texas

Forest Service, suggesting that it quickly became a secondary tower. In fact, the most recent references for the station date to 1959 and 1960, when the position was used for an "occasional look" on days when the fire danger was Class V or was needed for surveillance during McGee Bend clearing operations.

Overlapping "seen areas" and the increasing use of aircraft contributed to the early abandonment of the Hi Point Station. Above all, plans to remove the structure were hastened by the decision by the US Army Corps of Engineers to flood Ayish Bayou and the Angelina River valleys to create the Sam Rayburn Reservoir. This project had just been completed in 1965, and much of the forested land on three sides of the lookout had been drowned. When the little-used tower was inspected in June 1967, the assessment revealed that the door and lock to the cab were gone and the bottom section of ladder was missing. The woodwork in the cab and on the floor had to be replaced, and half of the ceiling was destroyed. More significantly, the first four diagonal braces had also been removed or stolen.

It took Chase and Bradford three trips to inspect Devils Lookout, and Burnside noted in a memorandum to the director during April 1968 that the tower is "presently almost surrounded by water and all the roadways leading to the tower are under water."[18] Conditions had dried the following month, and the inspection revealed that the tower, marked with Carnegie Steel Company stamps, was in satisfactory condition. The pair reported that approximately twenty-five braces required replacement due to electrolysis, and the woodwork on the stairway risers and in the interior of the cab needed replacement. Replacement window lights were also necessary, as were four new baseplates. Ten bullet holes in the roof would also need to be leaded. Fortunately, the inspection trip located a logging road that could be used to remove the lookout "only in very dry weather and after an extended dry period."[19]

Chase wrote to Dresser Industries, the manufacturer that acquired IDECO, to inquire about the availability of replacement parts for the Devils Lookout. When the company responded, Chase was informed that "the shop drawings we had of your towers have been changed so many times and part numbers have been changed equally as many times, that we would be afraid to manufacture them. . . . May I suggest that you use the enclosed list as a check for size along with the old parts, and have the members made locally."[20] Fire towers were becoming a thing of the past, and only Aermotor's Broken Arrow, Oklahoma, facility was in a position to fabricate the necessary components.

The tower at Devils Lookout was re-erected in Smithville after A. J. Rod donated one acre of land to the state. The International lookout was completed by O. C. Braly in March 1970, and two months later the Bastrop County Soil and Water Conservation District included the duty station and firefighting equipment on its annual tour.[21]

Crewleader R. T. Lewis appeared in a 1972 issue of *Have You Heard* to report that he never gets lonely in the Smithville Tower, as "his pet buzzards always drop by to keep him company."[22] Despite the fun, an incident occurred that year that strained relationships between the TFS and the landowner. The episode reached the president of Texas A&M University and stemmed from a crewman driving his personal vehicle across the property without permission on an interior road other than the agreed-upon easement to the tower. So stinging was the encounter that the director was forced to circulate a memorandum "in the interest of our image and effectiveness" requiring "a set of ground rules for the property." Violation of the rules, the memo read, would be "sufficient cause for immediate dismissal of any district personnel." The director closed the letter by noting that the actions were "not in keeping with the spirit or intent of this agency. However, at this time, it would appear to be the only recourse open."[23] It was not the official reason given to the Texas Parks and Wildlife Department, but within a year of the episode Director Kramer wrote to officials at nearby Buescher State Park petitioning managers for the use of a hillside southeast of Park Road 1 on which the service could relocate the three-year-old tower.[24]

The transfer never occurred, and by 1975 the primary means of fire detection were contract aerial surveys conducted from the Smithville airport. The towers, the *Bastrop Advertiser and Bastrop County News* reported, were used only for supplemental, periodic detection.[25] Though the tower was used briefly during the spring 1976 fire season, it was declared surplus by 1979. It was sold in 1980, and the Board of Regents at Texas A&M University quickly gave authority to "sell, exchange, convey, or transfer ownership" of the one-acre tract.[26] Following legislative approval, the site returned to the original land donor in 1982. The Depression-era tower, originally from the Sabine National Forest, stands in private ownership today.

Kramer noted the addition of the eight new lookouts in the *News* (table 11).[27] Three towers would be positioned in the Lost Pines near the communities of Smithville, Winchester, and Bastrop. Other towers in Titus, Smith, Henderson, and Anderson Counties would be erected at Wilhite, Chandler, Poynor, Lindale, and Montalba to protect 604,196 acres of forest.[28] Simultaneously, two "legacy" lookouts at Chita and Onalaska would be relocated,[29] probably because of flooding in the Trinity River bottomland forests that occurred during the construction of Lake Livingston.

Table 11. Lookout towers constructed between 1970 and 1979.

1970	1971	1972	1973	1974	1975	1976	1977	1978	1979
Bastrop	Lindale *from Moss Hill*			Texas Forestry Museum *from East River*					Beckville 2 *From Beckville 1*
Chandler									
Chita 2 *from Chita 1*									
Montalba									
Onalaska 2 *from Onalaska 1*									
Poynor *from Hi Point*									
Smithville *from Devils*									
Wilhite									
Winchester									

The original lease between the Longleaf Lumber Company and the TFS for the Onalaska Lookout has a handwritten note along the document's header indicating that the station would be staffed by "Stanford." Atma Stanford began working as a full-time patrolman at the Onalaska Duty Station in February 1935 and was a dedicated veteran of the TFS when he retired in July 1964.[30] As he neared retirement in 1963, a *Polk County Enterprise* announcement reminding residents to be careful with fire even urged everyone to do their part and show "appreciation to the men who sacrifice their time to protect our lives and property from the threat of fire."[31]

The lease for the relocated Onalaska Lookout was signed in August 1970, and the tower was reconstructed later that month. An announcement of the change was made in *Have You Heard* the following February.[32] Within a decade the tower was declared surplus, and it was one of three purchased by the St. Regis Paper Company. Though in private ownership, the lookout still stands in a heavily forested portion of Polk County.

The towers in the Lost Pines region were constructed without delay and were all operational by April 1970. The *Bastrop Advertiser and Bastrop County News* publicized the project's completion using language typical of the aging FCD's press releases. The article featured news of the crew, explained the operation of the tower, and highlighted the firefighting equipment that was invested in the county. But in recognition of the changing times, the piece concluded that "if you observe a wildfire in your area, call the Texas Forest Service."[33]

Things appeared to be running smoothly as other leases were signed in early 1970. Outwardly, this was also true at Lindale, where a lease was granted in May. With access secured, Braly prepared the site and poured the concrete for the lookout's piers. But Burnside was compelled to call him on July 29, 1970, and immediately followed his call with a letter on August 4, advising the contractor to cease all work because of a "flaw in the tower site lease."[34] In truth, it was less a flaw in the lease and more a family affair that required the Forest Service to "start over for another site,"[35] since some of the landowner's heirs refused to endorse the terms of the agreement. Ultimately, Braly had to bulldoze the recently cured piers and restore the area while Burnside scrambled to identify another location and negotiate a lease on which to erect the Lindale Tower. Finally, on February 2, 1971, a second location was procured. Burnside alerted Director Kramer and sheepishly asked, "Is this lease now sufficient before recording to initiate construction of the tower on this site?"[36] It was, and Braly diligently repoured piers for the Lindale Tower and completed construction by March 1971, making it the last tower to be erected in the state for fire detection purposes.[37] The lease was canceled only eleven years later in 1982.

While the towers at Lindale, Poynor, and Smithville were surplus, those at Bastrop, Chandler, Montalba, Winchester, and Wilhite were purchased "new" from the Dunlap Manufacturing Company in Tulsa, Oklahoma. A District 2 invoice for two towers, one of which was sent to Chandler, indicates that each was purchased for $4,707.

The Chandler Tower was completed by September 25, 1970,[38] and the tower was dedicated on November 10 when approximately thirty-five local landowners listened to remarks from Director Kramer, Senator Charles Wilson, and State Representative Bill Bass.[39] As plans were being made to phase out the fixed point lookout system a decade later, the lookout was retained for agency communications needs and remained part of the state system until it was transferred to Henderson County.[40]

The structure is the only standing Dunlap example, an Aermotor LS-40 look-alike. It is unclear from the available records if the tower was fabricated under contract with Aermotor during the period when the company was moving its manufacturing operations to Argentina in 1969 or if Dunlap simply created a comparable product that allowed it to undercut its competitors. It is noteworthy, however, that neither the drill holes nor the elongated "Aermotor, Broken Arrow, Okla." placard was discovered on any of the corner posts during a site visit.

Burnside retired May 31, 1971, soon after the Lindale Tower was finished. Though the Lost Pines lookouts had all been built on schedule, vandalism and misunderstandings strained relationships at some sites just as they had at Lindale. "Less than wholesome" relationships with other lessors were becoming more common. Challenges at the East River

Lookout forced Patrick Ebarb, the new head of the Fire Control Department, to write to the director in 1972 seeking permission to advertise the tower for sale so that the service could "unload it on someone" and avoid friction with the landowner.[41] More cordial requests were also being sent from lessors asking for towers to be removed or fenced to reduce area "night life."[42]

Ebarb was compelled to issue a memorandum in June 1975. In it, he pointed to "recent disgruntlement by a local tower site lessor" that prompted him to remind all district personnel that the lookout towers continued to be a "very real part of our total protection effort as a *secondary* detection system." He suggested that the towers be visited weekly to check against vandalism and neglect. "The continuance of most tower site leases," he added, "is contingent upon the site being used 'for fire protection purposes.'" Ebarb then advised each district to record their visits to "forestall voiding some tower site leases now in effect."[43]

Meanwhile, vandalism became a nuisance, and the TFS increasingly spent resources maintaining the towers. At Appleby, for example, the district forester wrote to Fire Control in 1975:

> At this point all tower windows have been broken and tower door removed by vandalism. The tower has been boarded up and trap door replaced in efforts to lessen weather damage to tower flooring. I believe that these repairs will not stop the problem at the tower. The adjacent landowner is currently putting in a road system to subdivide the area. This will further aggravate problems of tower repair by improved tower access. Should the Service ever need to revert to tower detection, Appleby tower detection area could easily be handled by other towers. For the above reasons I recommend Appleby tower be disposed of by including it on the sale of surplus property this year.[44]

Three years later, the District 4 forester wrote to the director informing him that "the Votaw Lookout Tower was set on fire during the early morning hours of December 25, 1977. All wood parts of the cabin and several ladder landings were destroyed. . . . The cost to repair would be considerable and the tower has been constantly subject to vandalism and trash dumping. I recommend that this tower be eliminated."[45]

Even in District 1 during 1979, the forester suggested that the East Mountain and Leesburg Lookouts be eliminated. This recommendation was based on the fact that the East Mountain Tower had "been a problem for as long as I have been with the TFS. It has been broken into on several occasions and equipment stolen or destroyed. Presently there is no equipment in the tower and almost all of the windows have been broken. Being located so close to a large urban area has resulted in a continuing vandalism problem."[46]

Other District 1 towers were deemed essential radio relays, despite heavy vandalism. In 1979, the district recommended retaining the Hallsville Tower, "even though vandalism has increased recently." The relay was "rarely bothered," the opinion noted, and the tower was "located on high ground with excellent range of vision."[47] The next year, however, the district suggested to Fire Control that the relay be moved to Mims Chapel "if at all possible and practical." The forester explained that vandalism had continued, and tower windows were repeatedly broken, the steps were burned, and there were shell holes in the cab. Also, all the gate chains had been destroyed or stolen. The Mims Chapel location was more secluded, off a public highway, behind a locked gate, and beyond a residence. Vandalism there "probably would be nil."[48]

At the same time, the towers were also becoming less useful in their primary function of protecting against fires. In 1976, for example, a TFS study indicated that no fires were reported from staffed lookout towers in four of the seven districts and that airplane spotters accounted for 43 percent of the fire calls.[49] The remaining alarms, the *Polk County Enterprise* explained, were sounded by the public or ground patrol units. Even in Bastrop County by 1975, a mere five years after the lookouts had been erected, the towers were being phased out, as made clear by a letter published by the *Bastrop Advertiser and Bastrop County News* between associate director of the Forest Service Bruce Miles and county judge Jack Griesenbeck. The judge had written over concerns that "the Texas Forest Service might withdraw their very worthwhile [fire] program from the Bastrop area" after supplying volunteer fire departments with firefighting equipment. In reply, Miles indicated that the TFS was there to stay, having recently "opened a new contract for aerial detection based at Smithville. This is our primary means of fire detection," he continued, "sometimes supplemented by periodic tower detection."[50]

Elsewhere, the District 6 forester had written to Ebarb regarding the Dodge Tower in Walker County. Changing technologies meant that the radio relay was eliminated from the tower by 1972. With the equipment removed and the lookout "superfluous to our detection needs," the forester requested permission to remove the structure. In the event of an emergency, the opinion continued, detection in the area could always be accomplished by sending personnel to the USFS lookout at Four Notch to achieve the same level of coverage.[51] The head of Fire Control foreshadowed changes in fire detection policy in his February 1972 reply, suggesting that the district delay any action on removing the tower "until we get a firm policy regarding detection systems."[52]

The decision would not be delayed long. The lookout was condemned in 1975, and the director approved sale of the tower "as is, where is."[53] Ultimately, an agreement between the TFS and the Nacogdoches Fire Department in March 1977 transferred ownership of the structure to that organization for training purposes.[54] Later that year, the tower was dismantled and the lease was canceled.

In a note thanking the lessor, District Forester James Blott wrote that the TFS appreciated "use of the land these past 44 years" and mentioned that "its use in fire protection has facilitated the future forest resource for all Texans." The forester described the condition of the site at the time and acknowledged that the tower piers were left in situ and could be removed, if necessary. As for the geodetic survey marker at the base of the lookout, Blott concluded, it "becomes a part of the landscape."[55]

Even today Patrick Ebarb warns, "Look Out, Lookout Towers!" and acknowledges that he considered them to be "anachronisms, transitions from an earlier age." Reminiscing about an early career assignment that entailed scuttling across District 1 to inventory twenty tower cabs, Ebarb jokes, "did nothing to change this negative opinion."[56]

Instead, the new head of Fire Control envisioned the future of Texas wildland fire response focused on visibility and aggressiveness. "Stationing fire crews at visible road hubs accessible to locations of historical fire occurrence," he explains, "made for quicker detection from aircraft and fast initial attack." He had tested this philosophy earlier in his career as a district forester, shuttering fifteen lookout towers within his area of responsibility "without official permission," relying alternatively "upon planes and the public" for fire detection.[57]

Ebarb's philosophy was consistent with earlier studies, identifying three deficiencies in the fixed point detection network. Importantly, he explains, the towers were vacant during "fire bursts" when crews were busy responding to fires. "If we could do without tower detection during periods of highest fire occurrence," he reasoned, the TFS could "certainly do without them during normal periods." Instead, with a supplement to aviation budgets, one winged "smokechaser" could survey one million acres in an hour, and nonproductive time decreased.[58]

His gamble worked. Average fire size decreased, and he won the support of the major timber companies and his staff. Nevertheless, Ebarb recalls that there were "several anecdotes circulated about visits from Director Kramer and his staff." The unsanctioned experiment caused Ebarb to exceed the budget for his plane allowance and ultimately resulted in a delegation flying from College Station in a $250,000 Aero Commander for condemnatory discussions. Ebarb pushed back and may have "peppered this challenge with a few epithets." Eventually, though, "they folded, made some small talk . . . and flew out of my life. It was a harbinger of things to come."[59]

Apparently, rumors that the network was becoming obsolete reached the executive vice president of the TFA in July 1972. By that time, Vice President Ed Wagoner wrote to Director Kramer, noting "considerable publicity locally concerning the possibility of discontinuing the use of fire towers for the detection of forest fires." This was partially confirmed, he felt, in speaking to a service employee who "informs me that you are entering a two-year trial period extending through October, 1974." At that time, he had heard, "you will make a decision regarding the use of aircraft as the sole means of fire detection."[60] Kramer replied quickly, saying that, despite rumors to the contrary, "the Texas Forest Service has no intention of using aircraft as a sole means of detection in the foreseeable future" and, perhaps added blushingly, that he did not know "how such rumors get started."[61]

Wagoner's ear was to the rail, though, and decommissioning the towers was not a rumor. Patrick Ebarb recalls that "the Fire Management Officer for the National Forests in Texas was a good friend who knew of my interest in getting out of the lookout tower business for good. He was of the same persuasion and was eager to shut down his 12 towers."[62] By April 1971, John Olson, the assistant forest supervisor for the TNFs, announced through an internal memorandum that an aerial detection feasibility study would begin during the 1972 fiscal year. Beginning October 1, he warned, funding would no longer be available for staffing lookout towers. Instead, a contract for aerial fire detection would be fulfilled by flights operated by the TFS. Olson also informed the district rangers in his announcement that telephone and electrical services would be disconnected to the guard stations, as the service had plans to remove the tower radios.

Olson sent a second memorandum to the district foresters that June. "Some clarification is necessary," he said, "to clear up any misunderstanding relative to removing telephone and electrical connections" from the lookouts. He then continued, "Here are the facts." Power would remain at only six towers, where it was necessary for key radio or repeater equipment. Because of this decision the lights would be left on at Jackson Hill, Dreka, Piney, Ratcliff, Nogalus, and Chambers Hill.[63]

Only a few years earlier Tommy Read began his 1967 *Silsbee Bee* article describing the view from the top of the Chambers Hill Lookout. His family's campsite on the shores of Red Hills Lake lay below him "like a blue gem in the deep green pine forest and rolling hills of upper Sabine County." Beyond the recreation area, "fleecy white clouds drifted lazily in the gentle breeze, giving the panorama of trees and hills a funny patchwork design."[64] All around were clearings of deep red earth and dozens of chicken ranches. Read could also see through the haze into Louisiana at a time when bottomland forests in the Sabine River valley were being flooded for the Toledo Bend Reservoir, a factor that may have led the USFS to expedite removal of the Devils Lookout.

The Chambers Hill site was within 2.5 miles of Milam and 5.5 miles of Geneva and may have been on the same hillside where the Milam Tree Cab was built in 1928. The station had always been an important primary lookout meant to be staffed whenever there was danger of fire, and it consistently appeared on interagency staffing schedules agreed on by the USFS and TFS throughout the 1940s and 1950s. The duty station persisted as a primary observation point into the 1970s, and detection schedules for the district required that the tower be staffed whenever the Buildup Index, which is a measure of fire danger conditions, rose during the regular fire season. Outside the peak fire period, an observer would still be posted in the tower whenever the fire weather was severe. Today, the abandoned relay house and the tower piers are enclosed by an isolation fence in a young pine stand, surrounded by a circular drive that once defined the compound. Just downhill to the west are the remains of a service garage.

Meanwhile, the TFS was making plans of its own. Ebarb's June memorandum acknowledged that some towers "chronically plagued" by vandalism should probably be removed. But any such recommendations, he cautioned, should "be compatible with the retention of a useable, skeletal tower system for your protection unit."[65] The towers might still be called back into service.

The year 1976 brought extreme fire danger and postponed any state decisions concerning the fate of the tower network. In the first two months of the year, the TFS fought 1,233 wildfires.[66] This was more than the figures reported for all of 1975, and the exceptional activity meant that the budget allocated for aerial detection was nearly depleted. Thus, in March, the decision was made that district foresters would be given the option to activate certain "key fire towers in high risk areas." It was strictly "a temporary measure due to the lack of funds," Ebarb said, but it might mean that two-thirds of the state's lookouts would be staffed.[67]

The lookouts held on that year, and the *Cherokeean* published a story in April, noting that vandalism of the towers was a serious risk to TFS personnel. Most of the damage, the article related, was from people shooting at the structures. Because of the large number of wildfires, it continued, the towers were being used in conjunction with fire patrol planes. Ebarb then sent a warning to readers, saying that forest workers should not be dodging bullets. "Fire fighting is hazardous enough without being shot at."[68] The destruction was a felony, and if any employee was shot on duty at the tower, much more serious charges would be filed.

Some towers were deemed important enough to preserve as late as 1978. It was that year that TFS received a letter from Texas Utilities Services advising them of the company's intention to strip-mine the tract of land leased by the agency for the Beckville Tower. The lignite mine was to be operational by 1980, the land agent wrote, and the company wished to work with the TFS to remove or relocate the structure.[69] Area Forester Joe Fox forwarded the letter to the associate director and outlined several courses of action, noting that the tower was in a strategic location as it could reference fires in Rusk, Panola, Harrison, and Gregg Counties. For that reason, it should be maintained, Fox thought, "even if our future plans are to reduce the present tower network to a skeleton system." Moreover, Fox acknowledged, the service had signed a radio relay agreement with the Texas Department of Public Safety in 1975 and the equipment installation at the site was nearly complete. In his judgment, the tower could be left in place, forcing the mine owner to avoid it during operations, or the company could relocate it to another suitable site in the Beckville area.[70]

Subsequent meetings resulted in an agreement to move the lookout at the company's expense, and an alternative site was selected in an area immediately adjacent to the Rock Hill water supply tank, about one-quarter mile from the historic position of the station. The relocation of the lookout and communication equipment began around June 15, 1979, and lasted until about August 1.[71] The plan called for the tower to be dismantled in three sections that would be transported to the new site by truck. Curiously, there had been some early discussion that these three pieces would be airlifted by helicopter.

Mr. Dependability and Other Remaining Old Hands

Emmett Turner was at the phone in 1951 coordinating the response of the DeKalb and Hooks Towers in Bowie County and directing firefighters to the blaze when his image was captured in the photograph. In it, Turner is keenly staring at the District 1 map in the dispatcher's office, but he must have known the details without looking. Emmett was born in 1912 and started his career as an emergency patrolman during the Depression, working first from a tree cab built on his father's land near Douglassville.[72] Later, the cab was replaced by the Douglassville Tower, one of sixteen built as the lookout network expanded into North Texas. Already an experienced firefighter by the beginning of World War II, Turner asked to be reassigned to the Hooks Duty Station when Robert Smith vacated his position.

White personally recommended him for the job, writing to Siecke that Turner was "a good boy and lives right close to the Douglassville tower, and I would recommend him for the position providing he can find a place to live somewhere near the tower and outside of the [military] reservation."[73] He accepted the transfer and temporarily left Douglassville to become one of five fire patrol officers stationed in Bowie County. The tower was located on the expansive arsenal, and Turner remembered wearing a special badge allowing him clearance to the property. During this period, he also staffed a "novel duty station" built off the side of the New Boston water tower (see fig. 12).

Later during the 1960s, Turner began occupying the Jefferson Lookout as a forest aide. He worked another decade, until 1975, before retiring with thirty-nine years of experience.[74] A *Have You Heard* article looked back on his career in 1972, saying that Emmett, "Mr. Dependability," was a man to be counted on and was always willing to tackle the job. With a lifetime commitment to the TFS, he was "probably the most experienced fire fighter in District 1."[75]

Turner's Aermotor MC-39 at Douglassville was sold in 1983. His Hooks Tower, also a MC-39, was moved off the arsenal in 1952 to Redwater, and that lease was terminated in 1996. At the time, District Forester Steve Adams wrote a note of thanks to the Redwater landowner, saying in part that the "tower site and tower have served Bowie County for many years and rendered a great service to its citizens."[76] But the tower became a nuisance by 2008 and was removed.

Fortunately, the Jefferson Lookout, also an Aermotor MC-39, now sits at the edge of Jefferson's downtown district. It was relocated and restored in 2003 and has been maintained for use as a cellphone tower. It also serves as an educational and historical resource that, in part, should recognize Jeff Hunter's pioneering role as one of the first African American lookouts in Texas and Emmett Turner's commitment and service to forestry.

10 Potshots or Pearls
The 1980s

FIRE TOWERS WERE OBSOLETE by the 1980s, both nationally and in Texas. Mark Thornton identified a number of reasons for the decline, including an increase in public fire reporting and the improved effectiveness and economy of aerial patrols.[1] These were important considerations, especially in view of the inflating costs of staffing, monitoring, and maintaining the towers. But Thornton also cited improved communication systems and more modern transportation routes as effective combatants against the spread of large fires. He also pointed to changes in fire policy decisions. Liability, wilderness designations, and smog were likewise identified by others to explain, in part, the rapid decline of the nation's expansive lookout system.

Bruce Miles replaced Kramer as director on October 29, 1980, and broadly outlined the TFS's fire detection strategy after reviewing several of his predecessor's pending files.[2] With increasing vandalism and maintenance costs, the plan was to reduce the number of Piney Woods lookouts and rely more heavily on aircraft detection. But, Miles noted, it was "critical" to keep an intact skeleton tower network in place to more adaptively respond to the increasing costs of aircraft fuel. The uncertainty created by the 1973 oil embargo and the Iranian Revolution in 1979 influenced decisions about which lookouts could be declared surplus, and one District 1 memorandum noted that "with the petroleum situation and the fact that we may have to revive the use of the towers in the district, I don't feel that we can have any less towers and really do the job of detection if called on to do so."[3]

Before Kramer's retirement, district-wide maps were updated with 15-mile detection radii around each tower, and those with closely overlapping "seen areas" were identified for possible elimination (fig. 46). By October 1979, twenty-five towers met these criteria and were declared surplus.[4] Elimination was imminent. Sixty-eight would be retained because of communication obligations and future use for skeletal detection (table 12). In the spring of 1980, shortly before retirement, Director Kramer sent the first instructions to Patrick Ebarb on how to dispose of the surplus. In his memorandum, he suggested that interested state or federal agencies be given the towers gratis, provided they assume responsibility for dismantling the structure and removing the piers from the ground. Otherwise, private parties or nonpublic organizations could be offered the towers on a negotiated basis, again assuming that they

Figure 46. The TFS began abandoning many duty stations by the end of the 1970s. District foresters developed "skeletal" lookout plans after studying local fire problems and determining where crew areas overlapped. Towers that were no longer necessary were proposed for elimination, such as those mapped here in Districts 4 and 5. Image courtesy of Texas A&M Forest Service.

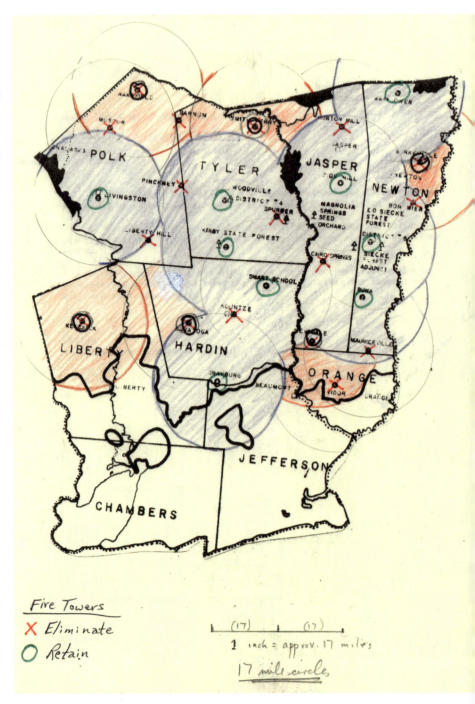

Table 12. Recommendations for the retention and elimination of lookout towers dated October 4, 1979. The report considered the future use of towers for a skeletal detection network and existing communication agreements and are ordered as they appeared. Note that seven towers had already been removed by the time this memorandum was written: Appleby (1975), Ariola (1974), Dodge (1977), East River (1974), Fern Lake (1976), Magnolia (1968), and Votaw (1978). By March 1981, the Barnum, Bird Mountain, Bon Weir, Cairo Springs, Elysian Fields, Kountze, Liberty Hill, Mauriceville, Meadows, Moscow, Pinckney, Salem, Smithville, Snap, Spurger, Vidor, and Weches lookouts had been sold on a negotiated basis. Some would revert to the TFS when the buyers failed to remove the structure.

	Retain	Retain	Eliminate
Area 1	Redwater Adams Hill DeKalb Negley Douglassville Union Hill Kildare Daingerfield	Wilhite Gilmont Latch Pine Mills Hallsville Jefferson Karnack Mims Chapel	East Mt. Leesburg Smithland Elysian Fields
Area 2	Deadwood Beckville Church Hill Mt. Enterprise Starrville Chandler Love's Lookout	Rusk Neches Montalba Lindale Elkhart Maydelle Arp	Snap Salem Meadows Poynor Bird Mt.
Area 3	Grapeland Chita Onalaska Central Cypress Lake Wolf Hill Cushing	Paxton Tabernacle Bronson Shady Grove Etoile Huntington Alto	Weldon Timpson San Augustine Weches
Area 4	Kirby Forest Woodville Livingston Mayflower Siecke Forest Buna Zion Hill Grayburg	Smart School Smith Ferry Wakefield Burkeville Evadale Kenefick Saratoga Horton Hill	Spurger Barnum Moscow Liberty Hill Pinckney Bon Wier Cairo Springs Mauriceville Vidor Kountze
Area 5	Fails Willow Springs Shepherd Splendora	Keenan Willis Bastrop Winchester	Jones Smithville

accepted the responsibility of dismantling the structure and disposing of the piers.[5] Some prioritization, Kramer thought, should be given to towers that were "eyesores." From October 1979 until October 1980, sixteen of the twenty-five lookouts were sold through negotiated agreements,[6] though the procedure hardly seems to have been successful. Together, ten lookouts were sold to just two individuals, and when a summary of the sales was written in March 1981, only four of the towers had actually been removed.

To expedite the process, the TFS decided that the public should bid on the remaining inventory of lookout towers. Invitations were announced in the local newspapers, and those interested in owning a fire tower could complete a bid form expressing the desired purchase price for one of the lookouts offered in the auction. The forms were mailed to the TFS along with a cashier's check or money order for 10 percent of the bid's value. On the closing date of the sale, the sealed envelopes were publicly opened at a prearranged time. The winner was the highest bidder, who then agreed to complete the remainder of the purchase within three days. Afterward, the winner had forty-five days (later extended to ninety days) to dismantle the lookout, remove the foundations, and dispose of any debris on the site.[7] When the work was satisfactorily completed, the state would then cancel the site lease and return the property to the lessor. In the event of two identical bids, the award for the tower was determined by drawing lots.[8] If the tower was not removed within the allotted period, the winner could ask for an extension or the bidder's payment to the TFS would be returned and the ownership of the structure would revert back to the state. Later, either a performance bond or a $500 security deposit was required.[9] Finally, if a tower was offered for sale and there were no bidders, then the TFS was allowed to sell the tower through a negotiated sale.

Emory Covington's lookout tower in Timpson, for instance, was advertised during the first auction of the eight remaining towers that had been proposed for elimination but had not been sold on a negotiated basis. Announcement of the sale was made in the local papers during November 1980. The winning bid, $500, was written on the bid form and placed in an envelope along with a $50 postal money order covering 10 percent of the purchase price. When the bid was opened and accepted on December 1, 1980, the winner paid the balance of the bill, as stipulated, within three days.

This particular transaction "stirred up a hornet's nest," however, when the landowner called in frustration that he was not made aware of the sale.[10] The owner even suggested that the tower was abandoned property that reverted back to him, the lessor.[11] Significantly, the owner argued that the lease for the property granted the TFS rights only to enter the property, and only they could remove the tower. Obviously, this interpretation had consequences for all of the state's fire towers. With the sale of the tower completed, however, the state was stuck, unable to facilitate delivery of the purchased goods. The state feared a breach of contract suit while still trying to negotiate with the lessor. Ultimately, the tower sale proceeded, but the experience changed the manner in which the TFS disposed of other towers, and all future bid announcements were preceded by letters to the landowners advising them of the Forest Service's intention to sell the tower.

Patrick Ebarb sent Director Miles a memorandum on the heels of the first bid round on January 12, 1981, requesting approval for the sale of additional towers.[12] Of the remaining sixty-eight towers, Ebarb recommended the disposition of an additional thirty-five and the retention of only thirty-three. Two days later, Miles replied to Ebarb approving the request, reminding him that the towers were to be placed on competitive bids, "regardless of the interest of tower site owners."[13]

Sometimes, local publicity temporarily ring-fenced towers. This was the case at Montalba, a new duty station erected during June 1970. Patrick Ebarb was forced to respond when the *Palestine Herald-Press* reported that a Sheriff's Office radio repeater was in danger and that "Anderson County may have to buy a fire tower to improve radio reception." Shielding the reputation of the agency, Ebarb noted that "the service 'would make every effort' to work with the County in cases where fire towers are used for radio communications."[14] This allowed the tower to stand until the summer of 1993, when it was removed and the lease canceled.

During the spring of 1981, the *News* published an article titled "Passing of an Era" that aimed, in part, to explain the disappearance of lookouts from the landscape. "Manning a fire tower," it reported, "was a boring job." A lookout would climb a tower and sit all day, "trying to stay alert to scan the horizon." It was a "chore on a warm, lazy day" and misery "while the chill February winds rattled the metal box." If these descriptions did little to sway the nostalgic, the article went on to justify why "the preservation of more than a limited number of towers for historical sites" could not be considered, citing that "in recent years the towers have become a nuisance. As prime targets for vandals and irresistible attractions for teenagers looking for a lark, the towers were defaced, burned, and partially destroyed in spite of fences, chains, and locked gates. The Service doesn't have enough manpower to defend these historical relics from an urban population looking for a hideaway to smoke 'pot' or to sunbathe nude."[15]

The dizzying spin of tower sales meant that some purchased lookouts remained standing past their removal deadline. This prompted the TFS to send certified letters to all buyers notifying them of their obligations, offering them an extension, and warning them that the purchase would be declared invalid and the money returned if the structures were still on-site at the end of the period.[16] In many cases, the purchaser's unfulfilled obligations resulted in the lookouts reverting back to the state before they were offered again in later bid rounds.

The sell-off was well under way when Kenneth Burton, assistant head of the Fire Control Department, wrote a memorandum to the districts in 1982 asking for an update concerning which towers had been discarded. "We don't want to advertise a tower for sale that 'ain't there,'" he said, but he hoped to firm up a date to "dispose of the remainder of these 'pearls.'"[17] Two months later, Burton again wrote to the districts alerting them that "about thirty some odd towers" would be retained in the state system for "various reasons." "Don't let them become eyesores," he continued, before recommending that the sites receive "regular periodic inspections."[18]

During the summer of 1983, eighteen additional towers were offered to the public as surplus.[19] Excluding recommendations for the Huntington and Bastrop Towers, sixteen lookouts on the announcement had been offered before. When Ebarb and Burton sent a memorandum on September 23 summarizing the August 2 sale,[20] only 55 percent of the listed towers received public bids and were sold. Table 13 inventories the towers that had not been listed as surplus by the state after the third bid round.

The Shady Grove Lookout, which was first listed as surplus in 1981, sold for $101.99 during the bid round. The station had long been an important link between the US and Texas Forest Services, and the 1949 memorandum of understanding between the two agencies indicated the tower should be staffed during the spring and fall fire seasons. The point was essential so that USFS personnel in the Nogalus and Piney Lookouts could coordinate with the state across the cooperative fire action boundary.

In a twist of fate, the purchaser wrote to the TFS in January 1983 indicating that he had made arrangements with the corporate timber owner to lease the property for approximately six months while the tower was dismantled. Plans proceeded slowly, however, and the state extended the project completion date on several occasions, the last of which expired on March 30, 1984. Finally, on March 21, 1984, TFS received copies of a new lease granting the tower owner occupancy of the property. The state's lease was canceled and the property reconveyed. Today, the lookout

Table 13. At the conclusion of the third bid round in 1983 only thirty-two state lookouts had not been declared surplus. This represents just one-third of the original network. Within the decade, many of these lookouts would be sold or transferred to private ownership through negotiated sales agreements.

Area 1	Redwater	Pine Mills
	Negley	Hallsville
	Gilmont	Jefferson
Area 2	Deadwood	Love's Lookout
	Beckville	Neches
	Church Hill	Montalba
	Chandler	Maydelle
	Rusk	
Area 3	Grapeland	Alto
	Tabernacle	
Area 4	Kirby Forest	Siecke Forest
	Woodville	Buna
	Mayflower	Grayburg
		Horton Hill
Area 5	Fails	Keenan
	Willow Springs	Willis
	Shepherd	Winchester

stands forgotten without window lights or a cabin floor off County Road 4150. Despite decades of neglect, the structure is in remarkable condition and meets the conditions of age and significance for inclusion on the National Registry.

After the third bid round, the remaining towers were slowly disposed of through negotiated sales. For example, Love's Lookout was acquired by the city of Jacksonville, Maydelle was purchased by a community water board, and the Horton Hill Tower was purchased by a local radio cooperative. Others like Willow Springs survived into the new century because of existing communication agreements, while a small minority decay silently as historic relics on state forests.

With the sales largely complete, the TFS began releasing the leased properties back to the landowners. The "Cancellation of Lease and Reconveyance of Title" for the Keenan Lookout site in Montgomery County is typical. Once the tower and tower piers had been removed from the leased acre, TFS personnel inspected the property and verified that the buyer had upheld the stipulations of the contract with the state as part of the sale. The language of the reconveyance then explained that the "above described property ceased to be used for the purpose for which it was leased" and advised the Montgomery County clerk that the Board of Regents of the Texas A&M University System declared the lease "null and void."[21]

11 Epilogue

THE IMPORTANCE OF LAND stewardship, timber management, forest restoration, and environmental protection has only increased throughout the one-hundred-year history of the Texas A&M Forest Service. Societal demands for affordability, home delivery, land and cultural resource protection, and spaces for outdoor recreation create multidimensional challenges for today's policy makers and land managers. Likewise, the development of Texas' fire control infrastructure coincided with a burst of state and federal activity brought about by the public's demand for resource protection during the bleakness of the Depression. During the 1930s, the solutions focused on tackling several societal issues ranging from conservation to unemployment. The program was coincident with the nation's budding conservation consciousness and the dominating midcentury management philosophy that all wildland fires had to be controlled.

Nationally, fixed point fire detection systems were part of that solution. They provided construction projects for unemployed laborers while protecting young, recently replanted forests. Soon, the lookouts and their staff became part of the community and the community's identity. Rangers hosted scouting groups, and the lookout grounds became the destination for local picnics and social events. Towers even became the setting for mischievous teenage rites of passage. In time they became recognizable symbols of forest protection, visible at intervals along the wayside during nearly every road trip through East Texas and the country.

The retrospective rebirth fire lookouts enjoy today is partially due to this nostalgia, as Americans escape to find moments of solitude in an increasingly connected world. In the ambience of literature, poetry, function, or personal legacy, the structures attract the curious and the adventurer. Peter Steer and Keith Miller described these attributes as an "*integrity of location*, design, *setting*, materials, workmanship, *feeling and association*" during their efforts to protect lookouts in the Southwestern Region and nominated 31 percent of the fire lookout towers and associated structures they studied for inclusion on the National Register. The pair argued that a structure's identity was directly associated with the development of the region's fire detection and protection history and, often, the CCC. More broadly, they felt that lookout compounds were also a product of the conservation movement and incipient public land management measures. Together, these factors made them "a significant contribution to the broad patterns of our history."[1]

While lookouts have become obsolete, the structures stand as a symbol of conservation and a tangible expression of our nation's ingenuity, human resources, and culture. Steer and Miller also recognized that tower compounds are "a finite, dwindling and nonrenewable cultural resource."[2] As such, at least some should be preserved. Using their evaluating criteria, the key attributes that make a lookout significant are age and integrity. Location and the lack of any structural modifications help define integrity.

Determining the amount of alterations can often be accomplished through site visits and a comparison of historic and recent photographs. While subjective, Steer and Miller's decision process focused on evaluating the window design and facade of each tower. The windows, they felt, were the major functional element linked to the lookout's purpose, and any alteration from the original style compromised the historical integrity of the design. Another important evaluation consideration included the crowding of a lookout by "non-fire detection" facilities such as communication towers.

Unfortunately, twenty-five years into the new century and on the eve of the one-hundredth anniversary of the CCC, only 21 percent of Texas' fixed point fire detection system survives. Of these towers, 65 percent are in private ownership with uncertain futures. Only twelve lookouts are in state ownership, and two are owned by the US government (table 14). Several have remained at the same vantage point for nearly a century. None have been given protection by management plans or through the National Register of Historic Places.

Changing technologies, maintenance expenditures, and liability are cited as the reasons for the rapid disposition of lookouts across the country. That most towers were not "stylistically or materially distinct in any meaningful way" only accelerated their demise.[3] Moreover, a positive feedback cycle developed when one lookout was removed from the network, creating unprotected acreage within the system that made triangulation and fire positioning more problematic. With "holes in the system," the value of standing towers diminished, and removal of the remaining lookouts was more easily justified. Lookouts with dual purpose, like those used for relays and fire protection, survived somewhat longer, until technology—again—made the structures obsolete.

Across the United States, volunteer groups, associations, and hiking clubs have increasingly and passionately revived lookout heritage and restored the tangible pieces of the lookout yard. Small donations and "pass the basket" contributions have successfully supplied the resources for restoration, but community involvement and skills sharing have

Table 14. Lookout towers owned by the TFS or USFS at the time of publication. Localities are arranged according to (1) quadrangle and (2) entity. Structures eligible for inclusion on the National Historic Register, based on the author's assessment, are identified.

Name	Entity	Latitude	Longitude	County	Quadrangle	Manufacturer	Model	Erected	Eligibility
Kirby State Forest	TFS	30.57435	-94.40708	Tyler	Beaumont	Aermotor	MC-40	1934	Yes
Siecke State Forest	TFS	30.63296	-93.82885	Newton	Beaumont	Aermotor	MC-39	1948	Yes
Woodville	TFS	30.73748	-94.43260	Tyler	Beaumont	EMSCO	TW-1	1934	Yes
Mayflower	TFS	31.10114	-93.75931	Newton	Palestine	EMSCO	TW-1	1934	Yes
Texas Forestry Museum	TFS	31.35069	-94.70506	Angelina	Palestine	Aermotor	MC-39	1974	No
Bronson	TFS	31.36042	-94.00208	Sabine	Palestine	EMSCO	TW-1	1934	Yes
Grapeland	TFS	31.46928	-95.48055	Houston	Palestine	Nashville Bridge Co.	—	1964	No
Tabernacle	TFS	31.66643	-94.23843	Shelby	Palestine	EMSCO	TW-1	1934	Yes
Neches	TFS	31.87346	-95.49713	Anderson	Palestine	EMSCO	TW-1	1936	Yes
Deadwood	TFS	32.16618	-94.16147	Panola	Tyler	Aermotor	MC-39	1936	Yes
Pine Mills	TFS	32.74020	-95.32620	Wood	Tyler	L. C. Moore	—	1960	No
Negley	TFS	33.74926	-95.05589	Red River	Texarkana	Aermotor	MC-39	1936	Yes
Liberty Hill	USFS	30.51808	-95.14494	San Jacinto	Beaumont	Aermotor	MC-39	1955	No
Ratcliff	USFS	31.39543	-95.13856	Houston	Palestine	Aermotor	MC-39	1936	Yes

provided the expertise and labor that allows America's next generation to discover our heritage firsthand. The model is easily replicated but begins with awareness and education.

The restoration of New York's Arab Mountain Lookout serves as an example of community involvement.[4] The tower was erected in 1918 and staffed through the 1980s. Deterioration and condemnation followed until 1997, when a citizens group called the Friends of Mt. Arab was established with an endorsement from the county's planning commission. The group's vision was to restore the lookout tower and the observer's cabin to a safe and useful condition to encourage public support and enjoyment of the lookout as a recreational site. The group's mission also included plans for interpretive programs on the area's human and natural history and a pledge for the continual maintenance and upkeep of the facilities.

The project gained momentum with support from the county legislature, local governments, and the state's Department of Environmental Conservation. In time, the Adirondack Architectural Heritage group agreed to handle the Friends' fund as a nonprofit extension of its mission in architectural preservation. Restoration of the tower began in 1998, and after $12,000 had been raised in 1999, the tower had been made safe for climbing.

Texas and Texans boast a unique identity and remarkable history. Still, the state has struggled at times to preserve spaces and structures that are significant reminders of the people and places that make up our past and provide us with the opportunity to reflect. This is exemplified by the delayed establishment of a forest conservation program, the late invitation for federal land purchases for park development, and the narrow definition applied to cultural resources in federal forests. To apply Foster's words in speaking of conservation in Texas one hundred years ago, "Public spirit has not as yet been sufficiently aroused" to recognize the significance of the Forest Service's lookout system and preserve at least a few examples for the future. As Steer and Miller point out, "No one can control situations in which lookouts are lost because of fire, vandalism or inclement weather." But, they concluded, most lookouts have been lost because of management decisions.[5]

This effort serves as an initial attempt to consolidate primary documents, field observations, and common design elements that will assist in better understanding the context and significance of Texas' lookout system. It is hoped that these are the first steps along a pathway for preserving some of our shared cultural resources. Most deserving of protection are the historical lookouts located on the Kirby State Forest, at the state's district forest offices in Woodville and Kirbyville, and the federal tower on the grounds of a former CCC campsite at Ratcliff. They exemplify several designs from different manufacturers, and two preserve toolboxes. Each is an educational resource, a symbol of conservation, and a tribute to the dedicated Texans that served to protect the state's future financial and cultural wealth.

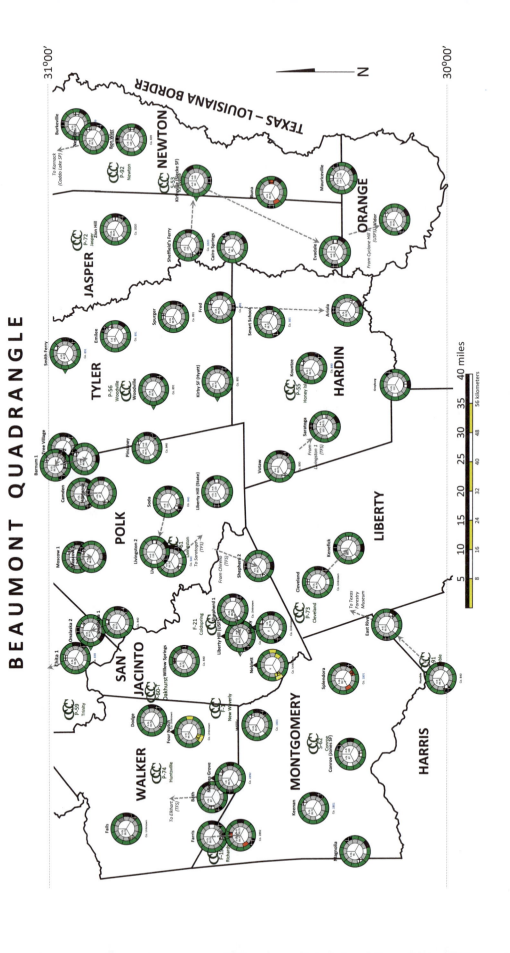

Appendix 1: Beaumont Quadrangle

EXPLANATION

A	Aermotor Co.	OL	Outside ladder
B	Bilby	IL	Inside ladder
C	Creosoted wood	IS	Inside stairway
D	Dunlap Manufacturing Corp.		
E	EMSCO Manufacturing Co.		
I	International Derrick & Equipment Co.		
L	L. C. Moore Manufacturing Co.		
M	McClintic-Marshall Corp.		
N	Nashville Bridge Co.		
P	BR Parkersburg		
U	Unknown		
CCC P-60-T Oakhurst	Historic Civilian Conservation Corps (CCC) Camp and Company (map only)		
838 840	Erecting CCC Company Hypothesized CCC Erecting Company	← — —	Lookout relocation (map only)

Name Ariola
Location 30.24500, -94.20889
Moved from Fred
Moved to
Quadrangle Beaumont
County Hardin
Entity TFS
District District 4
Lease date June 14, 1950
Survey R. C. Rogers A-46
NGS description None
Property number 5417   Tower/unit number 43 (408)
                       1   2   3
Erected 1950
Removed 1974
Model EMSCO TW-1
Height 99.75 feet
Observers Millard Wesley Palmer, crewleader (1950–64); Frank Hampshire, helper (1954); Floyd McDaniel (ca. 1958); Carroll L. Riley, crewman (1958–59); Charles St. Aubin, crewman (1960); Richard H. Ester, crewman/crewleader (1962–64); George Crowsom, crewman (1964); Warren P. Milstead, crewleader (1965); Robert W. "Buddy" Thompson, crewleader (1966); Ray E. Boyd, crewman (1965–66); Robert Antley (1967); Joseph A. Martin, crewman/crewleader (1961–71); George H. Hart, crewleader (1971)

223

Name	**Barnum, Location 1**
Location	30.99046, -94.64940
Moved from	Peach Tree Village
Moved to	Barnum, Location 2
Quadrangle	Beaumont
County	Tyler
Entity	TFS
District	District 4
Lease date	June 8, 1950
Survey	D. Millhome
NGS description	BL2138
Property number	5412	Tower/unit number	16 (411)
	1	2	3
Erected	1948
Removed	1961
Model	EMSCO TW-1
Height	99.75 feet
Observers	Arthur L. Williams (ca. 1944); Allen Griswold, crewleader/forest aide (1946–79)

Name	**Barnum, Location 2**
Location	30.92760, -94.63620
Moved from	Barnum, Location 1
Moved to	
Quadrangle	Beaumont
County	Tyler
Entity	TFS
District	District 4
Lease date	July 11, 1961
Survey	Gavino Aranio
NGS description	None
Property number	5412	Tower/unit number	16 (411)
	1	2	3
Erected	1961
Removed	1981
Model	EMSCO TW-1
Height	99.75 feet
Observers	Allen Griswold, crewleader/forest aide (1946–79)

Name	**Bath (Possum Walk)**		
Location	30.58872, -95.61067		
Moved from			
Moved to	Elkhart		
Quadrangle	Beaumont		
County	Walker		
Entity	TFS		
District	District 6		
Lease date	Unknown		
Survey	C. W. Bell		
NGS description	BL2272		
Property number	Unknown	Tower/unit number	3 (Unknown)
	1	2	3
Erected	1928		
Removed	1959		
Model	International		
Height	87 feet		
Observers	Alfred Chatham, lookout (ca. 1930); Edgar W. Sandel (1936); Eugene A. Heft, emergency patrolman (ca. 1942); James Arthur Owen, emergency patrolman (1942); Gus Randall, crewleader (ca. 1948); Linzie R. Stone, crewman (1958–59)		

I-OL 87

Name	**Bon Wier**		
Location	30.79833, -93.70806		
Moved from			
Moved to			
Quadrangle	Beaumont		
County	Newton		
Entity	TFS		
District	District 5		
Lease date	October 10, 1939		
Survey	T&NO RR—A473		
NGS description	None		
Property number	338	Tower/unit number	72 (508)
	1	2	3
Erected	1940	1958	
Removed	1958	1983	
Model	Creosoted wood	Nashville Bridge Co.	
Height	120 feet	122 feet	
Observers	John H. Newby, patrolman (1945); James A. Dougharty, crewleader (1951–74); Grady Castleberry, crewman (1959); A. J. Satterwhite (ca. 1963); Sam Marshall Poplin, crewman (1963–64); Thomas Carlton Herrin, crewman (1964); J. D. Miller Jr., crewman (1968–70)		

C-IS 120 | N-OL 122

Co. 839

Name	**Buna**		
Location	30.43750, -93.86472		
Moved from			
Moved to			
Quadrangle	Beaumont		
County	Newton		
Entity	TFS		
District	District 5		
Lease date	May 27, 1949		
Survey	H&TC RR A-204		
NGS description	None		
Property number	6014 1	Tower/unit number 2	76 (518) 3
Erected	1949		
Removed	After 1983		
Model	Aermotor MC-40		
Height	99.75 feet		
Observers	Aubrey E. Cole, crewleader (1948); Hardy Ray, senior crewleader (1951–52); Benjamin Franklin Jones, crewman (1953); Madison F. Bean, crewman (ca. 1954); Roy Albert Holmes, crewman (1954–56); Lewis Williams, crewleader at Vidor, moved to Buna (1956–59); Jim A. Walters, crewman (1956–67); C. Williams, crewman (1956); Paul J. West, crewman (1959); Alton Chamblee, crewleader (1963); Raymond Kelly (1964–66); Wesley A. Brockman, crewman/crewleader (1961–688)		

Name	**Burkeville**		
Location	30.93750, -93.66722		
Moved from			
Moved to			
Quadrangle	Beaumont		
County	Newton		
Entity	TFS		
District	District 5		
Lease date	September 8, 1939		
Survey	W. Wilson A-626		
NGS description	BK3234		
Property number	6020 1	Tower/unit number 2	71 (513) 3
Erected	1940		
Removed	1983		
Model	EMSCO TW-1		
Height	99.75 feet		
Observers	Jack C. Phelps, smokechaser (1942); Raymond M. Perkins, smokechaser (1942); Walter L. Smith, smokechaser (1942); Willie E. Greening (ca. 1956); T. M. Peterson, crewleader (1956); J. B. Smith Jr., crewman (1958–59); Oza Hall, crewleader (1945–71); Melbern Mitchell, crewman (1963); Garfield Johle Jr., crewman (1966); Orvis C. Woods, crewleader (1972)		

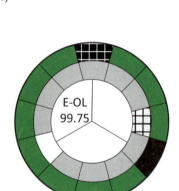
Co. 839

Name	**Cairo Springs**
Location	30.53606, -94.02177
Moved from	
Moved to	
Quadrangle	Beaumont
County	Jasper
Entity	TFS
District	District 5
Lease date	November 23, 1951
Survey	Lucretia Brown
NGS description	None
Property number	6023
	1
Erected	1952
Removed	1981
Model	EMSCO derrick
Height	96 feet
Observers	Homer F. Latham, crewman (1959); Ronnie C. Travis, crewman (1960); David Horn, crewleader (1963); Junious J. Arrant, crewleader (1964); Adouff Kirby, crewleader (1964–65); Leroy Walton, crewman (1969); Jodie S. Robbins, crewman/forestry aide (1971–72); William B. Hutchison, crewleader (1961–72)

Tower/unit number   Unknown (506)
2   3

E-OL 96

Name	**Camden**
Location	30.92204, -94.73337
Moved from	Hortense
Moved to	
Quadrangle	Beaumont
County	Polk
Entity	TFS
District	District 4
Lease date	June 8, 1950
Survey	H&TCC RR Co. No. 2
NGS description	None
Property number	Unknown
	1
Erected	1948
Removed	1956
Model	International
Height	87 feet
Observers	Barney S. Parrish, lookoutman (1949)

Tower/unit number   5 (Unknown)
2   3

I-OL 87

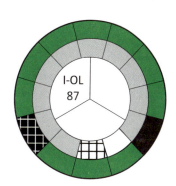

Beaumont Quadrangle   **227**

Name	**Chita, Location 1**			
Location	30.94318, -95.21198			
Moved from				
Moved to	Chita, Location 2			
Quadrangle	Beaumont			
County	Trinity			
Entity	TFS			
District	District 3			
Lease date	September 26, 1933			
Survey	BBB&C RR A-107			
NGS description	BL2168			
Property number	4875	Tower/unit number	31 (311)	
	1	2	3	
Erected	1934			
Removed	1970			
Model	EMSCO TW-1			
Height	99.75 feet			
Observers	J. W. Bowman; William (Willis) F. Loftin, crewleader (1942–66); James R. Dunaway, crewman, (1959); Philip N. Rowe, crewman/crewleader (1959–66); James M. Lawrence, crewman (1966); Hubert Dunaway, crewman (1966–67); James F. Parish, crewman (1970–71)			

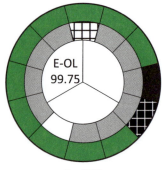

Co. 838

Name	**Cleveland (Oak Shade; Tarkington Prairie)**			
Location	30.32139, -94.99139			
Moved from				
Moved to	Kenefick			
Quadrangle	Beaumont			
County	Liberty			
Entity	TFS			
District	District 6			
Lease date	April 13, 1934			
Survey	B. Tarkington			
NGS description	None			
Property number	Unknown	Tower/unit number	27 (606)	
	1	2	3	
Erected	1934			
Removed	1960			
Model	EMSCO TW-1			
Height	99.75 feet			
Observers	George L. Hightower, patrolman/crewleader (1926–56); Robert Rice, emergency patrolman (1937–40); Francis M. Ott, smokechaser/emergency patrolman (1942–43); Alvin E. Franklin (ca. 1953); Jack Orton, crewman (1953–58); Hessie K. Parker, crewleader (1957–59); Jerry Douglas Thomas, crewman (1959)			

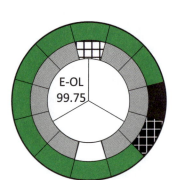

Co. Unknown

Name	Conroe (State Forest No. 2; W. G. Jones State Forest)		
Location	30.23849, -95.48874		
Moved from			
Moved to			
Quadrangle	Beaumont		
County	Montgomery		
Entity	TFS		
District	District 6		
Lease date	State property		
Survey	A. W. Springer		
NGS description	BL2211		
Property number	11126	Tower/unit number	7 (611)
	1	2	3
Erected	1931	1962	
Removed	1962	1996	
Model	International	Aermotor LS-40	
Height	87 feet	100 feet	
Observers	Coleman, patrolman; Jim I. Peoples, emergency lookoutman/AWS observer (1937–42); Surcy L. Peoples, crewleader (ca. 1931–52); Samuel T. Hoke, emergency patrolman (1937–40); Earl T. Johnson, CCC LEM/lookout/AWS observer (1937–43); Zebedee Rabon, emergency patrolman/crewleader (1941–ca. 1951); Lewis Tafelski (1945–52); Eugene D. Johns, smokechaser (1947–48); Andrew W. Blake (1962); Alcide Tomplait, crewleader (1963); Morris O. Wiggins, crewman/crewleader (1965–69); Monte Ray McDowell, crewman (1966–67); James L. Sunday, crewman (1959–67); Ivan B Aiken Jr., crewman (1968); Carlton Williams, crewman (1969)		

Name	Dodge		
Location	30.74587, -95.39027		
Moved from			
Moved to			
Quadrangle	Beaumont		
County	Walker		
Entity	TFS		
District	District 6		
Lease date	October 28, 1933		
Survey	C. M. Conrow A-137		
NGS description	BL2178		
Property number	6577	Tower/unit number	26 (614)
	1	2	3
Erected	1934		
Removed	1977		
Model	EMSCO TW-1		
Height	99.75 feet		
Observers	John Thigpen; Porter E. Key (1936–54); Jeff D. Crabb (1936); James W. Key, emergency patrolman (1942); Lewis P. Rush, emergency patrolman (1939–42); Stanford C. Key, emergency patrolman (1943–44); Kerry D. Dixon (1955–63); Ed Farris, crewleader (1955); Darrell D. Dickey, crewleader (1959); John R. Hall, crewman (ca. 1961–65); William Clayton Mathis, crewman (1965–66); Gerald Wayne Owens, crewman (1966–67); Lawrence E. "Bud" Swearingen, crewleader (1961–66); Charles H. Bartee, crewleader (1967–69); Thomas Walker, crewman (1967); Henry E. Farris, crewman (1967)		

Name	**East River**
Location	30.14066, -95.11407
Moved from	Humble
Moved to	Texas Forestry Museum
Quadrangle	Beaumont
County	Harris
Entity	TFS
District	District 6
Lease date	June 21, 1949
Survey	E. L. Branham
NGS description	BL2193
Property number	6572
	1
Erected	1949
Removed	1974
Model	Aermotor MC-39
Height	99.75 feet
Observers	Alcide J. Tomplait, crewleader (1956); Freeman, crewman (1955); Roscoe D. Munson, crewman (1958); Walter Mayo Yates, crewleader (1958–71); George Thomas Sallas, crewman/crewleader (1958–67); Juluas C. Sumner, forestry aide (1972)

Tower/unit number 70 (609)
2   3

Name	**Emilee**
Location	30.83852, -94.27811
Moved from	
Moved to	
Quadrangle	Beaumont
County	Tyler
Entity	TFS
District	District 4
Lease date	January 20, 1941
Survey	A. Barclay
NGS description	None
Property number	Unknown
	1
Erected	1941
Removed	1952
Model	Creosoted wood
Height	100 feet
Observers	Stephen Wyatt Hanks (1949)

Tower/unit number 73 (Unknown)
2   3

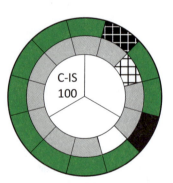

Co. 891

230   Appendix 1

Name	**Evadale**			
Location	30.27139, -94.03722			
Moved from	Kirbyville (Tower 1)			
Moved to				
Quadrangle	Beaumont			
County	Jasper			
Entity	TFS			
District	District 5			
Lease date	June 10, 1949			
Survey	F. L. DeSanque			
NGS description	None			
Property number	6018		Tower/unit number	66 (516)
	1	2		3
Erected	1949	1962		
Removed	1962	1983		
Model	Aermotor LS-40	Aermotor LS-40		
Height	100 feet	100 feet		
Observers	Willis Bernis Crosby (ca. 1952); Cecil Norman Hodges, crewman (1952); Frank L. Harris, crewleader (1956); Al A. Wicker, crewleader (1958); Fouler D. Garsee, crewleader (1959); Elton R. Stimits, crewman (1959); Berry Lee Lamert, crewleader (1960); William J. Smith, crewman (1962); William Terry, crewleader (ca. 1964); Sidney Ester (1963); Milton Eugene Reid, crewman (1962); Dwight Delaine Palmer, crewleader (1964–65); Walter A. Burrill (1964); Teddie Hagar, crewman (1964–65); Don Earl Harris, crewman (1965); Homer T. Young (ca. 1966); Ernest B. Harris, crewman (1966); Jimmy M. Latham, crewman (1966); Lewis Williams, crewleader (1965–66); Ronald R. Harmon, crewman (1967); Horace McCormick, crewman (1967); David L. Stanton, crewman (1970–71); Edward E. Cowart, crewleader (1966–71); Arthur R. Falkner, crewleader (1974)			

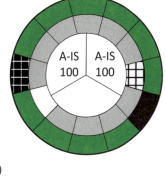

Name	**Fails (Mossy Grove)**			
Location	30.81111, -95.70389			
Moved from				
Moved to				
Quadrangle	Beaumont			
County	Walker			
Entity	TFS			
District	District 6			
Lease date	September 23, 1933			
Survey	LMC Guire			
NGS description	BL2265			
Property number	6575		Tower/unit number	25 (607)
	1	2		3
Erected	1934			
Removed	2008			
Model	EMSCO TW-1			
Height	99.75 feet			
Observers	Morris E. Cotton (1934); Alsey E. Cromeens (1934–56); Wayne D. Gaines, crewleader (1956–70); James W. Johnson, crewman (1957–58); Thomas G. Kelley, crewman (1968); James Bass, crewman (1962); Jerry L. Nichols, crewman (1962); Henry L. Nickerson, crewman (1965); Otha B. Fails, crewman (1971)			

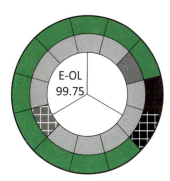

Co. Unknown

Beaumont Quadrangle **231**

Name	**Farris (West Sandy)**		
Location	30.58837, -95.72658		
Moved from			
Moved to	Pool		
Quadrangle	Beaumont		
County	Walker		
Entity	TFS		
District	District 6		
Lease date	Unknown		
Survey	H. Fares		
NGS description	BL2277		
Property number	Unknown	Tower/unit number	Unknown
	1	2	3
Erected	1934		
Removed	1936		
Model	EMSCO TW-1		
Height	99.75 feet		
Observers	Alfred Chatham (1934)		

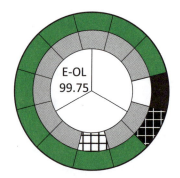

Co. Unknown

Name	**Four Notch**		
Location	30.64587, -95.42351		
Moved from			
Moved to			
Quadrangle	Beaumont		
County	Walker		
Entity	USFS		
District	Sam Houston NF, Raven District		
Lease date	Federal property		
Survey	J. H. Shepperd		
NGS description	BL2177		
Property number	Unknown	Tower/unit number	Unknown
	1	2	3
Erected	ca. 1937		
Removed	ca. 1977		
Model	L-1000		
Height	99.75 feet		
Observers	Alfred Chatham (1937–61)		

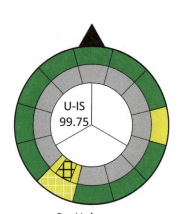

Co. Unknown

Name	**Fred**
Location	30.56652, -94.19422
Moved from	
Moved to	Ariola
Quadrangle	Beaumont
County	Tyler
Entity	TFS
District	District 4
Lease date	February 15, 1936
Survey	BBB&C Railway Section No. 10
NGS description	None
Property number 1	Unknown	Tower/unit number 2	43 (Unknown) 3
Erected	1936
Removed	1951
Model	EMSCO TW-1
Height	99.75 feet
Observers	None known

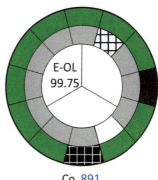

Co. 891

Name	**Grayburg**
Location	30.12070, -94.41805
Moved from	
Moved to	
Quadrangle	Beaumont
County	Hardin
Entity	TFS
District	District 4
Lease date	November 9, 1961
Survey	Mary Jones
NGS description	None
Property number 1	10062	Tower/unit number 2	Unknown 3
Erected	1962
Removed	1996
Model	EMSCO derrick
Height	122 feet
Observers	Walter Counce, crewleader (1962); Arthur McClusky and Carl Burke, crewmen (1962); John D. Showman, crewleader (1965); Pete Fabio and Bob Ray Allne, crewmen (1962); Buford Santos, crewman (1963); Kennith Walters, crewleader (1965); Charles S. Rowell, crewman (1965); Liston Riggs, crewman (1969); John Lamar Alpers, crewman (1970); Clarence P. Dugat, crewleader (1969–70)

Name	Hortense			
Location	30.87636, -94.73156			
Moved from				
Moved to	Camden			
Quadrangle	Beaumont			
County	Polk			
Entity	TFS			
District	District 4			
Lease date	Unknown			
Survey	J. Barrow			
NGS description	BL0723			
Property number	Unknown	Tower/unit number	5 (Unknown)	
	1	2	3	
Erected	1930			
Removed	1948			
Model	International			
Height	87 feet			
Observers	Arthur L. Williams, patrolman (ca. 1945); Alton B. David; Chesley A. Ogden, emergency smokechaser (1937); Joseph Carl Parrish, smokechaser/AWS observer (1937–46); Carlton L. Parrish (1944); B. B. Hickman (ca. 1944); Barney S. Parrish, lookoutman (1944–47)			

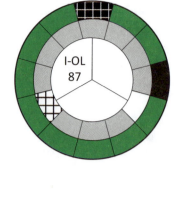

Name	Humble		
Location	30.00588, -95.26435		
Moved from			
Moved to	East River		
Quadrangle	Beaumont		
County	Harris		
Entity	TFS		
District	District 6		
Lease date	August 16, 1938		
Survey	W. B. Adams		
NGS description	None		
Property number	Unknown	Tower/unit number	70 (Unknown)
	1	2	3
Erected	1938		
Removed	1948		
Model	Aermotor MC-39		
Height	99.75 feet		
Observers	J. A. Koinm, patrolman (1938); Robert P. Jones, patrolman (1938); Alson G. Edgerton, alternate lookout (ca. 1942); Buster Ford, alternate lookout (ca. 1942); Hardy Ford (1938–42)		

Co. 840

Name	**Keenan**
Location	30.32972, -95.64365
Moved from	
Moved to	
Quadrangle	Beaumont
County	Montgomery
Entity	TFS
District	District 6
Lease date	March 17, 1934
Survey	Thomas Douglass
NGS description	BL2235
Property number	6573
Tower/unit number	30 (616)
Erected	1934
Removed	1981
Model	EMSCO TW-1
Height	99.75 feet
Observers	Morgan S. Cartwright, crewleader (1930–61); Robert C. Harris, crewman (1955–63); Leonard Singleton, crewleader/forest aide (1961–79); Arnold Sunday, crewman (1963); John W. Nicholson, crewman (1965); Joe Bly, crewman (1965–69); Bryan M. DuBose, crewman (1969); Oliver J. Payne, crewman (1970); Alma W. Vermillion, crewleader (1971)

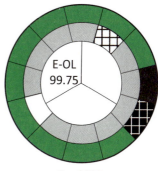

Co. 1801

Name	**Kenefick**
Location	30.23998, -94.89364
Moved from	Cleveland
Moved to	
Quadrangle	Beaumont
County	Liberty
Entity	TFS
District	District 6
Lease date	October 4, 1960
Survey	C. Burnet
NGS description	None
Property number	6574
Tower/unit number	Unknown
Erected	1960
Removed	1983
Model	EMSCO TW-1
Height	99.75 feet
Observers	Shirley J. Keith, crewman/crewleader (1962–75); Kenneth Page, crewleader (1964–67); William F. Gilley, crewman (1966–67); Donald A. Whelihan (1974)

Name	Kirby State Forest (State Forest No. 4; Hyatt State Forest)		
Location	30.57435, -94.40708		
Moved from			
Moved to			
Quadrangle	Beaumont		
County	Tyler		
Entity	TFS		
District	District 4		
Lease date	State property		
Survey	TC RR		
NGS description	BL2040		
Property number	5420	Tower/unit number	21 (416)
	1	2	3
Erected	1934		
Removed	Standing		
Model	Aermotor MC-40		
Height	99.75 feet		
Observers	Theordore Busselle; Enoch M. Pitts, patrolman/AWS observer (1937–43); William M. Hux, emergency patrolman (1940–41); Charles S. Price, emergency patrolman (1941); Thomas E. McGowan, smokechaser (1942); Chestley Byron Dickens (1943); Mabel L. Pitts, emergency lookout (1943); Louis C. Conner, crewman (1944–54); John A. Goings, crewman (1956); James B. Fountain, crewleader (1967); C. B. Wagnon, crewman (1967); Donald L. Powers, crewman (1966–67); Morris Blades, crewleader (1960–67)		

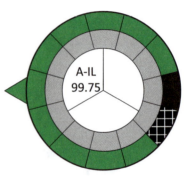

Co. 891

Name	Kirbyville State Forest (Siecke State Forest; State Forest No. 1)		
Location	30.63296, -93.82885		
Moved from	Sheffield's Ferry (Tower 2)		
Moved to	Evadale (Tower 1)		
Quadrangle	Beaumont		
County	Newton		
Entity	TFS		
District	District 5		
Lease date	State property		
Survey	J. T. Lewis		
NGS description	None		
Property number	6016	Tower/unit number	1 (510)
	1	2	3
Erected	1926	December 7, 1948	
Removed	1948	Standing	
Model	Aermotor LS-40	Aermotor MC-39	
Height	80 feet	120 feet	
Observers	W. H. Smith ("Uncle Hick") (1928); J. V. Sheffield, lookoutman (1930–37); R. J. Balthis; V. V. Bean; John W. Huffman, emergency patrolman/lookoutman (1930–44); Eddie L. Smith, emergency patrolman; Gordon Kelley, temporary emergency patrolman (1940); Jim Archie Walters, smokechaser (1942); William H. Richards, emergency patrolman (1942–43); Betty A. Huffman (1943); Grady P. Midkiff, patrolman (1945); Columbus E. Jones, emergency patrolman (1943–47); Floyd Bean, crewleader (1949); Harold Loyd Jones, crewman (1949); Buford D. Perry, crewman (1949); Grady P. Midkiff, lookoutman (ca. 1951); Cederic B. Powell, lookoutman (1951–52); D. L. Parmer (1953–59); Mathas N. "Bo" Nichols, crewleader (1967); Charles D. Powell, crewleader/forestry aide (1960–73)		

Name	**Kountze**			
Location	30.33639, -94.37111			
Moved from				
Moved to				
Quadrangle	Beaumont			
County	Hardin			
Entity	TFS			
District	District 4			
Lease date	October 19, 1933			
Survey	H. Stamper			
NGS description	BL2102			
Property number	5415	Tower/unit number	24 (419)	Co. 890
	1	2	3	
Erected	1934			
Removed	1981			
Model	EMSCO TW-1			
Height	99.75 feet			
Observers	Willie C. Spears, patrolman (1940); Gilbert B. Richardson, AWS observer/patrolman (1940–42); Woods, smokechaser (1943); Frank B. Anderson (ca. 1946); M. F. Jordan, patrolman (1946–47); Herman Lamar Jones, crewleader (1949); Clarence Williford (1952); Harold F. Peck, crewleader (1952–63); James C. Fowler, crewman (ca. 1953); Norman Noah Neff, crewman (1953–54); Ver Ben Jordan and Robert Lee Smith, crewmen (1955); Joseph Daigle (ca. 1955); Harold M. Holland, crewman (1955); John E. Wheeler (ca. 1955); David Martin (ca. 1958); John Dee Wilkins, crewman (1958); John D. Castilaw, crewman (1958–60); John S. Read Jr., crewman (1960); Alfred L. Langston, crewman (1961); Vestes Moss, crewman (1961); Luther E. Henderson, crewman (1962–63); Jeppie McCormick, crewman (1964); Richard E. Smith, crewman (1966); John Wheeler, crewleader (1961–66); Robert Antley, crewleader (1966); Ira Sykes Jr. (1968)			

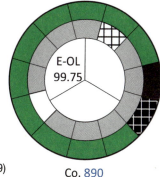

Name	**Liberty Hill (State)**		
Location	30.57423, -94.72635		
Moved from			
Moved to			
Quadrangle	Beaumont		
County	Polk		
Entity	TFS		
District	District 4		
Lease date	Unknown		
Survey	J. W. Moreland		
NGS description	BL2143		
Property number	11127	Tower/unit number	4 (413)
	1	2	3
Erected	1929	1963	
Removed	1963	1983	
Model	International	Aermotor LS-40	
Height	114 feet	120 feet	
Observers	William Nixon Cain (ca. 1932–49); Roy H. Gay, crewleader (ca. 1937–65); Charlie Lilly, crewman (1954); Ottice J. Moore, crewman (1958–69); Neal Elliott, crewleader (1965); Landron B. Herring, crewleader (1966–68); Charles R. Cain, crewleader (1968–69); Hugh Horace Ricks, crewman (1970); William B. Hullin, crewleader (1970); Edwin E. Theeck, crewleader (1971); Walter L. Wilkerson, crewman (1971)		

Name	**Liberty Hill (USFS)**		
Location	30.51808, -95.14494		
Moved from	Shephard (Tower 1, Location 1)		
Moved to			
Quadrangle	Beaumont		
County	San Jacinto		
Entity	USFS		
District	Sam Houston NF, Big Thicket District		
Lease date	Federal property		
Survey	G. Taylor		
NGS description	BL2172; BL2173		
Property number	Unknown	Tower/unit number	Unknown
	1	2	3
Erected	1938	1955	
Removed	1955	Standing	
Model	EMSCO TW-1	Aermotor MC-39	
Height	99.75 feet	99.75 feet	
Observers	None known		

Name	**Livingston, Location 1**		
Location	30.69338, -94.93038		
Moved from			
Moved to	Saratoga (Tower 2)		
Quadrangle	Beaumont		
County	Polk		
Entity	TFS		
District	District 4		
Lease date	Unknown		
Survey	J. Boulter		
NGS description	BL2148		
Property number	Unknown	Tower/unit number	18 (409)
	1	2	3
Erected	1934		
Removed	1957		
Model	Aermotor MC-40		
Height	99.75 feet		
Observers	William H. Parker (1929–47); Arthur L. Williams (ca. 1926–57); Lexa L. Smith, alternate lookout (ca. 1942); Edgar Williams, alternate lookout (ca. 1942); Maggie Mae V. Standard, alternate lookout (ca. 1942); Wilburn Rowe (ca. 1952); Billy Frank Platt, crewleader (1952); R. V. Jones, crewman (1953); Louis D. Clifton, crewman/crewleader (1954–57); Samuel A. Kirpatrick Jr., crewman (1957–68)		

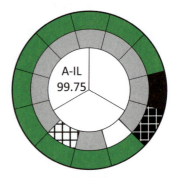

Name	**Livingston, Location 2**		
Location	30.73547, -94.90426		
Moved from	Soda		
Moved to			
Quadrangle	Beaumont		
County	Polk		
Entity	TFS		
District	District 4		
Lease date	September 23, 1957		
Survey	I&GN RR A-681		
NGS description	None		
Property number	5410	Tower/unit number	Unknown (409)
	1	2	3
Erected	1957		
Removed	1993		
Model	Aermotor MC-39		
Height	120 feet		
Observers	L. D. Clifton, crewleader (1957); Samuel A. Kirpatrick Jr., crewleader (1957–68); Joseph H. Matthews (1960–67); Bordal P. Ogden, crewman/forest aide (1967–79); Rufus Hickman (1976)		

Name	**Magnolia**		
Location	30.22097, -95.77050		
Moved from			
Moved to			
Quadrangle	Beaumont		
County	Montgomery		
Entity	TFS		
District	District 6		
Lease date	December 17, 1951		
Survey	W. T. Dunlavy		
NGS description	BL2247		
Property number	Unknown	Tower/unit number	Unknown (612)
	1	2	3
Erected	1952		
Removed	1968		
Model	EMSCO derrick		
Height	96 feet		
Observers	Bannon E. Damuth, crewman/crewleader (1954–73); Jimmy Carter, crewman (1956); J. C. Flemming, crewman (1956); Henry A. Martin, crewman (1948–64); George B. Hosford, crewman (1965); James L. Peacock, crewman (1969)		

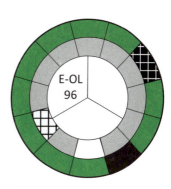

Name	**Mauriceville (Deweyville)**		
Location	30.24517, -93.82062		
Moved from			
Moved to			
Quadrangle	Beaumont		
County	Newton		
Entity	TFS		
District	District 5		
Lease date	December 12, 1951		
Survey	James S. Russell		
NGS description	None		
Property number	11128	Tower/unit number	Unknown (517)
	1	2	3
Erected	1952	1963	
Removed	1963	1980	
Model	International	Aermotor LS-40	
Height	114 feet	120 feet	
Observers	L. C. "Pete" Matthews, crewleader (1952–73); Clarence M. Teal, crewman, 1959; James O. Goddard (ca. 1961); Calvin Edward Hutson, crewman (1961–65); Monroe L. Wroten, crewman (1967–76)		

Name	**Mercy**		
Location	30.43847, -95.12086		
Moved from	Haleyville, Alabama (Tower 2)		
Moved to			
Quadrangle	Beaumont		
County	San Jacinto		
Entity	USFS		
District	Sam Houston NF, Big Thicket District		
Lease date	Federal property		
Survey	I. Lowry		
NGS description	BL2184		
Property number	Unknown	Tower/unit number	Unknown
	1	2	3
Erected	1938	1949	
Removed	1949	1971	
Model	Unknown	IDECO	
Height	Unknown	99.75 feet	
Observers	None known		

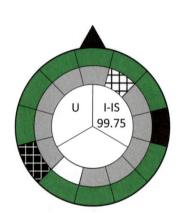

Co. Unknown

Name	**Moores Grove**			
Location	30.54345, -95.55762			
Moved from				
Moved to				
Quadrangle	Beaumont			
County	Walker			
Entity	USFS			
District	Sam Houston NF, Raven District			
Lease date	Federal property			
Survey	H. Applewhite			
NGS description	BL2274			
Property number	Unknown	Tower/unit number	Unknown	
	1	2	3	
Erected	1939			
Removed	1971			
Model	IDECO (L-1000)			
Height	99.75 feet			
Observers	W. E. Parrish (ca. 1942)			

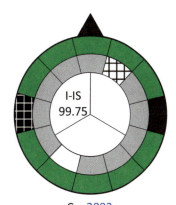

Co. 2892

Name	**Moscow, Location 1**			
Location	30.93584, -94.91582			
Moved from				
Moved to	Moscow, Location 2			
Quadrangle	Beaumont			
County	Polk			
Entity	TFS			
District	District 4			
Lease date	November 29, 1933			
Survey	Phillip Dixon			
NGS description	BL2160			
Property number	5423	Tower/unit number	17 (417)	
	1	2	3	
Erected	1934			
Removed	1961			
Model	Aermotor MC-40			
Height	99.75 feet			
Observers	Brodie H. Jones, emergency patrolman/patrolman (1937–44); Barney B. Tarver (ca. 1942); Billie Jack Jones, lookoutman (1943–44); Billy Lafferty, crewleader (ca. 1955); Billy Stamford, crewman (1953); Brunett Broom, crewman/crewleader (1955); Eugene Cockrell, crewleader (1956); Maxie Cooper, crewleader (1956); William L. Brooks, crewman (1958)			

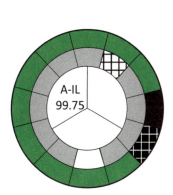

Co. 840

Beaumont Quadrangle **241**

Name	**Moscow, Location 2**				
Location	30.90192, -94.91568				
Moved from	Moscow, Location 1				
Moved to					
Quadrangle	Beaumont				
County	Polk				
Entity	TFS				
District	District 4				
Lease date	May 25, 1961				
Survey	Andrew J. Ford				
NGS description	None				
Property number	5423	Tower/unit number	17 (417)		
	1	2	3		
Erected	1961				
Removed	1981				
Model	Aermotor MC-40				
Height	99.75 feet				
Observers	Rufus Hickman, crewleader (1963–66)				

A-IL
99.75

Name	**Neblett**				
Location	30.43263, -95.23002				
Moved from					
Moved to					
Quadrangle	Beaumont				
County	San Jacinto				
Entity	USFS				
District	Sam Houston NF, Big Thicket District				
Lease date	Unknown				
Survey	R. O. Lusk				
NGS description	BL2191				
Property number	Unknown	Tower/unit number	Unknown		
	1	2	3		
Erected	ca. 1938				
Removed	ca. 1948				
Model	Unknown				
Height	Unknown				
Observers	None known				

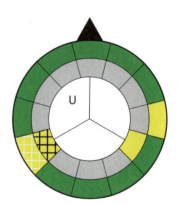

U

Co. Unknown

Name	**Newton**
Location	30.89038, -93.70708
Moved from	
Moved to	Karnack, Location 1
Quadrangle	Beaumont
County	Newton
Entity	TFS
District	District 5
Lease date	November 13, 1933
Survey	T&NO RR—5
NGS description	BK3233
Property number	Unknown
Erected	1934
Removed	1942
Model	EMSCO TW-1
Height	99.75 feet
Observers	Wilson

Tower/unit number 39 (Unknown)

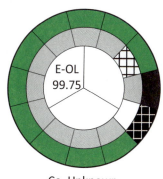

Co. Unknown

Name	**Onalaska, Location 1**
Location	30.83448, -95.10654
Moved from	
Moved to	Onalaska, Location 2
Quadrangle	Beaumont
County	Polk
Entity	TFS
District	District 4
Lease date	September 26, 1933
Survey	I&GN RR—A668
NGS description	BL0914
Property number	4868
Erected	1934
Removed	1970
Model	EMSCO TW-1
Height	99.75 feet
Observers	Atma D. Stanford, crewleader (1934–65); Elliott E. Sheffield, crewman/crewleader (1959–71); Rollen M. Tanner, crewman (1967); Jessie Richard Oliver, crewman (1963–67)

Tower/unit number 32 (324)

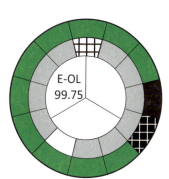

Co. 840

Name	**Onalaska, Location 2**
Location	Withheld
Moved from	Onalaska, Location 1
Moved to	
Quadrangle	Beaumont
County	Trinity
Entity	TFS
District	District 3
Lease date	August 17, 1970
Survey	H&TC RR Block 23, A-304
NGS description	None
Property number	4868
Tower/unit number	32 (324)
Erected	1970
Removed	Standing (sold 1981)
Model	EMSCO TW-1
Height	99.75 feet
Observers	Elliott E. Sheffield, crewman/crewleader (1959–71)

Name	**Peach Tree Village**
Location	30.97220, -94.60355
Moved from	
Moved to	Barnum, Location 1
Quadrangle	Beaumont
County	Tyler
Entity	TFS
District	District 4
Lease date	January 17, 1934
Survey	I&GN RR A-707
NGS description	None
Property number	Unknown
Tower/unit number	16 (Unknown)
Erected	1934
Removed	1948
Model	EMSCO TW-1
Height	99.75 feet
Observers	Arthur L. Williams, patrolman (1937–45); Elbert C. Hughes, smokechaser (1937–41); John A. Richardson, emergency patrolman (1937–40); Travis A. Barnes, smokechaser (1942–43); Frank W. Ripley, emergency smokechaser (1940–41); Herman Winn Hopkins, emergency smokechaser/AWS observer/lookout (1942–47)

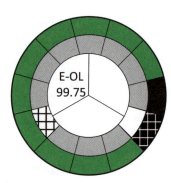

Co. 891

Name	Pinckney (Kiam; County Line)		
Location	30.75674, -94.60190		
Moved from			
Moved to			
Quadrangle	Beaumont		
County	Polk		
Entity	TFS		
District	District 4		
Lease date	January 17, 1934		
Survey	J. B. Woods		
NGS description	BL2133; BL2134; BL2135		
Property number	5411	Tower/unit number	63 (410)
	1	2	3
Erected	1934	1937	
Removed	1937	1981	
Model	Aermotor MC-40	Aermotor MC-39	
Height	99.75 feet	120 feet	
Observers	Edd A. Rhodes, smokechaser (1936); William Albert Holder, smokechaser/AWS observer/crewleader (1936–65); Sidney Emil Revia, crewman (1951–57); Paul G. Schlatlman Jr., crewman (1968); Leon Dickens, crewleader/forest aide (1965–79)		

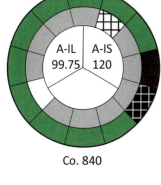

Co. 840

Name	Pool		
Location	30.51506, -95.72379		
Moved from	Farris (Tower 1)		
	Polkville Lookout, Mississippi (Tower 2)		
Moved to			
Quadrangle	Beaumont		
County	Montgomery		
Entity	USFS		
District	Sam Houston NF, Raven District		
Lease date	Federal property		
Survey	S. T. Moore		
NGS description	BL2280		
Property number	Unknown	Tower/unit number	Unknown
	1	2	3
Erected	1937	1963	
Removed	1963	ca. 1980	
Model	L-1000 (EMSCO TW-1)	Aermotor MC-39	
Height	99.75 feet	99.75 feet	
Observers	Kimbro (ca. 1954–61); Eddie Ray Stewart, crewman (1960); Ralph Gilley, crewman (1962–68); Edwin R. Arnold, crewleader (1966); Roy Arnold, crewleader (1967); George G. (Gus) Randall, crewleader (1971)		

Co. 2892

Name	**Saratoga**			
Location	30.30039, -94.53954			
Moved from	Livingston, Location 1 (Tower 2)			
Moved to				
Quadrangle	Beaumont			
County	Hardin			
Entity	TFS			
District	District 4			
Lease date	June 14, 1950			
Survey	J. Coit			
NGS description	BL2109			
Property number	5409		Tower/unit number	77 (418)
	1		2	3
Erected	1950		1961	
Removed	1961		1983	
Model	Bilby Survey		Aermotor MC-40	
Height	102 feet		99.75 feet	
Observers	Clarence Williford, crewleader (1955); Joseph Martin, crewman (1955–62); Hugh Bevil Means, crewleader (1960); Buford Santos, crewman/crewleader (1960–62); Mose F. Ryan, crewleader (1965); Wilbur W. Collins, crewleader (1965); George H. Hart, crewleader (1971); James Ryan, forest aide (1975)			

Name	**Sheffield's Ferry**			
Location	30.64963, -94.01583			
Moved from				
Moved to	Siecke State Forest			
Quadrangle	Beaumont			
County	Jasper			
Entity	TFS			
District	District 5			
Lease date	September 14, 1937			
Survey	James H. Blount			
NGS description	None			
Property number	Unknown		Tower/unit number	66 (Unknown)
	1		2	3
Erected	1937			
Removed	1948			
Model	Aermotor MC-39			
Height	120 feet			
Observers	J. V. Sheffields (ca. 1937)			

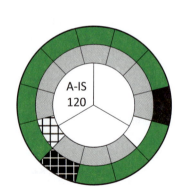

Co. 1820

Name	**Shepherd, Location 1**
Location	30.53472, -95.07056
Moved from	
Moved to	Liberty Hill (USFS)
Quadrangle	Beaumont
County	San Jacinto
Entity	TFS/USFS
District	Sam Houston NF, Big Thicket District
Lease date	Unknown
Survey	J. Hardin
NGS description	BL1106
Property number 1	Unknown
Erected	1934
Removed	1938
Model	EMSCO TW-1
Height	99.75 feet
Observers	None known

Tower/unit number 2: Unknown
3: Unknown

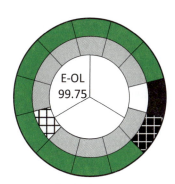

Name	**Shepherd, Location 2**
Location	30.46865, -94.94559
Moved from	Chireno
Moved to	
Quadrangle	Beaumont
County	San Jacinto
Entity	TFS
District	District 6
Lease date	August 21, 1960
Survey	H. White
NGS description	None
Property number 1	4879
Erected	1960
Removed	1988
Model	Aermotor MC-40
Height	99.75 feet
Observers	Jack Orton, crewman (1953–58); John B. Ragsdale, crewman (1961); Edward Shrader, crewman (1962); H. K. Parker, crewleader (1962); Bruce Thomas, crewleader (1962–71); Jack Long, crewman (1966); Forest Young, forest aide (1970)

Tower/unit number 2: Unknown
3: Unknown

Beaumont Quadrangle **247**

Name	**Smart School**				
Location	30.44333, -94.23778				
Moved from					
Moved to					
Quadrangle	Beaumont				
County	Hardin				
Entity	TFS				
District	District 4				
Lease date	July 16, 1938				
Survey	A. A. Burrill				
NGS description	None				
Property number	5421	Tower/unit number	69 (412)	Co.	891
	1	2	3		
Erected	1938				
Removed	1984				
Model	Aermotor MC-40				
Height	99.75 feet				
Observers	Ashley Gore, smokechaser/crewman (1938–62); Abner W. Bell, crewleader (ca. 1953–59); Edgar Coleman Hunt, crewleader (1960); Virgil Eason, crewleader (1959–64); Kenneth J. Keith, crewman (1961); Elmo Davis, crewman (1963); R. B. Watts, crewman (1964); Raymond Hooks, crewman/crewleader (1961–64); Charles Kent, crewman (1966); Leroy Ballard, crewman (1966); Wiley B. Kelly, crewman/crewleader (1967–70); Edgar L. Brookins, crewman (1970)				

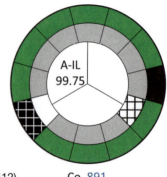

Name	**Smith Ferry (Colmesneil)**				
Location	30.96810, -94.32722				
Moved from					
Moved to					
Quadrangle	Beaumont				
County	Tyler				
Entity	TFS				
District	District 4				
Lease date	December 15, 1933				
Survey	G&BN Company Section 44				
NGS description	BL2049				
Property number	Unknown	Tower/unit number	42 (414)	Co.	891
	1	2	3		
Erected	1934				
Removed	Standing (sold 1981)				
Model	EMSCO TW-1				
Height	99.75 feet				
Observers	Noah A. Platt, crewleader/educational officer (1934–67); Richard J. Owens, lookout (1937–49); Billy M. Nowlin, crewleader (1952–54); George S. Downing, crewman (1957); Elza Oates, crewman (1962); Otis Sanders, crewman (1963); Kenneth Graham, crewman/crewleader (1963–67); Leonard B. Conner, crewleader (1964–65); Gene C. Russell, crewman/crewleader (1965–66); Billy Ross Harkness, crewleader (1966–67); Jeff Griffin, crewleader (1967); John Durham, crewman (1967); Wesley P. McDonald, crewman (1969); Kimmie D. Hensarling, crewman (1974)				

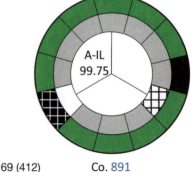

Name	Soda
Location	30.70151, -94.75584
Moved from	
Moved to	Livingston, Location 2
Quadrangle	Beaumont
County	Polk
Entity	TFS
District	District 4
Lease date	February 1, 1937
Survey	J. Falcon
NGS description	BL2144

Property number	Unknown		Tower/unit number	64 (Unknown)	
	1			2	3

Erected	1937
Removed	1957
Model	Aermotor MC-39
Height	120 feet
Observers	Arthur L. Williams, patrolman (ca. 1930–47); James J. Howard (ca. 1948); Herman Winn Hopkins, crewman (1947–48)

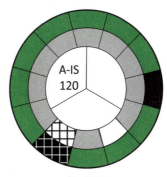

Co. 840

Name	Splendora (Ben Milam)
Location	30.25442, -95.27061
Moved from	
Moved to	
Quadrangle	Beaumont
County	Montgomery
Entity	TFS
District	District 6
Lease date	October 20, 1933
Survey	Walker Co. School Land
NGS description	BL2215

Property number	6579		Tower/unit number	29 (613)	
	1			2	3

Erected	1934
Removed	After 1984
Model	EMSCO TW-1
Height	99.75 feet
Observers	Alfred J. DeFoor, crewleader/crewman (1949–73); Cap Page (ca. 1955); Harvey L. Smith, crewleader (1958–59); Linzie Stone, crewman (1962); John Barbee, crewleader (1962); Wilmer Cook, crewman (1963–73)

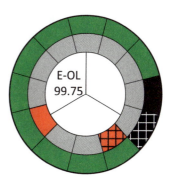

Co. 1801

Beaumont Quadrangle 249

Name	**Spurger**			
Location	30.68863, -94.22235			
Moved from				
Moved to				
Quadrangle	Beaumont			
County	Tyler			
Entity	TFS			
District	District 4			
Lease date	November 25, 1933			
Survey	Joseph Collier			
NGS description	None			
Property number	5414 1	Tower/unit number	23 (415) 2	Co. 891 3
Erected	1934			
Removed	1980			
Model	EMSCO TW-1			
Height	99.75 feet			
Observers	Tharp R. Jordan, patrolman (1944); Jerry P. Jordan, lookoutman (1944–47); Willie L. Holland, lookoutman/patrolman/crewleader (1944–51); A. J. Spurlock, crewleader (1942–70); Millard W. Palmer, crewman (1948–49); Waldrep (ca. 1950); Thomas Baten Jordan, crewman (1951); Pat Alvin Riley, crewleader (1951); George McInnis (1954); Oliver Grissom, crewman (1962)			

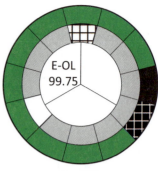

Name	**Vidor (Terry)**			
Location	30.12368, -93.94210			
Moved from	Cyclone Hill (Tower 1)			
Moved to				
Quadrangle	Beaumont			
County	Orange			
Entity	TFS			
District	District 5			
Lease date	June 9, 1958			
Survey	G. A. Patillo			
NGS description	None			
Property number	11059 1	Tower/unit number	Unknown (515) 2	3
Erected	1958			
Removed	1981			
Model	EMSCO TW-1			
Height	99.75 feet			
Observers	J. H. Walker, crewman (1956); Joseph M. Smith (ca. 1956); Lewis Williams, crewleader (ca. 1956); Cola Chalmer Henley, crewman/crewleader (1956); E. J. Turpin, crewman II (1958); Royce L. Fancher, crewleader (1959); Wilbur M. Eaves, crewman (1959); Joseph Frederick, crewman (1962); Jefferson Byrd, crewman (1963); Clarence Keels, crewman (1966); Joseph Singleton, crewleader (1961–66); John H. Shelby, crewleader (1970); Herman Joyce, crewleader (1971–72); Milton C. Clark, forest aide (1972); Carl E. Boatman, forest aide (1974)			

Name	**Votaw (Rye)**			
Location	30.41517, -94.65524			
Moved from				
Moved to				
Quadrangle	Beaumont			A-IL
County	Hardin			99.75
Entity	TFS			
District	District 6			
Lease date	January 17, 1934			
Survey	T. H. Lewis			
NGS description	BL2113			
Property number	5419	Tower/unit number	19 (618)	Co. 890
	1		2	3
Erected	1934			
Removed	1978			
Model	Aermotor MC-40			
Height	99.75 feet			
Observers	Perry L. Moye (ca. 1925–39); Newton A. Perkins, emergency patrolman (1938); William Lloyd Kelley, crewleader (1950–52); William Tanton, crewman (ca. 1952); Ulys Laverne Knight, crewman (1952); John H. Poole, crewman (1953); Dow T. Clifton, crewleader (1954–59); Chester F. Young, crewman (1959); Ernest Munson, crewleader (1962); Johnnie L. Currie, crewman/crewleader (1966–69); James Rhodes, crewman (1966); Roy Lowe, crewman (1967)			

Name	**Willis**			
Location	30.47684, -95.40403			
Moved from				
Moved to				
Quadrangle	Beaumont			E-OL
County	Montgomery			99.75
Entity	TFS			
District	District 6			
Lease date	January 27, 1934			
Survey	De La Garza			
NGS description	BL2231			
Property number	6576	Tower/unit number	28 (610)	Co. 1801
	1		2	3
Erected	1934			
Removed	1982			
Model	EMSCO TW-1			
Height	99.75 feet			
Observers	Bob M. Williams, patrolman (1934–55); Clyde Thomas Parrish, seasonal crewman (1949); Henry A. Martin, senior lookout (ca. 1949); James Edward Hosea, senior lookout (1951–52); Bernice L. Clifton, crewman/crewleader (1955–68); Paul DeFoor (1958–59); Alton Robinson, crewman (1961–64); Ralph W. Gilley, crewman/crewleader (1961–69); Paul Buchanan, crewman (1966–67); Alvin L. Hardy, crewman (1971)			

Name	**Willow Springs**
Location	30.66147, -95.21966
Moved from	
Moved to	
Quadrangle	Beaumont
County	Sam Jacinto
Entity	TFS/USFS
District	District 6
Lease date	October 23, 1933
Survey	I&GN No. 10
NGS description	BL1126
Property number	6578
Tower/unit number	73 (608)
Co.	840
Erected	1933
Removed	2000
Model	EMSCO TW-1
Height	99.75 feet
Observers	T. J. Williamson, patrolman (1949); James Edward Hosea, seasonal crewman (1949–51); Tom B. Winfrey, crewleader (ca. 1939–74); Lloyd Bryant, crewman (1959–67); Tommy Dirden Jr. (1970); Frank Phlegm Jr., crewman (1971); Arnold D. Sunday, forestry aide (1972); Ocie Norman, crewleader (1974); Tucker Miller Jr., crewman (1975)

E-OL 99.75

Name	**Woodville**
Location	30.73748, -94.43261
Moved from	
Moved to	
Quadrangle	Beaumont
County	Tyler
Entity	TFS
District	District 4
Lease date	November 6, 1933
Survey	John Nowlin
NGS description	BL2045
Property number	5413
Tower/unit number	22 (406)
Co.	891
Erected	1934
Removed	Standing
Model	EMSCO TW-1
Height	99.75 feet
Observers	Allen M. Riley (1937); Jesse E. Hyde, emergency patrolman/AWS observer (1936–43); Gordon, smokechaser (1943); Noah A. Platt, crewleader (ca. 1954); Vernon Walton Holton, crewleader (1954); G. K. McKee, crewman (1958); Robert L. Comte, crewleader (1958–68); Gerald Spurlock, crewman (1963); Allan Wilson, crewman (1964–66); Arvist L. Franklin, crewman (1967); James H. DeRamus, crewman (1968); Leroy Hamilton, crewman (1968)

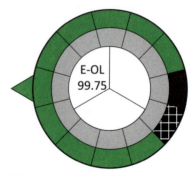

E-OL 99.75

Name	**Zion Hill**
Location	30.83719, -93.97072
Moved from	
Moved to	
Quadrangle	Beaumont
County	Jasper
Entity	TFS
District	District 5
Lease date	September 29, 1936
Survey	T. M. Stone
NGS description	BK3241
Property number	6021
Tower/unit number	38 (512)
Erected	1934
Removed	1983
Model	EMSCO TW-1
Height	99.75 feet
Observers	Ider J. Kelley, crewleader (1951–56); Eddie L. Smith, temporary replacement, from Kirbyville (1940); Thomas E. Rutledge, patrolman (1943); Robert E. L. Jones, emergency patrolman/patrolman/crewleader (1943–50); Herbert Henry Hall, crewman/crewleader (1951–64); Legran J. Bradshaw, crewleader (1957–72); John D. Showman, crewleader (1961); Kimble Singletary, crewman (1965); Chester E. Stowell [Stovall], crewman (1971)

E-OL 99.75
Co. 1820

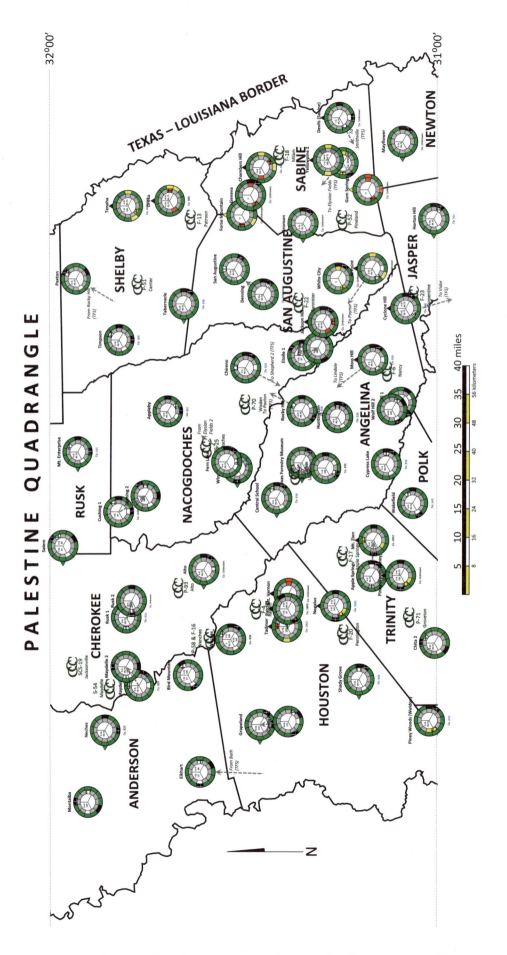

Appendix 2: Palestine Quadrangle

EXPLANATION

A	Aermotor Co.	OL	Outside ladder	
B	Bilby	IL	Inside ladder	
C	Creosoted wood	IS	Inside stairway	
D	Dunlap Manufacturing Corp.			
E	EMSCO Manufacturing Co.			
I	International Derrick & Equipment Co.			
L	L. C. Moore Manufacturing Co.			
M	McClintic-Marshall Corp.			
N	Nashville Bridge Co.			
P	BR Parkersburg			
U	Unknown			
CCC P-60-T Oakhurst	Historic Civilian Conservation Corps (CCC) Camp and Company (map only)			
838 / 840	Erecting CCC Company / Hypothesized CCC Erecting Company	←---	Lookout relocation (map only)	

Name	Alto (Primrose)
Location	31.61532, -95.02222
Moved from	
Moved to	
Quadrangle	Palestine
County	Cherokee
Entity	TFS
District	District 2
Lease date	October 20, 1933
Survey	J. Durst
NGS description	BY2716
Property number	4289

Tower/unit number 10 (209)
1 2 3

Erected	1934
Removed	Standing (sold 1986)
Model	Aermotor MC-40
Height	99.75 feet
Observers	Alec Black, patrolman (1934); Fred F. Neeley, emergency patrolman/patrolman (1934–39); Nolan T. Johnson, emergency patrolman (1939–42); Nesbett Foster, patrolman (1939–46); William P. James, emergency patrolman/patrolman (1942–47); J. A. Benge (ca. 1952); Floy M. Creel, crewleader (1953–67); James F. Landrum, crewleader (1965–66); Charles McGaughey, crewman (1966)

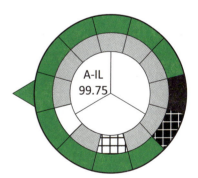

Co. Unknown

Name	**Appleby**		
Location	31.70898, -94.55645		
Moved from			
Moved to			
Quadrangle	Palestine		
County	Nacogdoches		
Entity	TFS		
District	District 3		
Lease date	October 18, 1935		
Survey	M. Galan		
NGS description	BY1148		
Property number	4881	Tower/unit number	48 (325)
	1	2	3
Erected	1936		
Removed	1975		
Model	Aermotor MC-39		
Height	99.75 feet		
Observers	M. D. Blanton (1937); Herbert L. Polk, emergency patrolman (1937–40); Clarence E. Greening, crewleader (1959); Wilburn H. Olds, crewman (1952–62); John D. Hancock (1962); George Lavon Hartt, crewman (1964); General Gabe Stewart, crewman/crewleader (1964–67); Murvey C. Corley, crewman (1967); Laroy Skeeters, crewman (1967); David Swift, forest aide (1975)		

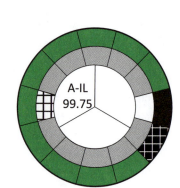

Co. 891

Name	**Apple Springs**		
Location	31.18267, -95.04218		
Moved from			
Moved to	Mt. Zion		
Quadrangle	Palestine		
County	Trinity		
Entity	TFS/USFS		
District	Davy Crockett NF, Trinity District		
Lease date	Unknown		
Survey	I&GN RR A-358		
NGS description	None		
Property number	Unknown	Tower/unit number	Unknown
	1	2	3
Erected	1934		
Removed	1939		
Model	L-1000 (Aermotor MC-40)		
Height	99.75 feet		
Observers	None known		

Co. 838

Name	**Bird Mountain**		
Location	31.65611, -95.33333		
Moved from			
Moved to			
Quadrangle	Palestine		
County	Anderson		
Entity	TFS		
District	District 2		
Lease date	May 1, 1931		
Survey	E. Kennedy A-452		
NGS description	BY2721		
Property number	11123	Tower/unit number	2 (207)
	1	2	3
Erected	1928	1963	
Removed	1963	1982	
Model	International	Aermotor LS-40	
Height	87 feet	100 feet	
Observers	Coy E. Cliburn, patrolman (1937–42); W. A. Cliburn Jr., emergency patrolman (ca. 1937); Roy M. Watkins, lookoutman (1937–40); Lewis H. Jordan, smokechaser (1937–39); Albert H. Rich, emergency patrolman, smokechaser (1937–49); Lawrence P. Johnston (1942–57); Joe C. Rich, emergency lookoutman/lookoutman/smokechaser/crewman (1931–56); Raymond Tillman, crewleader (1964–65); Robert E. Tucker, crewman (1970); Ira C. Bowman, crewleader (1957; 1970–71)		

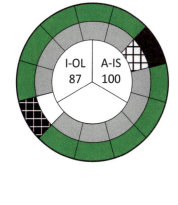

Name	**Bronson**		
Location	31.36042, -94.00208		
Moved from			
Moved to			
Quadrangle	Palestine		
County	Sabine		
Entity	TFS		
District	District 5		
Lease date	August 24, 1935		
Survey	J. H. Kirby A-410		
NGS description	BY2630		
Property number	6017	Tower/unit number	34 (511)	Co. Unknown
	1	2	3
Erected	1934		
Removed	Standing		
Model	EMSCO TW-1		
Height	99.75 feet		
Observers	William F. Fullen, patrolman (1943); Price Rhame Jr., emergency patrolman (1943); Cleo C. Horn, patrolman (1943–44); Noice Walter Horn (1943–44); William F. Parmer, crewmember (1944–45); Denies L. Parmer, crewleader (ca. 1944); Laurie L. Ford, patrolman (1944); Johnie A. Page, emergency patrolman/patrolman (1944–46); Glenon B. Page, temporary emergency patrolman (1945–46); Carl B. Partin, crewleader (1952–72); John Horton, crewman (1948–58); Aubrey E. Cole, crewleader (1958); Maurice Daniel, crewman (1959); Noah Ash Kincel, crewman (1961); Benton Parton, crewleader (1964); Paul A. Harvey, crewman (1961–64); Vernon Wright, crewman (1971); I. T. Strickland, crewleader (1972)		

Name	**Candy Hill (Latexo)**				
Location	31.39889, -95.47056				
Moved from					
Moved to	Grapeland				
Quadrangle	Palestine				
County	Houston				
Entity	TFS				
District	District 3				
Lease date	September 13, 1958				
Survey	S. F. Wall				
NGS description	None				
Property number	Unknown	Tower/unit number	Unknown (317)		
	1	2	3		
Erected	1958				
Removed	1964				
Model	Nashville Bridge Co.				
Height	122 feet				
Observers	Grady G. Woolley, crewleader (1956–59); Dewey B. Sims, crewman (1957–64)				

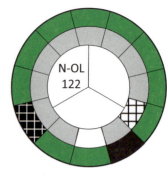

N-OL 122

Name	**Central School (O'Quinn)**				
Location	31.42769, -94.81134				
Moved from					
Moved to					
Quadrangle	Palestine				
County	Angelina				
Entity	TFS				
District	District 3				
Lease date	October 11, 1933				
Survey	J. D. Gann				
NGS description	BY2664				
Property number	4878	Tower/unit number	13 (309)		
	1	2	3		
Erected	1934				
Removed	Standing (sold 1982)				
Model	Aermotor MC-40				
Height	99.75 feet				
Observers	T. J. O'Quinn (ca. 1934–ca. 1940); Grover Cleveland Birdsong, patrolman/crewleader (1937–58); Alton Allen, crewleader (1959); Thomas R. Allen, crewman (1960); Archie Edwards, crewman (1959); Levi Dunn, crewman (1961); Lonnie P. Ricks, crewman (1932–72)				

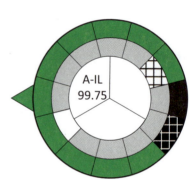

A-IL 99.75

Co. 838

Name	**Chambers Hill**
Location	31.46754, -93.83663
Moved from	
Moved to	
Quadrangle	Palestine
County	Sabine
Entity	USFS
District	Sabine NF, Yellowpine District
Lease date	Federal property
Survey	I. Powell
NGS description	BX2990

Property number	N/A	Tower/unit number	N/A		
	1		2		3

Erected	ca. 1936
Removed	Unknown
Model	Unknown
Height	99.75 feet
Observers	Rita Chance, lookout (1957–58); W. Owens, alternate lookout (1957–58); Oreta Chance, lookout (1959); Cecil Craig, alternate lookout (1959)

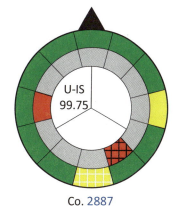

Co. 2887

Name	**Chireno**
Location	31.50825, -94.42678
Moved from	
Moved to	Shepherd, Location 2
Quadrangle	Palestine
County	Nacogdoches
Entity	TFS
District	District 3
Lease date	August 9, 1933
Survey	J. A. Chirino
NGS description	BY2620

Property number	Unknown	Tower/unit number	20 (Unknown)		
	1		2		3

Erected	1934
Removed	1960
Model	Aermotor MC-40
Height	99.75 feet
Observers	Glenn Mettauer; Harold K. Greer, emergency patrolman (ca. 1942); Hollie Terry Posey, emergency patrolman (1942–43)

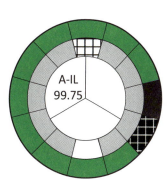

Co. 838

Palestine Quadrangle **259**

Name	**Chita, Location 2**			
Location	31.01442, -95.23913			
Moved from	Chita, Location 1			
Moved to				
Quadrangle	Palestine			
County	Trinity			
Entity	TFS			
District	District 3			
Lease date	August 17, 1970			
Survey	H. Bond			
NGS description	None			
Property number	4875	Tower/unit number	31 (Unknown)	
	1	2	3	
Erected	1970			
Removed	Sold 1981			
Model	EMSCO TW-1			
Height	99.75 feet			
Observers	James F. Parish, crewman (1971)			

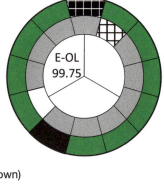

Name	**Cushing, Location 1 (Button Mountain)**			
Location	31.83592, -94.85356			
Moved from	Possibly moved from Flower Mt.? (Cushing)			
Moved to				
Quadrangle	Palestine			
County	Nacogdoches			
Entity	TFS			
District	District 2			
Lease date	September 16, 1935			
Survey	J. S. Roberts			
NGS description	None			
Property number	Unknown	Tower/unit number	61 (221)	
	1	2	3	
Erected	1936			
Removed	1962			
Model	International			
Height	87 feet			
Observers	J. N. Slatterwhite, patrolman (1936); John Acrey, emergency patrolman (1936); Oscar G. Mullins, crewleader (1959); Carl E. Mullins (1960–67)			

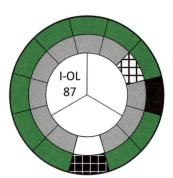

Co. Unknown

Name	**Cushing, Location 2**		
Location	31.76750, -94.80694		
Moved from			
Moved to			
Quadrangle	Palestine		
County	Nacogdoches		
Entity	TFS		
District	District 2		
Lease date	June 26, 1962		
Survey	A. Oliver		
NGS description	None		
Property number	11124	Tower/unit number	61 (221)
	1	2	3
Erected	1962		
Removed	Sold 1983		
Model	Aermotor LS-40		
Height	100 feet		
Observers	Johnny Halsted, crewman (1964–67); Aubrey Strahan, crewman (1967)		

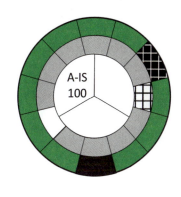

Name	**Cyclone Hill (Kimmey)**		
Location	31.08804, -94.25263		
Moved from	Vilas, Florida (?) (Tower 2)		
Moved to	Cabin 2 at Texas Forestry Museum		
Quadrangle	Palestine		
County	Angelina		
Entity	TFS/USFS		
District	Angelina NF, Angelina District		
Lease date	Unknown		
Survey	A. Ashley		
NGS description	BY2634		
Property number	N/A	Tower/unit number	N/A
	1	2	3
Erected	1934	1956	
Removed	1956	1977	
Model	EMSCO TW-1	Aermotor MC-39	
Height	99.75 feet	99.75 feet	
Observers	Rudolphus Hopson, lookout (ca. 1950)		

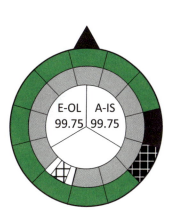

Co. 838

Palestine Quadrangle   **261**

Name	**Cypress Lake**			
Location	31.14167, -94.71167			
Moved from				
Moved to				
Quadrangle	Palestine			
County	Angelina			
Entity	TFS			
District	District 3			
Lease date	October 11, 1933			
Survey	Ben Ford			
NGS description	BY0906			
Property number	4877	Tower/unit number	15 (322)	
	1	2	3	
Erected	1934			
Removed	1983			
Model	Aermotor MC-40			
Height	99.75 feet			
Observers	Clifford C. DuBose, crewleader (1956–59); Charles Morris, crewman (1963); Wilmer L. "Bo" Richardson, crewman/crewleader (1963–76); Edwin L. Weisinger, crewman (1965); Ray Hambrick, crewman (1965); David Dial, crewman (1965); James D. Jeans, crewman (1967); John P. Morris, crewman (1968); Winfred Watson, forest aide (1975)			

A-IL
99.75

Co. 838

Name	**Denning**			
Location	31.46359, -94.21039			
Moved from				
Moved to	San Augustine			
Quadrangle	Palestine			
County	San Augustine			
Entity	TFS			
District	District 3			
Lease date	April 5, 1934			
Survey	SP RR A-274			
NGS description	None			
Property number	Unknown	Tower/unit number	35 (Unknown)	
	1	2	3	
Erected	1934			
Removed	1944			
Model	EMSCO TW-1			
Height	99.75 feet			
Observers	Robert W. Lacy (ca. 1943); Joe Bailey Neely, patrolman (1943–44); Giles Overton Perry, emergency patrolman (1934–44)			

E-OL
99.75

Co. Unknown

Name	**Devils (Sabine)**			
Location	31.26653, -93.68867			
Moved from				
Moved to	Smithville			
Quadrangle	Palestine			
County	Sabine			
Entity	USFS			
District	Sabine NF, Yellowpine District			
Lease date	Unknown			
Survey	J. S. Chiveral A-187			
NGS description	BX2963			
Property number	T-726	Tower/unit number	Unknown	
	1	2	3	
Erected	1936			
Removed	1968			
Model	T-1000 (International)			
Height	99.75 feet			
Observers	None known			

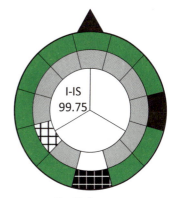

Co. Unknown

Name	**Dreka**			
Location	31.70188, -93.93426			
Moved from	Arkansas (Tower 2)			
Moved to				
Quadrangle	Palestine			
County	Shelby			
Entity	TFS/USFS			
District	Sabine NF, Tenaha District			
Lease date	Unknown			
Survey	John Latham			
NGS description	BX3048			
Property number	Unknown	Tower/unit number	Unknown	
	1	2	3	
Erected	ca. 1934	1955		
Removed	1955	Unknown		
Model	L-1000 (OL)	McClintic-Marshall		
Height	99.75 feet	99.75 feet		
Observers	J. L. Mattox, lookout (1957); Kelly O'Rear, alternate lookout (1957); S. J. Lynch, lookout (1958–59); Jim Vaughn, alternate lookout (1958); Elwood Kay, alternate lookout (1959)			

Co. 880

Palestine Quadrangle **263**

Name	**Elkhart**		
Location	31.62167, -95.61891		
Moved from	Bath		
Moved to			
Quadrangle	Palestine		
County	Anderson		
Entity	TFS		
District	District 2		
Lease date	November 20, 1958		
Survey	William Frost		
NGS description	None		
Property number	11112	Tower/unit number	Unknown (222)
	1	2	3
Erected	1959	1962	
Removed	1962	1993	
Model	International	Aermotor LS-40	
Height	87 feet	80 feet	
Observers	Lawrence Johnson, crewleader (1953–56); James W. Bridges, crewleader (1958–69); Zack E. Sheridan, crewman (1959–69); William E. Dunnam, crewman (1969)		

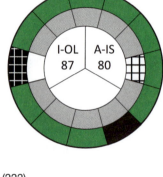

Name	**Etoile (Camp Pershing), Location 1**		
Location	31.36111, -94.38861		
Moved from			
Moved to	Etoile, Location 2		
Quadrangle	Palestine		
County	Nacogdoches		
Entity	TFS		
District	District 3		
Lease date	Unknown		
Survey	I&GN RR A-652		
NGS description	BY0471		
Property number	11125	Tower/unit number	6 (308)
	1	2	3
Erected	1931	1963	
Removed	1963	1984	
Model	International	Aermotor LS-40	
Height	87 feet	100 feet	
Observers	J. D. (Jim) Crawford, crewleader (1937–57); James W. Spurgeon, crewman (1957–61); Robert Morgan (1961); James C. Morton, crewman (1962); Claude Daniel, crewman (1963); E. O. Lowery, crewman/crewleader (1944–65); Mervis Lowery, crewman/crewleader (1965–70); Ellis A. Wooten, crewman (1966–71); Charles E. Brookshire, crewleader (1971); Wilbert L. Burran, crewman (1972)		

Name	**Etoile, Location 2**
Location	31.33156, -94.36711
Moved from	Etoile, Location 1
Moved to	
Quadrangle	Palestine
County	Nacogdoches
Entity	Private
District	District 3
Lease date	N/A
Survey	I&GN RR A-652
NGS description	None
Property number 1	11125
Erected	1984
Removed	Standing
Model	Aermotor LS-40
Height	100 feet
Observers	N/A

Tower/unit number 2: 6 (308)
Tower/unit number 3:

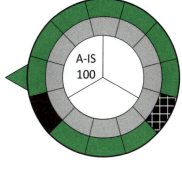

A-IS 100

Name	**Fern Lake (Mitchell)**
Location	31.54160, -94.70538
Moved from	
Moved to	
Quadrangle	Palestine
County	Nacogdoches
Entity	TFS
District	District 3
Lease date	November 12, 1933
Survey	L. Watkins
NGS description	BY2683
Property number 1	4880
Erected	1934
Removed	1976
Model	Aermotor MC-40
Height	99.75 feet
Observers	Charles F. Williams, crewman (1957); Leonard L. Smith, crewman (1958); Gilbert L. Henson, crewman (1959); Raymond R. Barker, crewleader (1951–71); Robert Haney, crewman (1959–71)

Tower/unit number 2: 12 (321)
Tower/unit number 3:

A-IL 99.75

Co. 838

Palestine Quadrangle **265**

Name	**Forse Mountain**				
Location	31.52693, -93.98049				
Moved from					
Moved to					
Quadrangle	Palestine				
County	Sabine				
Entity	Unknown				
District	Sabine NF, Tenaha District				
Lease date	Unknown				
Survey	E. A. Long				
NGS description	None				
Property number	Unknown	Tower/unit number	Unknown		
	1	2	3		
Erected	ca. 1937				
Removed	ca. 1939				
Model	Unknown				
Height	Unknown				
Observers	None known				

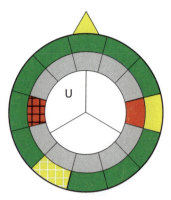

Co. Unknown

Name	**Geneva**				
Location	31.49366, -93.91890				
Moved from					
Moved to					
Quadrangle	Palestine				
County	Sabine				
Entity	TFS				
District	District 3				
Lease date	Unknown				
Survey	J. I. Pifirmo				
NGS description	None				
Property number	Unknown	Tower/unit number	Unknown		
	1	2	3		
Erected	1934				
Removed	ca. 1938				
Model	Unknown				
Height	Unknown				
Observers	None known				

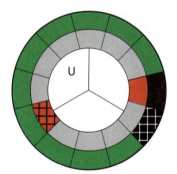

Co. Unknown

Name	**Grapeland**
Location	31.46928, -95.48055
Moved from	Candy Hill
Moved to	
Quadrangle	Palestine
County	Houston
Entity	TFS
District	District 3
Lease date	October 9, 1963
Survey	George Poe A-823
NGS description	None
Property number	337 Tower/unit number Unknown (317)
	1 2 3
Erected	1964
Removed	Standing
Model	Nashville Bridge Co.
Height	122 feet
Observers	James Williams, crewman (1964); Elza Shaw, crewleader (1965–67); James E. Shaw, crewleader (1966); Hilton Daniels, crewleader (1966); Justin Kennedy, crewman/crewleader (1964–71); Alpha Shipper, crewman (1967)

Name	**Gum Springs**
Location	31.19167, -93.90167
Moved from	
Moved to	
Quadrangle	Palestine
County	Sabine
Entity	Unknown
District	Sabine NF, Yellowpine District
Lease date	Unknown
Survey	R. E. Bass
NGS description	None
Property number	Unknown Tower/unit number Unknown
	1 2 3
Erected	ca. 1936
Removed	ca. 1938
Model	Unknown
Height	Unknown
Observers	None known

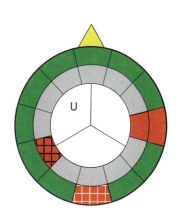

Co. Unknown

Name	**Hi Point (Ziegler Hill)**
Location	31.17833, -94.13333
Moved from	
Moved to	Poynor
Quadrangle	Palestine
County	San Augustine
Entity	USFS
District	Angelina NF, Angelina District
Lease date	Unknown
Survey	SP RR
NGS description	BY2633
Property number 1	T-126
Tower/unit number 2	Unknown
3	
Erected	ca. 1937
Removed	1968
Model	EMSCO TW-1
Height	99.75 feet
Observers	None known

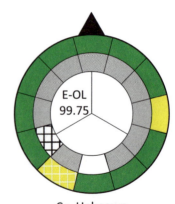

Co. Unknown

Name	**Horton Hill**
Location	31.01627, -93.98726
Moved from	
Moved to	
Quadrangle	Palestine
County	Jasper
Entity	TFS
District	District 5
Lease date	November 13, 1933
Survey	H&TC RR A-312
NGS description	BX2983
Property number 1	6022
Tower/unit number 2	36 (509)
3	
Erected	1934
Removed	Standing (sold 1984)
Model	EMSCO TW-1
Height	99.75 feet
Observers	William E. Horn, patrolman (1937–44); Neal W. Kincel, emergency patrolman/AWS observer/lookout (1944); Steven I. Coleman (ca. 1943); Carrie Reese, emergency patrolman/AWS observer (1943); Robert E. Lee Jones, crewleader (1945–47); Ralph McBride, crewleader (1953–59); Austin Whitehead, crewman (1965); Joe McNair, crewleader (1961–67); Gilbert Hill, crewman (1968); Jim Powell, crewleader (1967–68); Marvin Poindexter, crewman (1967); William H. Forse (ca. 1967); Issac Rayburn Hampshire, crewman (1967); John W. Rhame, crewman (1971); James A. Whitehead, crewman (1952–71); Charles K. MaGree, forest aide (1974)

Co. 893

Name	**Huntington**		
Location	31.27068, -94.57426		
Moved from			
Moved to			
Quadrangle	Palestine		
County	Angelina		
Entity	TFS		
District	District 3		
Lease date	November 15, 1933		
Survey	J. Bradshaw		
NGS description	BY2640		
Property number	4870	Tower/unit number	14 (319)
	1		2	3
Erected	1934		
Removed	1983		
Model	Aermotor MC-40		
Height	99.75 feet		
Observers	Charlie Forrest, towerman (1949–53); Charles A. Modisett, crewleader (1957); R. V. Hayes, crewman (1959–61); Lacy L. Laird (ca. 1961); Woodie Eugene Page, crewman (1961); Rayford T. Barrett, crewleader (1959–61); Tommy Anderson, crewman (1962); John H. Snelson, crewman (1957–60); Kenneth L. Treadaway, crewleader (1965); Harold Chamblee Sr., crewleader (1967)		

A-IL
99.75

Co. 838

Name	**Jackson Hill**		
Location	31.30605, -94.29207		
Moved from			
Moved to			
Quadrangle	Palestine		
County	San Augustine		
Entity	USFS		
District	Angelina NF, Angelina District		
Lease date	Unknown		
Survey	M. C. Flourney		
NGS description	BY2637		
Property number	Unknown	Tower/unit number	Unknown
	1		2	3
Erected	1934		
Removed	After 1977		
Model	L-1000 (Aermotor MC-39?)		
Height	99.75 feet		
Observers	None known		

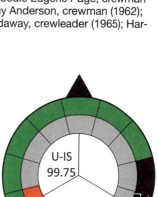

U-IS
99.75

Co. Unknown

Palestine Quadrangle **269**

Name	**Lufkin**			
Location	31.30304, -94.72432			
Moved from				
Moved to				
Quadrangle	Palestine			
County	Angelina			
Entity	TFS			
District	District 3			
Lease date	Unknown			
Survey	McKinney & Williams A-463			
NGS description	None			
Property number	Unknown	Tower/unit number	67 (Unknown)	
	1	2	3	
Erected	1937			
Removed	1957			
Model	Creosoted wood			
Height	100 feet			
Observers	Hayward Metts (1941); Claud C. Purvis, CCC/smokechaser (1933–57); Edwin E. Reynolds, senior crewleader (1951–53); O. F. Reel, crewman (1958)			

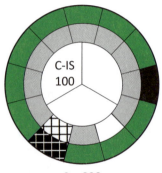

Co. 838

Name	**Maydelle, Location 1 (Fairchild State Forest; State Forest No. 3)**			
Location	31.78400, -95.36500			
Moved from				
Moved to	Maydelle, Location 2			
Quadrangle	Palestine			
County	Cherokee			
Entity	TFS			
District	District 2			
Lease date	State property			
Survey	W. M. Evans			
NGS description	BY2737			
Property number	4288	Tower/unit number	8 (208)	
	1	2	3	
Erected	1934			
Removed	1960			
Model	Aermotor MC-40			
Height	99.75 feet			
Observers	Ray Sides (1943–71); Don Austin (1953); John N. Hassell, crewman (1956); Sam Ezell, crewman (1958–67); Everett C. Sherman, crewleader (1958); Olan M. Ferguson, crewleader (1959)			

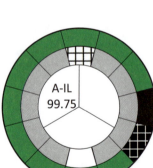

Co. 833

Name	**Maydelle, Location 2**			
Location	31.81966, -95.30935			
Moved from	Maydelle, Location 1			
Moved to				
Quadrangle	Palestine			
County	Cherokee			
Entity	TFS			
District	District 2			
Lease date	September 2, 1960			
Survey	L.B. Parish			
NGS description	None			
Property number	4288	Tower/unit number	8 (208)	
	1		2	3
Erected	1960			
Removed	Standing (transferred 1989)			
Model	Aermotor MC-40			
Height	99.75 feet			
Observers	None known			

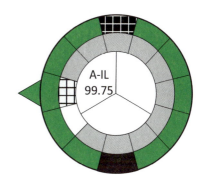

Name	**Mayflower**			
Location	31.10114, -93.75931			
Moved from				
Moved to				
Quadrangle	Palestine			
County	Newton			
Entity	TFS			
District	District 5			
Lease date	October 23, 1933			
Survey	T&NO RR A-1102			
NGS description	BX2979			
Property number	6019	Tower/unit number	37 (514)	
	1		2	3
Erected	1934			
Removed	Standing			
Model	EMSCO TW-1			
Height	99.75 feet			
Observers	Charlie A. Weeks (ca. 1943); Andrew E. Knighton, patrolman (1943); Henry A. Marshall, crewleader (1953); James T. Furlow, crewman (1958–59); Melvin R. Kerr, crewman (1968); Orlean G. Miller (1953–69)			

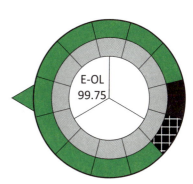

Co. Unknown

Name	**Montalba (Wilkerson Mountain)**
Location	31.91906, -95.70855
Moved from	
Moved to	
Quadrangle	Palestine
County	Anderson
Entity	TFS
District	District 2
Lease date	June 6, 1970
Survey	E. E. Lowder A-489
NGS description	None
Property number 1	Unknown
Tower/unit number 2	Unknown 3
Erected	1970
Removed	1993
Model	Dunlap T-10
Height	100 feet
Observers	Charles A. Seagler, crewleader (1970); Arvis Waldon Sr., crewman (1970)

D-IS 100

Name	**Moss Hill (Zavalla)**
Location	31.17763, -94.40421
Moved from	
Moved to	Lindale
Quadrangle	Palestine
County	Angelina
Entity	TFS/USFS
District	Angelina NF, Angelina District
Lease date	Unknown
Survey	HT&B RR
NGS description	BY2583
Property number 1	1–28
Tower/unit number 2	Unknown 3
Erected	1934
Removed	1968
Model	Aermotor MC-40
Height	99.75 feet
Observers	None known

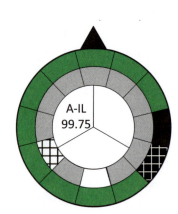

A-IL 99.75

Co. 838

Name	**Mt. Enterprise**			
Location	31.94574, -94.67852			
Moved from				
Moved to				
Quadrangle	Palestine			
County	Rusk			
Entity	TFS			
District	District 2			
Lease date	August 28, 1935			
Survey	L. Williams			
NGS description	BY2679			
Property number 1	4292	Tower/unit number 2	47 (213) 3	
Erected	1936			
Removed	Standing (sold 1981)			
Model	Aermotor MC-39			
Height	99.75 feet			
Observers	Robert S. Flanagan (1935–50); Robert Collier, emergency patrolman (1936); John Allen Red, emergency patrolman (1941–42); Jess H. Phillips, emergency patrolman (1942–43); Eli C. Jones, emergency patrolman (1943–45); James B. Flanagan (1958); James C. Moore, crewman (1962); David W. Matlock (ca. 1955); Calvin W. Jackson, crewman (1962); Ralph C. Greene, crewman (1970); James Ross Jr., crewleader (1965–70); Albert C. Shumate, crewman (1967)			

A-IS 99.75
Co. 833

Name	**Mt. Vernon**[1]			
Location	31.39630, -95.09480			
Moved from				
Moved to				
Quadrangle	Palestine			
County	Houston			
Entity	USFS			
District	Davy Crockett NF, Neches District			
Lease date	Federal property			
Survey	J. Henley			
NGS description	None			
Property number 1	Unknown	Tower/unit number 2	Unknown 3	
Erected	Unknown			
Removed	Unknown			
Model	Unknown			
Height	Unknown			
Observers	None known			

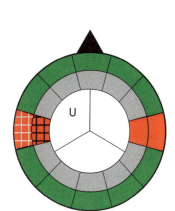

U
Co. Unknown

Name	**Mt. Zion**
Location	31.17748, -94.93712
Moved from	Apple Springs
Moved to	
Quadrangle	Palestine
County	Trinity
Entity	USFS
District	Davy Crockett NF, Trinity District
Lease date	Federal property
Survey	P. Dixon
NGS description	BY2662

Property number	Unknown	Tower/unit number	Unknown
	1		2		3

Erected	1939
Removed	ca. 1970
Model	L-1000 (Aermotor MC-40)
Height	99.75 feet
Observers	None known

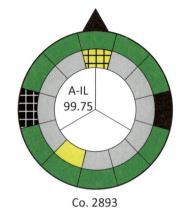

Co. 2893

Name	**Neches**
Location	31.87346, -95.49713
Moved from	
Moved to	
Quadrangle	Palestine
County	Anderson
Entity	TFS
District	District 2
Lease date	August 27, 1935
Survey	A. R. Stevens
NGS description	BY2740

Property number	4293	Tower/unit number	45 (206)
	1		2		3

Erected	1936
Removed	Standing
Model	EMSCO TW-1
Height	99.75 feet
Observers	John F. Lewis, patrolman (until 1942); Marvin Leeroy Perkins, patrolman/emergency patrolman (1942–47); Frank R. Petri, emergency patrolman (1939–42); Cleo Perkins, crewleader (1942–73); William J. Furnish, emergency patrolman (1947); Lawrence D. Yancey, crewman (1959); Rogers Tulsem, crewleader (1959)

Co. 833

Name	**Nogalus**		
Location	31.26806, -95.12944		
Moved from	Ozark NF, Arkansas (Tower 2)		
Moved to			
Quadrangle	Palestine		
County	Houston		
Entity	USFS		
District	Davy Crockett NF, Trinity District		
Lease date	Federal property		
Survey	I&GN RR		
NGS description	BY2743		
Property number	Unknown	Tower/unit number	Unknown
	1	2	3
Erected	1939	1956	
Removed	1956	ca. 1978	
Model	L-2400 (Creosoted wood)	Aermotor MC-39	
Height	120 feet	99.75 feet	
Observers	R. L. Ivie, forest guard (1943)		

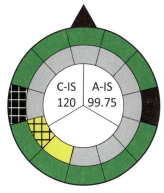

Co. 2893

Name	**Paxton**		
Location	31.94816, -94.16489		
Moved from	Rocky Hill		
Moved to			
Quadrangle	Palestine		
County	Shelby		
Entity	TFS		
District	District 3		
Lease date	June 27, 1951		
Survey	J. D. Freeland A-233		
NGS description	BY2608		
Property number	4867	Tower/unit number	65 (323)
	1	2	3
Erected	1951		
Removed	Sold 1981		
Model	Aermotor MC-39		
Height	99.75 feet		
Observers	Oren O. Ross, crewman (1959); John U. Majors (1959); John H. Jones, crewleader (1957–68); Frank Crawford, crewleader (1958–60); Edwin Wall Hooker, crewman (1960–63); Howard H. Townsend, crewman (1963–67); Andrew O. Scholar, crewman (1968)		

Palestine Quadrangle **275**

Name	**Piney Woods (Piney)**		
Location	31.10826, -95.04127		
Moved from	From Key Bridge Lookout, South Carolina (Tower 2)		
Moved to			
Quadrangle	Palestine		
County	Trinity		
Entity	USFS		
District	Davy Crockett NF, Trinity District		
Lease date	Federal property		
Survey	J. L. Boden		
NGS description	None		
Property number	Unknown	Tower/unit number	Unknown
	1	2	3
Erected	1939	1959	
Removed	1959	ca. 1978	
Model	L-2300 (Creosoted wood)	Unknown	
Height	100 feet	99.75 feet	
Observers	None known		

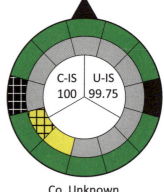

Co. Unknown

Name	**Piney Woods (Weldon)**		
Location	31.03083, -95.46667		
Moved from			
Moved to			
Quadrangle	Palestine		
County	Walker		
Entity	TFS		
District	District 3		
Lease date	Unknown		
Survey	L. R. Clapp A-155		
NGS description	BY2754		
Property number	4874	Tower/unit number	41 (320)
	1	2	3
Erected	1934		
Removed	After 1998		
Model	EMSCO TW-1		
Height	99.75 feet		
Observers	Herbert F. Hare (1936–56); Oscar P. Conner (1940–70); Low Keels, crewleader (1957–67); James A. McFarland, crewman (1957–59); Elmer Little, crewman (1963–67); J. T. Jordan, crewman (1971); Jack K. Carlton, crewman/crewleader (1959–71)		

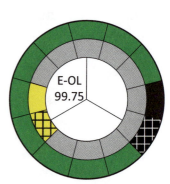

Co. 839

Name	**Ratcliff**			
Location	31.39543, -95.13856			
Moved from				
Moved to	Cushing (?)			
Quadrangle	Palestine			
County	Houston			
Entity	TFS/USFS			
District	Davy Crockett NF, Neches District			
Lease date	Verbal agreement			
Survey	I&GN RR A-604			
NGS description	BY2745			
Property number	Unknown	Tower/unit number	Unknown	
	1	2	3	
Erected	1928	1936		
Removed	1936	Standing		
Model	International	Aermotor MC-39		
Height	87 feet	99.75 feet		
Observers	Charles T. McCurdy, lookout (1947); Clayton Ashby (1947)			

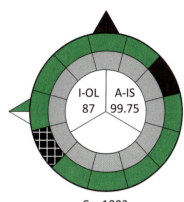

Co. 1803

Name	**Rocky Hill (Ewing)**			
Location	31.36130, -94.55348			
Moved from				
Moved to	Paxton			
Quadrangle	Palestine			
County	Angelina			
Entity	TFS			
District	District 3			
Lease date	August 26, 1937			
Survey	I. Wray A-825			
NGS description	BY2641			
Property number	Unknown	Tower/unit number	65 (Unknown)	
	1	2	3	
Erected	1937			
Removed	1951			
Model	Aermotor MC-39			
Height	99.75 feet			
Observers	Sam Vardeman, smokechaser (1931–45); Rube A. Pinner, emergency smokechaser (1940–44)			

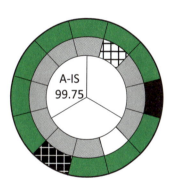

Co. 838

Palestine Quadrangle **277**

Name	**Rusk, Location 1**			
Location	31.81696, -95.16485			
Moved from				
Moved to	Rusk, Location 2			
Quadrangle	Palestine			
County	Cherokee			
Entity	TFS			
District	District 2			
Lease date	November 16, 1933			
Survey	J. Morgan A-579			
NGS description	None			
Property number	4290	Tower/unit number	9 (Unknown)	
	1	2	3	
Erected	1934			
Removed	1939			
Model	Aermotor MC-40			
Height	99.75 feet			
Observers	Ned E. Harris (1936); John DeFoor (1936–37)			

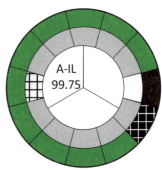

Co. 833

Name	**Rusk, Location 2**			
Location	31.80444, -95.11667			
Moved from	Rusk, Location 1			
Moved to				
Quadrangle	Palestine			
County	Cherokee			
Entity	TFS			
District	District 2			
Lease date	Unknown			
Survey	S. Halbert A-344			
NGS description	BY2706			
Property number	4290	Tower/unit number	9 (210)	
	1	2	3	
Erected	1939			
Removed	1999			
Model	Aermotor MC-40			
Height	99.75 feet			
Observers	Jesse Layton Jones, senior crewleader (1943–56); Andrew H. Riggs, crewman (1949–54); Bertis Lee Watson, crewleader (1956–63); R. S. Dyess (ca. 1960); Roy Kennedy, crewman (1955–63); Jack L. Berry, crewman (1966); Bolton "Bo" M. Hansen, crewleader (1964–71)			

Co. Unknown

Name	**Salem**		
Location	31.98134, -94.95282		
Moved from			
Moved to			
Quadrangle	Palestine		
County	Rusk		
Entity	TFS		
District	District 2		
Lease date	October 12, 1951		
Survey	R. Marlow A-569		
NGS description	BY2700		
Property number	12060	Tower/unit number	78 (212)
	1	2	3
Erected	1951	1965	
Removed	1965	Standing (sold 1982)	
Model	International	Aermotor LS-40	
Height	87 feet	80 feet	
Observers	Russell Burd (1952); Allen T. Cleaver, crewleader (1953–73); Henry Roberts, crewman (1957)		

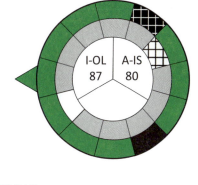

Name	**San Augustine**		
Location	31.53436, -94.13627		
Moved from	Denning		
Moved to			
Quadrangle	Palestine		
County	San Augustine		
Entity	TFS		
District	District 3		
Lease date	August 16, 1944		
Survey	William Garrett		
NGS description	BY2616		
Property number	4871	Tower/unit number	35 (312)
	1	2	3
Erected	1944		
Removed	1980		
Model	EMSCO TW-1		
Height	99.75 feet		
Observers	Tom C. Renfroe, patrolman (1946); Allen P. Travis (ca. 1944–58); J. L. Finn Jr. (ca. 1955); Willie Ray Horne, crewman (1955–56); Houston Hinton, crewleader (1959); Mrs. Gilcrease, alternate lookout (1959); Horace J. Whitehead, crewman (1956–67); Ervin Davis, crewleader (1964–67); Horace D. Butler, crewman (1968–71); Robert D. Johnson, crewman/crewleader (1967–68); Clifton Holloway, crewman (1969); Kenneth Wayne Moye, crewman (1970); Garland O. Evett, crewman (1971)		

Name	**Shady Grove (Augusta)**			
Location	31.21401, -95.34807			
Moved from				
Moved to				
Quadrangle	Palestine			
County	Houston			
Entity	TFS			
District	District 3			
Lease date	September 26, 1933			
Survey	R. S. De Las Coy A-25			
NGS description	BY2753			
Property number	4876	Tower/unit number	40 (315)	
	1	2	3	
Erected	1934			
Removed	Standing (sold 1983)			
Model	EMSCO TW-1			
Height	99.75 feet			
Observers	Friday; Vera Mae Woolley, lookout (1957–59); Thomas H. Buller, emergency patrolman/crewleader (1937–58); Marshall Paxton Lively, crewleader (1959–67); Jessie A. Wooley, crewman (1958–67); Lucky B. Williams, crewman (1967); Stanley Freeman, crewman (1969)			

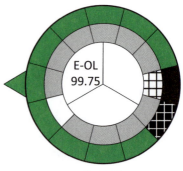

Co. 838

Name	**Tabernacle (Choice)**			
Location	31.66643, -94.23843			
Moved from				
Moved to				
Quadrangle	Palestine			
County	Shelby			
Entity	TFS			
District	District 3			
Lease date	Unknown			
Survey	W. M. Perry			
NGS description	None			
Property number	4872	Tower/unit number	33 (313)	
	1	2	3	
Erected	1934			
Removed	Standing			
Model	EMSCO TW-1			
Height	99.75 feet			
Observers	Daniel O. Mask (1936); James W. Kimbro (1936); James A. Kimbro, crewleader (1935–64); Lettie Hughes, emergency lookout (1943); Euel L. Hopkins, smokechaser/crewman (1944–48); Tommy Franklin Hughes, crewman (1949); Nealy J. Kimbro, lookout (1949); Archie L. Anderson, crewman/crewleader (1958–ca. 1962); Lonnie L. Carmichael, crewman (1962–73)			

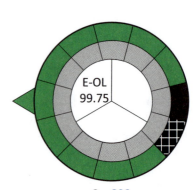

Co. 838

Name	Tadmor[2]		
Location	31.39938, -95.19202		
Moved from			
Moved to			
Quadrangle	Palestine		
County	Houston		
Entity	USFS		
District	Davy Crockett NF, Neches District		
Lease date	Federal property		
Survey	W. M. Conner		
NGS description	None		
Property number	Unknown	Tower/unit number	Unknown
	1	2	3
Erected	ca. 1939		
Removed	Unknown		
Model	CT-1 (?)		
Height	100 feet		
Observers	None known		

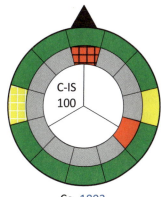

Co. 1803

Name	Tenaha		
Location	31.81267, -93.95071		
Moved from	Fullerton Lookout, Kisatchie NF, Louisiana (Tower 2)		
Moved to			
Quadrangle	Palestine		
County	Shelby		
Entity	USFS		
District	Sabine NF, Tenaha District		
Lease date	Federal property		
Survey	I. McDaniel A-467		
NGS description	BX3057		
Property number	Unknown	Tower/unit number	Unknown
	1	2	3
Erected	ca. 1938	1962	
Removed	1962	ca. 1977	
Model	L-2300 (IS)	Aermotor MC-39	
Height	100 feet	99.75 feet	
Observers	Kelly O'Rear, lookout (1957); Jim Vaughn, alternate lookout/lookout (1957–59); Elwood Kay, alternate lookout/lookout (1958–59)		

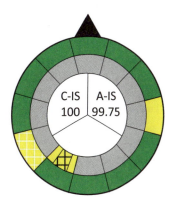

Co. Unknown

Palestine Quadrangle   **281**

Name	Texas Forestry Museum				
Location	31.35069, -94.70506				
Moved from	East River				
Moved to					
Quadrangle	Palestine				
County	Angelina				
Entity	TFS				
District	District 3				
Lease date					
Survey	J. L. Quinalty				
NGS description	None				
Property number	6572		Tower/unit number	N/A	
	1		2	3	
Erected	1974				
Removed	Standing				
Model	Aermotor MC-39				
Height	99.75 feel				
Observers	N/A				

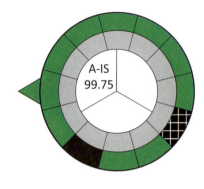

Name	Timpson (Oak Grove)				
Location	31.83542, -94.34591				
Moved from					
Moved to					
Quadrangle	Palestine				
County	Shelby				
Entity	TFS				
District	District 3				
Lease date	September 17, 1935				
Survey	Emily Irish A-901				
NGS description	BY2623				
Property number	4873		Tower/unit number	44 (316)	Co. 880
	1		2	3	
Erected	1936				
Removed	1981				
Model	EMSCO TW-1				
Height	99.75 feet				
Observers	Flanagan (1935); Emory E. Covington (1936–68); W. M. Adams (1936); Lee Edward Hollaway, crewman (1949); John H. Jones, crewman (1959); Doyle F. Ramsey, crewman/crewleader (1961–68); Charles J. Hughes, crewman (1968); Hoye Thompson, crewman (1968)				

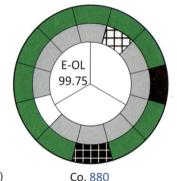

Name	**Wakefield**				
Location	31.07389, -94.82611				
Moved from					
Moved to					
Quadrangle	Palestine				
County	Polk				
Entity	TFS				
District	District 4				
Lease date	December 18, 1937				
Survey	Houston CSL A-270				
NGS description	BY2659				
Property number	5422	Tower/unit number	68 (407)		
	1		2	3	
Erected	1938				
Removed	Sold 1982				
Model	Aermotor MC-39				
Height	99.75 feet				
Observers	Billy M[ac?] Cockrell, crewman (1957–58); S. A. Williams, crewleader (1958); Wallace Dickson, crewman (1956); Simeon Williams, crewleader (1937–63); Maxie Cooper, crewleader (1963–67); James Liljequist, crewman (1967); Robert F. White, crewman (1967); John F. Stout, crewman (1967); Jerry E. Nowlin, crewman (1968)				

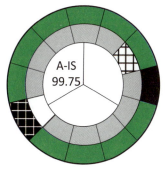

Co. 840

Name	**Weches**				
Location	31.54493, -95.22765				
Moved from					
Moved to					
Quadrangle	Palestine				
County	Houston				
Entity	TFS				
District	District 3				
Lease date	September 2, 1933				
Survey	H. Ware A-1240				
NGS description	BY2718				
Property number	35	Tower/unit number	11 (307)		
	1		2	3	
Erected	1934	1938	1958		
Removed	1938	1958	1981		
Model	Unknown (MC-40?)	Creosoted wood	Nashville Bridge Co.		
Height	99.75 feet	100 feet	105 feet		
Observers	Henry P. Cutler, crewleader (1953–58); Doyle W. Pyle, crewman (1959–67); Robert J. "Dooley" Rodgers, crewman/crewleader (1956–67)				

Co. 838

Palestine Quadrangle **283**

Name	**White City**				
Location	31.26917, -94.16306				
Moved from					
Moved to					
Quadrangle	Palestine				
County	San Augustine				
Entity	TFS/USFS				
District	Angelina NF, Angelina District				
Lease date	Unknown				
Survey	C. Chaplin				
NGS description	None				
Property number	Unknown	Tower/unit number	Unknown		
	1	2	3		
Erected	1934				
Removed	ca. 1938				
Model	Unknown				
Height	Unknown				
Observers	None known				

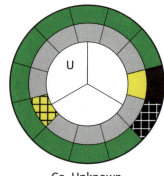

Co. Unknown

Name	**Winston Ranch**			
Location	31.52664, -94.71474			
Moved from	Elysian Fields (Tower 2)			
Moved to				
Quadrangle	Palestine			
County	Nacogdoches			
Entity	Private			
District	N/A			
Lease date	Private property			
Survey	A. Bermea			
NGS description	None			
Property number	N/A	Tower/unit number	N/A	
	1	2	3	
Erected	1999			
Removed	Standing			
Model	Aermotor LS-40			
Height	100 feet			
Observers	N/A			

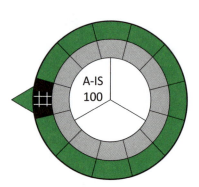

Name	**Wolf Hill, Location 1**			
Location	31.10875, -94.53024			
Moved from				
Moved to				
Quadrangle	Palestine			
County	Angelina			
Entity	TFS			
District	District 3			
Lease date	May 8, 1950			
Survey	East Texas Railroad #45 A-233			
NGS description	BY2654			
Property number	Unknown 1	Tower/unit number	75 (Unknown) 3	
Erected	1948			
Removed	1958			
Model	OL Creosoted wood with steel cab			
Height	45 feet			
Observers	Charles Modisett, crewleader (1947–64)			

Name	**Wolf Hill, Location 2**
Location	31.11944, -94.55111
Moved from	
Moved to	
Quadrangle	Palestine
County	Angelina
Entity	TFS
District	District 3
Lease date	September 15, 1958
Survey	East Texas Railroad #49 A-242
NGS description	None
Property number	336 1
Tower/unit number	75 (310) 3
Erected	1958
Removed	1984
Model	Nashville Bridge Co.
Height	105 feet
Observers	Charles Modisett, crewleader (1947–64); John H. Snelson, crewman (1957–59); Doyle Snelson, crewleader/forest aide (1963–74); Vestal Cryer (ca. 1964); William Glenn Havard, crewman (1962–64); Charlie Carson Havard, crewman (1961–64)

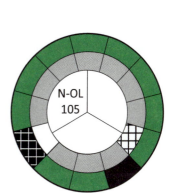

Palestine Quadrangle

Name	**Yellowpine, Location 1**		
Location	31.25794, -93.82197		
Moved from			
Moved to	Elysian Fields (Tower 1)		
Quadrangle	Palestine		
County	Sabine		
Entity	TFS/USFS		
District	District 3, Sabine NF, Yellowpine District		
Lease date	Verbal agreement		
Survey	A. W. Walters A-223		
NGS description	None		
Property number	Unknown	Tower/unit number	Unknown
	1	2	3
Erected	1928		
Removed	ca. 1939		
Model	International		
Height	87 feet		
Observers	W. F. Fullen (ca. 1928)		

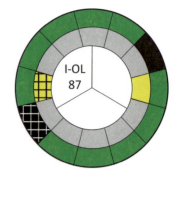

Name	**Yellowpine, Location 2**		
Location	31.28204, -93.81191		
Moved from	Mississippi (Tower 2)		
Moved to			
Quadrangle	Palestine		
County	Sabine		
Entity	USFS		
District	Sabine NF, Yellowpine District		
Lease date	Federal property		
Survey	TC RR A-383		
NGS description	BX2994		
Property number	Unknown	Tower/unit number	Unknown
	1	2	3
Erected	ca. 1939	1953	
Removed	1953	2014	
Model	CT-4 (Creosoted wood)	McClintic-Marshall	
Height	100 feet	99.75 feet	
Observers	Albert Jones (ca. 1950–53)		

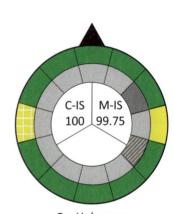

Appendix 3: Tyler Quadrangle

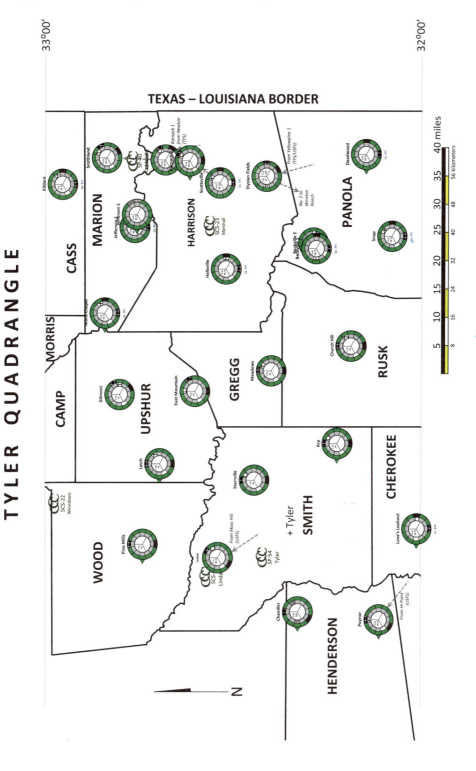

EXPLANATION

A	Aermotor Co.	OL	Outside ladder
B	Bilby	IL	Inside ladder
C	Creosoted wood	IS	Inside stairway
D	Dunlap Manufacturing Corp.		
E	EMSCO Manufacturing Co.		
I	International Derrick & Equipment Co.		
L	L. C. Moore Manufacturing Co.		
M	McClintic-Marshall Corp.		
N	Nashville Bridge Co.		
P	BR Parkersburg		
U	Unknown		
P-60-T Oakhurst	Historic Civilian Conservation Corps (CCC) Camp and Company (map only)		
838 / 840	Erecting CCC Company / Hypothesized CCC Erecting Company	←---	Lookout relocation (map only)

Name	Arp
Location	32.24016, -95.02506
Moved from	
Moved to	
Quadrangle	Tyler
County	Smith
Entity	TFS
District	District 2
Lease date	July 12, 1960
Survey	William Irons
NGS description	None
Property number	10527
Tower/unit number	Unknown
Erected	1960
Removed	Standing (sold 1981)
Model	L. C. Moore K-Type
Height	94 feet
Observers	Howard Williams, crewleader (1953–72); Neal G. Parker, crewman (1958–72)

288   Appendix 3

Name	**Beckville (Tatum), Location 1**		
Location	32.29589, -94.44057		
Moved from			
Moved to	Beckville, Location 2		
Quadrangle	Tyler		
County	Panola		
Entity	TFS		
District	District 2		
Lease date	August 29, 1935		
Survey	W. Hamilton		
NGS description	CR1134		
Property number	4286	Tower/unit number	51 (217)
	1	2	3
Erected	1936		
Removed	1979		
Model	Aermotor MC-39		
Height	99.75 feet		
Observers	Joseph B. Akins, patrolman (1936–61); William C. Pruitt, emergency patrolman (1936–37); Albert Bennett, emergencyman (1942); Walter Phelps; Richard W. Brevard, crewman/crewleader (1958–67); Arthur M. Ross, crewman (1961–64); Jerry Houston Hughes, crewleader (1969–70); Johnny R. Metcalf, crewleader (1967–70); Jerry Hill, crewleader (1970); James D. Curry, crewman (1969–70); Charles D. Hillin, crewman (1964–71); Charles R. Peavey, crewleader (1970–72)		

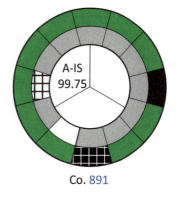

Co. 891

Name	**Beckville, Location 2**		
Location	32.29934, -94.43776		
Moved from	Beckville, Location 1		
Moved to			
Quadrangle	Tyler		
County	Panola		
Entity	TFS		
District	District 2		
Lease date	Unknown		
Survey			
NGS description	None		
Property number	4286	Tower/unit number	51 (217)
	1	2	3
Erected	1979		
Removed	1993		
Model	Aermotor MC-39		
Height	99.75 feet		
Observers	None known		

Tyler Quadrangle **289**

Name	**Chandler**
Location	32.34035, -95.52880
Moved from	
Moved to	
Quadrangle	Tyler
County	Henderson
Entity	TFS
District	District 2
Lease date	April 7, 1970
Survey	Juan M. Martinez
NGS description	None
Property number	Unknown
Tower/unit number	Unknown
Erected	1970
Removed	Standing (transferred 2019)
Model	Dunlap T-10
Height	100 feet
Observers	Macon L. Carver, crewman (1970); Donald L. Deniell, crewleader (1971); James Draper, crewman (1975)

Name	**Church Hill**
Location	32.20425, -94.73204
Moved from	
Moved to	
Quadrangle	Tyler
County	Rusk
Entity	TFS
District	District 2
Lease date	June 21, 1960
Survey	William T. Davis
NGS description	None
Property number	10526
Tower/unit number	Unknown
Erected	1960
Removed	1982
Model	L. C. Moore K-Type
Height	94 feet
Observers	James Wooley, crewman (1961); Spencer Cooper, crewman (1962); John Grant, crewleader (1963); Barton Jarrell, crewleader (1963–70); Claude M. Mullins, crewleader (1971); William E. Hudson, crewman (1971); Gary Davidson, forest aide (1975)

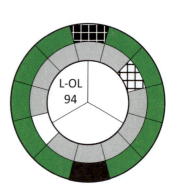

Name	**Deadwood**			
Location	32.16618, -94.16147			
Moved from				
Moved to				
Quadrangle	Tyler			
County	Panola			
Entity	TFS			
District	District 2			
Lease date	October 30, 1935			
Survey	B. M. Lewin A-397			
NGS description	CR1116			
Property number	4295	Tower/unit number	50 (216)	
	1		2	3
Erected	1936			
Removed	Standing			
Model	Aermotor MC-39			
Height	99.75 feet			
Observers	Sam Colburn (1936–43); Hudson Randolph, crewman (1949); Mitchel Emmett LaGrone, crewman/crewleader (1949–69); William A. Dickerson, crewman (1953–58); Albert N. Dickerson, crewman (1959); Jimmy Owen Powell, crewman/crewleader (1959–66); Jimmy Powell, crewman/crewleader (1962); Dillard L. Johnson, crewman (1969); Clarence W. Barnett, crewleader (1969)			

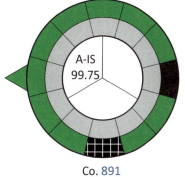

Co. 891

Name	**East Mountain**			
Location	32.60772, -94.86856			
Moved from				
Moved to				
Quadrangle	Tyler			
County	Upshire			
Entity	TFS			
District	District 1			
Lease date	December 9, 1961			
Survey	William King			
NGS description	None			
Property number	10062	Tower/unit number	Unknown (123)	–
	1		2	3
Erected	1962			
Removed	1980			
Model	EMSCO derrick			
Height	87 feet			
Observers	Leon B. Patrick, crewleader (1957–67); Bennie Wood, crewman (1962); Michael Kirk Kennimer, crewman (1964); Lester Ayars, crewman (1967); Bev Davis, crewman/crewleader/forest aide (1966–76); Louis Powell, forest aide (1975–76); Winton L. Sanders, forest aide (1969–75)			

Name	**Elysian Fields**		
Location	32.42194, -94.22528		
Moved from	Yellowpine (Tower 1)		
Moved to	Winston Ranch (Tower 2)		
Quadrangle	Tyler		
County	Harrison		
Entity	TFS		
District	District 1		
Lease date	September 16, 1953		
Survey	Henry Harper		
NGS description	CR1109		
Property number	11122	Tower/unit number	Unknown (117)
	1	2	3
Erected	1953	1963	
Removed	1963	1999	
Model	International	Aermotor LS-40	
Height	87 feet	100 feet	
Observers	Jimmy Mercer, crewman (1957); J. R. Hill Jr., crewman (1958); Harold J. Mann, crewman (1959); Leroy Wilson, crewman (1962); Artis Williams, crewman (1966); R. R. Fonville, crewleader (1963–76); Louie Randle, crewman (1963); Roosevelt Walton Jr., crewman (1971)		

Name	**Gilmont**		
Location	32.80268, -94.87887		
Moved from			
Moved to			
Quadrangle	Tyler		
County	Upshire		
Entity	TFS		
District	District 1		
Lease date	December 19, 1951		
Survey	James Wagstaff		
NGS description	CR1224		
Property number	3779	Tower/unit number	Unknown (108)
	1	2	3
Erected	1952		
Removed	Sold 1983		
Model	EMSCO derrick		
Height	87 feet		
Observers	Carl Stewart, crewleader (1956–76); Kenneth Ray Fountain (ca. 1957); Gene Carroll Nix, crewman (1956–65); Colman L. Palmer, crewman (1961); Stephen Chandler, crewman (1962); Donald Ehrlish, crewman (1963); Michael B. Irons, crewman (1969); John R. Harris, crewman (1966–69); Paul Chandler, crewman (1970); George Shakelford, forest aide (1970); Travis Powell, crewman (1974–75); Grady King (1976)		

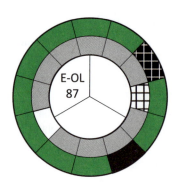

Name	**Hallsville**
Location	32.53012, -94.50304
Moved from	
Moved to	
Quadrangle	Tyler
County	Harrison
Entity	TFS
District	District 1
Lease date	September 11, 1935
Survey	Clery Grillett
NGS description	CR1190
Property number	3775
	1
Erected	1936
Removed	1983
Model	Aermotor MC-39
Height	99.75 feet

Tower/unit number 53 (120)
2 3

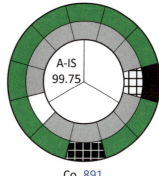

Co. 891

Observers: Junius Perry, patrolman; Sanford K. Sparks, patrolman (1936–ca. 1942); John C. Petty, emergency patrolman (1936–42); Luners Mack Perry, patrolman (1943–46); Danial Guffey, crewleader (1948–63); William (Billy) Haynes, crewleader (1963–67); Nolan Welty, crewman (1964); William C. McCay, crewman (1965); Emanuel W. Phillips, crewman/crewleader (1966–75); Perry Young, crewman (1970–75)

Name	**Jefferson, Location 1**
Location	32.76609, -94.37769
Moved from	
Moved to	Jefferson (Location 2)
Quadrangle	Tyler
County	Marion
Entity	TFS
District	District 1
Lease date	October 17, 1935
Survey	W. M. Archer
NGS description	CR1091
Property number	3773
	1
Erected	1936
Removed	2003
Model	Aermotor MC-39
Height	99.75 feet

Tower/unit number 54 (109) Co. 891
2 3

Observers: Floyd A. Bramlett; Whitfield (ca. 1937); James P. Neely, emergency patrolman (1943); George C. Reed, emergency patrolman/lookout (1937–48); Jeff Hunter, crewman (1952); W. M. Walker (ca. 1956); R. A. Powell, crewman (1956–7); Ivan W. Foster (ca. 1958); James E. Jones, crewman (1958); Louie D. Dodson (ca. 1957); Henry Lee Price, crewleader (1957–59); Harold Gene Harvey, crewman (1959–61); B. Wonston Lemmon (1961); W. B. Powell, crewman (1962); Fred Edward Hawkins, crewman (1962); Hillary Smith, crewman (1963); Robert Taylor, crewman (1964); Emmett Turner, crewleader (1964–68); James H. Morris, crewman (1968–69); Alan D. Bolick, forest aide (1974)

Name	**Jefferson, Location 2**		
Location	32.75952, -94.34229		
Moved from	Jefferson, Location 1		
Moved to			
Quadrangle	Tyler		
County	Marion		
Entity	Private		
District	District 1		
Lease date	N/A		
Survey	A. Urquhart		
NGS description	None		
Property number	3773	Tower/unit number	54 (109)
	1	2	3
Erected	2003		
Removed			
Model	Aermotor MC-39		
Height	99.75 feet		
Observers	N/A		

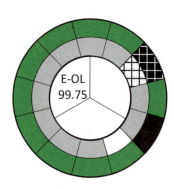

Name	**Karnack, Location 1**		
Location	32.68364, -94.17863		
Moved from	Newton		
Moved to			
Quadrangle	Tyler		
County	Harrison		
Entity	TFS		
District	District 1		
Lease date	Unknown		
Survey	William R. D. Ward		
NGS description	None		
Property number	Unknown	Tower/unit number	39 (Unknown)
	1	2	3
Erected	1942		
Removed	1952		
Model	EMSCO TW-1		
Height	99.75 feet		
Observers	Lawrence H. Bramlett (1942–57); Joseph Bison, crewman (1944); Wade W. Futrell, crewman (1944)		

Name	**Karnack, Location 2**
Location	32.62116, -94.18207
Moved from	Scottsville
Moved to	
Quadrangle	Tyler
County	Harrison
Entity	TFS
District	District 1
Lease date	March 6, 1961
Survey	W. R. Ward
NGS description	None
Property number	3774

1	2	3
	Tower/unit number	39 (Unknown)

Erected	1961
Removed	Standing (sold 1981)
Model	Aermotor MC-39
Height	99.75 feet
Observers	Lawrence Bramlett, crewleader (1942–57); Leonard Alcorn, crewleader (1963); William McCoy, crewleader (1963); Pete Coltharpe (1965–70); Wallace J. Stanley, crewleader (1970–72); Ricky Lawless, forest aide (1975); Tony Reese, forest aide (1975); Ottis Mickelboro, forest aide (1975); Joseph E. Moore (ca. 1976)

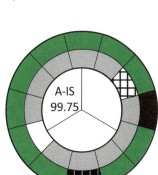

Co. 891

Name	**Kildare**
Location	32.94888, -94.25096
Moved from	
Moved to	
Quadrangle	Tyler
County	Cass
Entity	TFS
District	District 1
Lease date	September 13, 1935
Survey	H. Reams
NGS description	CR1084
Property number	3768

1	2	3
	Tower/unit number	56 (115)

Erected	1936
Removed	Sold 1982
Model	Aermotor MC-39
Height	99.75 feet
Observers	Mott (ca. 1936); A. Z. Melton (1940–ca. 1946); Thomas Dooley, crewman (1967); Charley Hendrick, crewleader (1966–70); Donald Williams, forest aide (1975); Kenneth Waters, forest aide (1975)

Tyler Quadrangle

Name	**Latch**		
Location	32.69691, -95.08512		
Moved from			
Moved to			
Quadrangle	Tyler		
County	Upshire		
Entity	TFS		
District	District 1		
Lease date	July 2, 1960		
Survey	Nacogdoches Co. School Land, Block 9		
NGS description	None		
Property number	10525 1	Tower/unit number 2	Unknown 3
Erected	1960		
Removed	Standing (sold 1982)		
Model	L. C. Moore K-Type		
Height	94 feet		
Observers	J. D. Warren, crewman (1962); Ollie V. Schuler, crewleader (1962–63); William Gunn, crewman (1962); Clarence Fieldmen, crewman (1964); Millard F. Jones, crewman (1967); Stephen E. Borden, crewleader (1966); Gerald D. Barton, crewleader (1967); Max Blackstone, crewman/crewleader (1966–76); John Harris, crewman (1970–75); Wayne Stafford Jr., crewman (1976)		

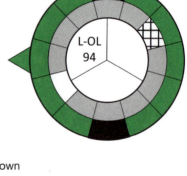

Name	**Lindale**		
Location	32.54598, -95.36681		
Moved from	USFS Moss Hill		
Moved to			
Quadrangle	Tyler		
County	Smith		
Entity	TFS		
District	District 2		
Lease date	February 2, 1971		
Survey	John S. McCoy		
NGS description	None		
Property number	Unknown 1	Tower/unit number 2	Unknown 3
Erected	1971		
Removed	1982		
Model	Aermotor MC-40		
Height	99.75 feet		
Observers	George M. Stripling, crewleader (1971); Herschell Bridges, forest aide (1975)		

Name	**Love's Lookout**
Location	32.03015, -95.28060
Moved from	
Moved to	
Quadrangle	Tyler
County	Cherokee
Entity	TFS
District	District 2
Lease date	August 31, 1935
Survey	T. Queuedo
NGS description	CR1278
Property number	4291
	1
Erected	1936
Removed	Standing (transferred 1993)
Model	Aermotor MC-39
Height	99.75 feet
Observers	William M. Smith (1936–43); Clyde C. Tidwell (1943–59); Allen Henderson, emergency patrolman (ca. 1943); James P. Neely, emergency patrolman (1943–45); Satterwhite (1953); William R. Kennedy, crewleader (1959); Jasper Carroll, crewleader (1965–66); V. D. Pierce, crewman/crewleader (1965–71); Charlie McElyes, crewman (1966)

Tower/unit number 46 (211)
2 3

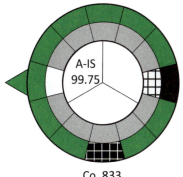

Name	**Meadows**
Location	32.41194, -94.80806
Moved from	
Moved to	
Quadrangle	Tyler
County	Gregg
Entity	TFS
District	District 2
Lease date	August 2, 1960
Survey	G. W. Floyd
NGS description	None
Property number	10528
	1
Erected	1960
Removed	1981
Model	L. C. Moore K-Type
Height	94 feet
Observers	Billy D. Mullins (ca. 1961); Leon B. Patrick (1961); Joe Clyde Horton, crewleader (1962); Guy G. Dunn, crewman (1962); Joseph Gibson, crewleader (1963–75); Virgil R. Thompson, crewman (1965); R. L. McKinnon (ca. 1965); John Cole Price, crewman (1965); Orland Gibson, crewleader (1966–68)

Tower/unit number Unknown
2 3

Name	**Mims Chapel (Ero)**			
Location	32.83750, -94.63417			
Moved from				
Moved to				
Quadrangle	Tyler			
County	Marion			
Entity	TFS			
District	District 1			
Lease date	September 12, 1935			
Survey	W. H. Gilbert			
NGS description	CR1179			
Property number	3769	Tower/unit number	55 (114)	
	1	2	3	
Erected	1936			
Removed	Standing (sold 1982)			
Model	Aermotor MC-39			
Height	99.75 feet			
Observers	H. G. Fite; C. J. Wicker (1946); Gully Cowsert Jr., crewleader (1962–63); Paul Cleere, crewleader (1963); Cleo Roberson, crewman (1959–68); Bobby Starkey, crewman (1966); Henry L. Gilliam, crewman (1967); Alfred Oneal Police, crewman (1967); Elmer Braddock, crewleader (1969–72); Garland Faulk, crewman (1971); Marion Berry, crewman (1970–71); Terry E. Wicker, forest aide (1974)			

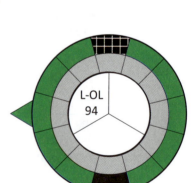

Name	**Pine Mills**			
Location	32.74020, -95.32620			
Moved from				
Moved to				
Quadrangle	Tyler			
County	Wood			
Entity	TFS			
District	District 2			
Lease date	July 28, 1960			
Survey	John P. Rossier			
NGS description	None			
Property number	10530	Tower/unit number	Unknown	
	1	2	3	
Erected	1960			
Removed	Standing			
Model	L. C. Moore K-Type			
Height	94 feet			
Observers	Cleo Calvin Hall, crewman (1962–63); Robert Vaughn, crewman (1963); Clarence Blackmon, crewman (1967); Joyce M. Whittiker, crewman (1964); Thurman M. Dobbs, crewman (1964); Stafford S. Shamburger, crewleader (1959–69)			

Name	**Poynor (Fincastle)**
Location	32.12956, -95.55005
Moved from	USFS Hi Point
Moved to	
Quadrangle	Tyler
County	Henderson
Entity	TFS
District	District 2
Lease date	April 2, 1970
Survey	W. W. Hawkins A-347
NGS description	None
Property number	13958

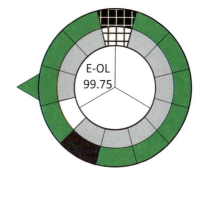

1		2	3
Erected	1970		
Removed	Standing (sold 1980)		
Model	EMSCO TW-1		
Height	99.75 feet		
Observers	John W. Hardman III, crewleader (1970); J. W. Largent, crewman (1971); Haskell Crossley, crewleader (1971–73); Guy Lambright, forest aide (1975–76); Jack Crossley (1976)		

Tower/unit number — Unknown

Name	**Scottsville**
Location	32.54410, -94.24550
Moved from	
Moved to	Karnack, Location 2
Quadrangle	Tyler
County	Harrison
Entity	TFS
District	District 1
Lease date	Unknown
Survey	Henry Blossom
NGS description	None
Property number	Unknown

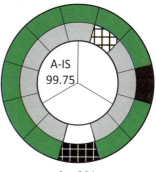

Co. 891

1		2	3
Erected	1936		
Removed	1961		
Model	Aermotor MC-39		
Height	99.75 feet		
Observers	Rose, patrolman (1941); Lawrence H. Bramlett, crewleader (1942–57); Lawrence E. Dameron, crewleader (1958); Joseph R. Johnson, crewleader (1958); Ervin B. Horton, crewman (1958); Tucker O. Rogers, crewleader (1959); James A. Baker, crewman (1959)		

Tower/unit number — 52 (110)

Name	**Smithland**
Location	32.83425, -94.17305
Moved from	Karnack, Location 1
Moved to	
Quadrangle	Tyler
County	Marion
Entity	TFS
District	District 1
Lease date	February 20, 1952
Survey	A. Fitzgerald
NGS description	CR1062
Property number	3778 Tower/unit number Unknown (113)
	1 2 3
Erected	1952
Removed	1981
Model	EMSCO TW-1
Height	99.75 feet
Observers	Major Martin, crewman (1962–67); Earl Belcher, crewleader (1963); Travis D. Dempsey, crewleader (1964); Arthur D. Wright, crewleader (1967–68)

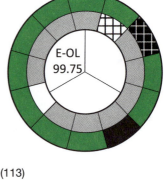

Name	**Snap**
Location	32.09666, -94.40702
Moved from	
Moved to	
Quadrangle	Tyler
County	Panola
Entity	TFS
District	District 2
Lease date	September 18, 1935
Survey	J. A. Lane
NGS description	CR1127
Property number	4292 Tower/unit number 49 (215)
	1 2 3
Erected	1936
Removed	1981
Model	Aermotor MC-39
Height	99.75 feet
Observers	Emery Marshall, patrolman (1936–43); William L. Kelley, emergency patrolman (1937); William Bush, emergency patrolman/patrolman/crewleader (1937–57); A. C. Roe, crewman (1951); James Bush, crewleader (1957); Stephen B. Lum, crewman (1957–68); Mason Bush, crewleader (1963); Thomas Robinson, crewman (1968); Tracy H. Wedgeworth, crewleader (1965–68)

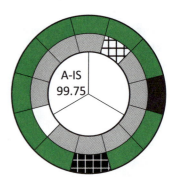

Co. 891

Name	**Starrville (Winona)**		
Location	32.45357, -95.13280		
Moved from			
Moved to			
Quadrangle	Tyler		
County	Smith		
Entity	TFS		
District	District 2		
Lease date	June 17, 1960		
Survey	Matthew Maise		
NGS description	None		
Property number	10529	Tower/unit number	Unknown
Erected	1960		
Removed	Sold 1981		
Model	L. C. Moore K-Type		
Height	94 feet		
Observers	Ray Lowery, crewleader (1953); Frank Myers, crewman (1961); John C. Weaver, crewman/crewleader (1963–66); Billy Gene Roberts, crewleader/forest aide (1966–76); Elijah Melton, forest aide (1975); Dale Geddie, forest aide (1976)		

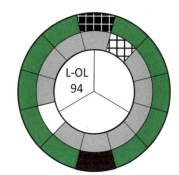

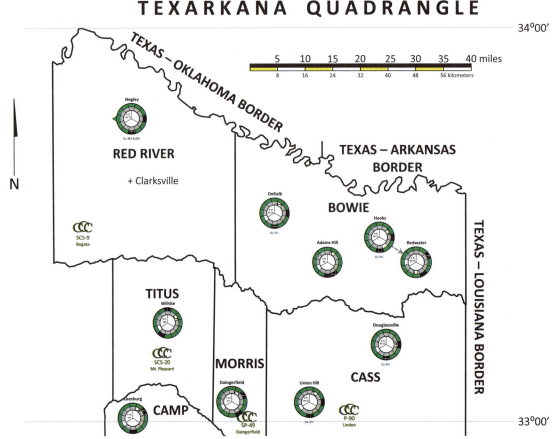

Appendix 4: Texarkana Quadrangle

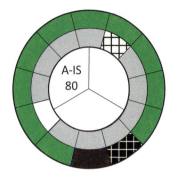

Name	**Adams Hill**
Location	33.38994, -94.45500
Moved from	
Moved to	
Quadrangle	Texarkana
County	Bowie
Entity	TFS
District	District 1
Lease date	September 9, 1965
Survey	J. Eskell
NGS description	None
Property number 1	12248
Tower/unit number 2	
Unknown 3	
Erected	1965
Removed	Sold 1981
Model	Aermotor LS-40
Height	80 feet
Observers	Clinton Boyd Ivey, crewleader (1966–69); Lester Tidwell, crewman (1966–67); Jackie Maroney, crewman (1968–69); Loyd W. Barrett, crewman (1966); Charles Foster, crewman (1969)

Name	**Daingerfield**
Location	33.03918, -94.74802
Moved from	
Moved to	
Quadrangle	Texarkana
County	Morris
Entity	TFS
District	District 1
Lease date	January 2, 1952
Survey	H. S. Proctor
NGS description	DM1263
Property number 1	3780
Tower/unit number 2	
Unknown (116) 3	
Erected	1952
Removed	Transferred 1981
Model	EMSCO derrick
Height	87 feet Unknown
Observers	Connor (1952), crewleader; Carl T. Moursund, crewleader (1955–69); C. L. Springfield (1956); John Gholston, crewman (1959); Roy Dee Skipper, crewleader (1970); Billie R. Carpenter, crewman (1970); Charles O. Barnwell, crewleader (1971); Roger D. Shaddix, crewman (1971)

Name	**DeKalb**		
Location	33.51394, -94.61701		
Moved from			
Moved to			
Quadrangle	Texarkana		
County	Bowie		
Entity	TFS		
District	District 1		
Lease date	March 2, 1936		
Survey	W. L. Browning		
NGS description	DM1292		
Property number	3771	Tower/unit number	60 (111)
	1	2	3
Erected	1936		
Removed	Sold 1981		
Model	Aermotor MC-39		
Height	99.75 feet		
Observers	C. W. Lann, patrolman (1937); Jim Howell, emergency patrolman (1938); James H. Howell, emergency patrolman (1937–42); John K. Wray, patrolman (1941–69); J. E. Moreland, emergencyman (1942); Roy Wiser, crewman (1957); Hercle Lee Cooper, crewleader (1969); Troy Morine, crewman (1969); Owen V. Finley, crewman (1959–1969)		

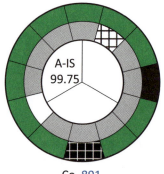

Co. 891

Name	**Douglassville**		
Location	33.18519, -94.27639		
Moved from			
Moved to			
Quadrangle	Texarkana		
County	Cass		
Entity	TFS		
District	District 1		
Lease date	September 4, 1935		
Survey	J. Baty		
NGS description	DM1229		
Property number	3777	Tower/unit number	58 (118)
	1	2	3
Erected	1936		
Removed	Sold 1983		
Model	Aermotor MC-39		
Height	99.75 feet		
Observers	Almer M. Allen, patrolman (1940); Lewis L. Hunt, emergency patrolman (1941–42); Emmett Turner, patrolman (1946); T. Warrington, patrolman (1946); William Johnson Swint, crewleader (1947–61); Joe Earl Tucker, crewman (1960); Fleming Waddill, crewman/crewleader (1958–66); Billy Tucker, crewleader (1964); Alton Dell McWaters, forest aide (1972); James Thompson, forest aide (1974–75)		

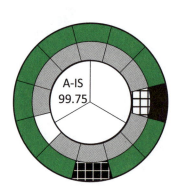

Co. 891

Name	**Hooks**
Location	33.45218, -94.29495
Moved from	
Moved to	Redwater
Quadrangle	Texarkana
County	Bowie
Entity	TFS
District	District 1
Lease date	October 17, 1935
Survey	G. Collum
NGS description	DM1241
Property number 1	Unknown
Tower/unit number 2	
3	59 (Unknown)
Erected	1936
Removed	1952
Model	Aermotor MC-39
Height	99.75 feet
Observers	R. V. Smith, patrolman (1941); Emmett Turner, patrolman (1942); John P. Pipes, crewman (1943–44); Thomas J. Clary, crewman (ca. 1944); Warner H. Dodd, crewman (ca. 1944); Wallace D. Grimes, crewman (ca. 1944); Chester A. Johnson, crewman (1944); Johnnie W. Powers, lookoutman (1944); Lawrence Hayes, emergencyman

A-IS
99.75

Co. 891

Name	**Leesburg**
Location	33.00542, -95.05547
Moved from	
Moved to	
Quadrangle	Texarkana
County	Camp
Entity	TFS
District	District 1
Lease date	April 24, 1961
Survey	S. D. Thomas
NGS description	None
Property number 1	11403
Tower/unit number 2	
3	Unknown
Erected	1962
Removed	1981
Model	Parkersburg derrick
Height	94 feet
Observers	Robert Redding, crewleader (1961–63); Jeff Hagen, crewleader (1963–71); Vance Vanderburg, crewman/crewleader (1966–76)

P-OL
94

Texarkana Quadrangle   **305**

Name	**Negley**				
Location	33.74926, -95.05589				
Moved from					
Moved to					
Quadrangle	Texarkana				
County	Red River				
Entity	TFS				
District	District 1				
Lease date	March 27, 1936				
Survey	J. C. Brown				
NGS description	DM1327				
Property number	3722	Tower/unit number	62 (107)	Co. 891 & 833	
	1		2	3	
Erected	1936				
Removed	Standing				
Model	Aermotor MC-39				
Height	99.75 feet				
Observers	William J. Leatherwood (1936–51); Oscar Gallander, emergency patrolman (ca. 1940); Hassel Johnston, emergency patrolman (1941–46); Pat H. Gandy, crewleader (1951–64); Henry M. Shipp, crewman (1952–56); Juanita Westbrook, seasonal lookout (1953–54); Jake Roberts, crewman (1956); Jerry Westbrook, crewman/crewleader (1961–73); George White, crewman (1964–68); Thomas Ray Stringer, crewman (1970); A. B. Hall, crewleader (1973)				

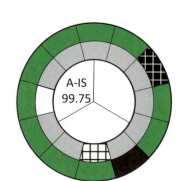

Name	**Redwater**				
Location	33.39333, -94.18250				
Moved from	Hooks				
Moved to					
Quadrangle	Texarkana				
County	Bowie				
Entity	TFS				
District	District 1				
Lease date	May 23, 1952				
Survey	J. Jackson				
NGS description	None				
Property number	3770	Tower/unit number	Unknown (106)		
	1		2	3	
Erected	1952				
Removed	Sold 1996				
Model	Aermotor MC-39				
Height	99.75 feet				
Observers	John Dee McDougal, crewman (1960); William Banks, crewman (1963); Mitch Missildine, crewleader (1963); Coy Corn, crewman (1961–66); John L. Sutton, crewman/crewleader (1966–67); Edward Neff, crewman (1966); Ray Neff, crewman/forest aide (1966–76)				

Name	**Union Hill (Mitchell; Hughes Springs)**
Location	33.04139, -94.51056
Moved from	
Moved to	
Quadrangle	Texarkana
County	Cass
Entity	TFS
District	District 1
Lease date	September 4, 1935
Survey	W. B. Wilson
NGS description	DM1259
Property number 1	Unknown
Tower/unit number 2	
3	57 (119)
Erected	1936
Removed	Sold 1981
Model	Aermotor MC-39
Height	99.75 feet
Observers	Earl T. Strickland, patrolman/lookout (1942–57); Melvin Finley, crewleader (1951); Oren Newsom (ca. 1955); Leonard Pruitt, crewman/crewleader (1955–63); Fred Edwards, crewman (1958); Vernon E. Hicks (ca. 1961); Howard Wayne Ray, crewman (1960); William Rawle Jackson, crewman (1961); Otis Blackburn, crewman/crewleader (1962–63); Oscar McGary, crewman (1963); John Bockman, crewleader (1963–64); Berryman Driskell, crewleader (1966); Billy Roy Skeleton, crewman (1967–68); Marion A. Hamilton, crewleader (1967–69); Curtis E. Maloney, crewman (1969–70); William T. Ball, crewman (1971)

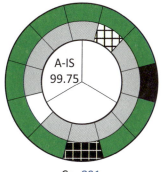

Co. 891

Name	**Wilhite**
Location	33.24002, -94.94330
Moved from	
Moved to	
Quadrangle	Texarkana
County	Titus
Entity	TFS
District	District 1
Lease date	January 28, 1970
Survey	Washington Gray
NGS description	None
Property number 1	Unknown
Tower/unit number 2	
3	Unknown
Erected	1970
Removed	Sold 1982
Model	Dunlap T-10
Height	100 feet
Observers	Billy C. Hammond, crewleader (1970–71); Dallas L. Brush (1970)

LOST PINES REGION

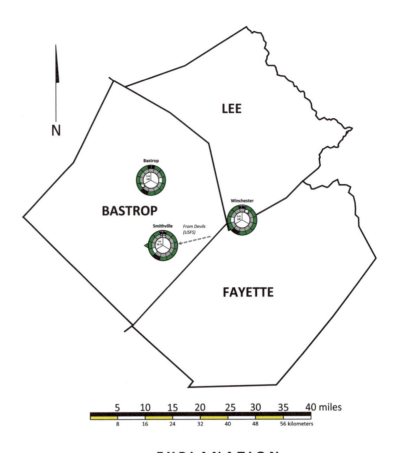

EXPLANATION

A	Aermotor Co.	OL	Outside ladder
B	Bilby	IL	Inside ladder
C	Creosoted wood	IS	Inside stairway
D	Dunlap Manufacturing Corp.		
E	EMSCO Manufacturing Co.		
I	International Derrick & Equipment Co.		
L	L. C. Moore Manufacturing Co.		
M	McClintic-Marshall Corp.		
N	Nashville Bridge Co.		
P	BR Parkersburg		
U	Unknown		

CCC P-60-T Oakhurst	Historic Civilian Conservation Corps (CCC) Camp and Company (map only)		
838 Erecting CCC Company 840 Hypothesized CCC Erecting Company		←---	Lookout relocation (map only)

Appendix 5: Lost Pines

Name	**Bastrop**
Location	30.17398, -97.24602
Moved from	
Moved to	
Quadrangle	Austin
County	Bastrop
Entity	TFS
District	District 7
Lease date	March 12, 1970
Survey	Bastrop Town Tract
NGS description	None
Property number	13851
	1
Erected	1970
Removed	1983
Model	Dunlap T-10
Height	100 feet
Observers	Harold D. Farley, crewleader (1969); Benjamin Devereux Jr., crewman (1969); Calvin L. Dixon, crewman (1971)

Tower/unit number Unknown (711)
2 3

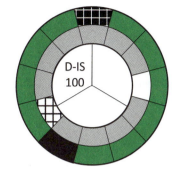

Name	**Smithville**
Location	30.00711, -97.21261
Moved from	USFS Devils
Moved to	
Quadrangle	Seguin
County	Bastrop
Entity	TFS
District	District 7
Lease date	January 15, 1970
Survey	J&J Stewart
NGS description	None
Property number	13849
	1
Erected	1970
Removed	Standing (sold 1980)
Model	International
Height	99.75 feet
Observers	Shelby Harton, crewleader (1969–76); Fred Anderson, crewman (1969); R. T. Lewis, crewman/crewleader (1970–72); Orville Beggs, crewman (1975); Ronald Leath, forest aide (1975)

Tower/unit number Unknown
2 3

Name	**Winchester**		
Location	30.06944, -96.96694		
Moved from			
Moved to			
Quadrangle	Austin		
County	Lee		
Entity	TFS		
District	District 7		
Lease date	February 2, 1970		
Survey	F. Taylor		
NGS description	None		
Property number	13850	Tower/unit number	Unknown
	1	2	3
Erected	1970		
Removed	1991		
Model	Dunlap T-10		
Height	100 feet		
Observers	Carl Newton Froehlich, crewleader (1969); George Turman, forest aide (1971); James C. Ephraim, crewman/crewleader (1969–72)		

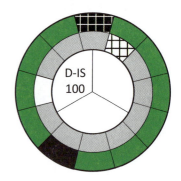

310   Appendix 5

Appendix 6: Pier Dimensions

AERMOTOR

LS-40
Inside Stairway

Height feet	a feet	h (b x 2) feet
22.00	10.063	13.271
35.00	12.094	16.146
47.00	14.109	19.000
60.00	15.911	21.542
73.00	17.917	24.375
80.00	18.922	25.802
93.00	20.938	28.656
100.00	21.943	30.073

LX-24

Height feet	a feet	h (b x 2) feet
21.00	10.469	14.096
36.00	12.844	17.456
48.00	15.182	20.763
62.00	17.521	24.070
75.00	19.865	27.385
80.00	20.677	28.466

MC-39
Inside Stairway

Height feet	a feet	h (b x 2) feet
33.00	11.984	16.948
45.75	13.906	19.667
59.25	15.948	22.552
79.50	19.021	26.901
86.25	20.031	28.328
99.75	22.063	31.203

MC-40
Platform Ladder & Straight Ladder

Height feet	a feet	h (b x 2) feet
45.75	13.906	19.667
59.25	15.938	22.542
79.50	19.021	26.901
86.25	20.042	28.344
99.75	22.073	31.219

INTERNATIONAL DERRICK

Straight Ladder Type

Height feet	a feet	h (b x 2) feet
72.75	17.984	25.432
79.50	19.021	26.901
86.25	20.042	28.344
99.75	22.073	31.214
120.00	25.135	35.547

Inside Stairway

Height feet	a feet	h (b x 2) feet
30.00	11.531	16.307
41.25	13.229	18.708
54.00	15.156	21.443
67.50	17.193	24.313
82.50	19.458	27.521
99.75	22.063	31.198
120.00	25.120	35.526

NASHVILLE BRIDGE COMPANY
Straight Ladder Type

Height feet	a feet	h (b x 2) feet
122.00	24.260	32.178

PLAQUEMINES

Height feet	a feet	h (b x 2) feet
22.00	10.063	13.271
35.00	12.094	16.146
47.00	14.109	19.000
60.00	15.911	21.542
73.00	17.917	24.375
80.00	18.922	25.802
93.00	20.938	28.656
100.00	21.943	30.073

USDA CT-1
Inside Stairway

Height feet	a feet	b feet	h feet
99.00	18.750	18.844	~26.500
119.00	21.177	21.281	~30.250

Note: Length a and b are not equal and tower is not square. Length h is variable.

Notes

Introduction

1. Carolyn Francis Hyman, "A History of the Texas National Forests" (master's thesis, University of Texas at Austin, 1948).

2. W. Ellis Groben, *Acceptable Plans, Forest Service Administrative Buildings* (USDA Forest Service, 1938), https://www.fs.usda.gov/eng/facilities/documents/acceptable_plans.pdf.

3. Philip Conners, *Fire Season: Field Notes from a Wilderness Lookout* (New York: HarperCollins, 2011).

4. Louis Untermeyer, *New Enlarged Pocket Anthology of Robert Frost Poems*, 4th ptg. (New York: Washington Square Press, 1962).

5. John Suiter, *Poets on the Peaks: Gary Snyder, Philip Whalen and Jack Kerouac in the Northern Cascades* (New York: Counterpoint Press, 2002), 268.

Chapter 1. The Breaking Wave: Forestry in Texas

1. William H. McWilliams and Roger G. Lord, *Forest Resources of East Texas*, Resource Bulletin SO-136 (New Orleans: USDA Forest Service, Southern Forest Experiment Station, 1988).

2. USDA Forest Service, *Standard Lookout Structure Plans* (1938), i, https://www.fs.usda.gov/eng/facilities/documents/1938_USDA-FS_StdLookout Plans.pdf.

3. Loren L. Bishop, "Texas National Forests," *Texas Geographic Magazine* 1.2 (1937): 1–15.

4. Robert S. Maxwell and Robert D. Baker, *Sawdust Empire: The Texas Lumber Industry, 1830–1940* (College Station: Texas A&M University Press, 1983).

5. Robert S. Maxwell and James W. Martin, *A Short History of Forest Conservation in Texas, 1880–1940*, Bulletin 20 (Nacogdoches: Stephen F. Austin State University, 1970), 7.

6. McWilliams and Lord, *Forest Resources of East Texas*.

7. William L. Bray, *Forest Resources of Texas*, Bulletin 47 (Washington, DC: USDA Bureau of Forestry, 1904), 7–71.

8. Robert S. Maxwell, "One Man's Legacy: W. Goodrich Jones and Texas Conservation," *Southwestern Historical Quarterly* 77.3 (1974): 355–80.

9. Maxwell and Martin, *Short History of Forest Conservation*, 1.

10. Maxwell and Martin, *Short History of Forest Conservation*, 2.

11. Ronald F. Billings, *A Century of Forestry, 1914–2014* (Sussex, VA: Donning, 2014).

12. Thomas C. Croker Jr., "The Longleaf Pine Story," *Journal of Forest History* 23.1 (1979): 32–43.

13. Herman H. Chapman, "Is the Longleaf Type a Climax?," *Ecology* 13.4 (1932): 328–34.

14. H. H. Chapman, "A Method of Studying Growth and Yield of Longleaf Pine Applied in Tyler Co., Texas," *Proceedings of the Society of American Foresters*

4.2 (1909): 207–20; Joe C. Truett and Daniel W. Lay, *Land of Bears and Honey: A Natural History of East Texas* (Austin: University of Texas Press, 1984).

15. Chapman, "A Method of Studying Growth and Yield of Longleaf Pine," 214.

16. Herman H. Chapman, "Hogs vs. Longleaf Pine Forests," *Texas Forest News* 16.6 (1936): 2.

17. Thad Sitton and James H. Conrad, *Nameless Towns: Texas Sawmill Communities, 1880–1942* (Austin: University of Texas Press, 2008).

18. Chapman, "Hogs vs. Longleaf Pine Forests," 2.

19. W. Goodrich Jones, "Texas Timber Resources," in Maxwell and Martin, *Short History of Forest Conservation in Texas*, 43–55.

20. James W. Cruikshank and I. F. Eldredge, *Forest Resources of Southeastern Texas*, Miscellaneous Publication 326 (Washington, DC: USDA Forest Service, 1939).

21. McWilliams and Lord, *Forest Resources of East Texas*; Maxwell and Baker, *Sawdust Empire*.

22. Croker, "The Longleaf Pine Story"; Maxwell and Martin, *Short History of Forest Conservation*.

23. Kevin Guthrie, Rebecca Barlow, and John S. Kush, "Restoring an Ecosystem with Silvopasture: A Short (leaf) Story," *Ecological Restoration* 34.1 (2016): 16–19.

24. McWilliams and Lord, *Forest Resources of East Texas*; Maxwell and Martin, *Short History of Forest Conservation*.

25. Maxwell and Martin, *Short History of Forest Conservation*.

26. Jones, "Texas Timber Resources," 51.

27. Maxwell and Martin, *Short History of Forest Conservation*; Bruce Kershner, Daniel Mathews, Gil Nelson, Richard Spellenberg, Terry Purinton, Andrew Block, Gerry Moore, and John W. Thieret, *National Wildlife Federation Field Guide to Trees of North America* (New York: Sterling Publishing, 2008).

28. Maxwell and Martin, *Short History of Forest Conservation*.

29. Guthrie, Barlow, and Kush, "Restoring an Ecosystem with Silvopasture."

30. Cruikshank and Eldredge, *Forest Resources of Southeastern Texas*, 7.

31. I-Kuai Hung, Daniel Unger, Yanli Zhang, Jeff Williams, Jason Grogan, Dean Coble, and Jimmie Yeiser, "Forest Landscape Change in East Texas: 1974–2009," *Southeastern Naturalist* 15.9 (2016): 1–15.

32. Conners, *Fire Season*.

33. William Oates, letter to the author, April 13, 2022.

34. W. Goodrich Jones, "Texas Timber Resources," in Maxwell and Martin, *Short History of Forest Conservation*, 50.

35. Bishop, "Texas National Forests."

36. Billings, *A Century of Forestry*.

37. Sitton and Conrad, *Nameless Towns*; Bishop, "Texas National Forests."

38. Maxwell and Baker, *Sawdust Empire*.

39. Laurence C. Walker, "Texas Forests: Their Fall and Rise: David Anderson Recalls the Texas Situation," unpublished interview with D. A. Anderson, Stephen F. Austin State University, 1980.

40. "Aldridge Sawmill," Texas beyond History, October 2, 2021, https://texasbeyondhistory.net/aldridge.

41. T. Lindsay Baker, *More Ghost Towns of Texas* (Norman: University of Oklahoma Press, 2005).

42. "Aldridge Sawmill."

43. Jones, "Texas Timber Resources," 47.

44. Maxwell, "One Man's Legacy."

45. Maxwell, "One Man's Legacy."

46. Maxwell, "One Man's Legacy."

47. Maxwell, "One Man's Legacy."

48. David L. Chapman, "An Administrative History of the Texas Forest Service, 1915–1975" (PhD diss., Texas A&M University, 1981).

49. Maxwell, "One Man's Legacy," 358.

50. Maxwell, "One Man's Legacy," 358.

51. Billings, *A Century of Forestry*.

52. Chapman, "Administrative History of the Texas Forest Service."

53. Chapman, "Administrative History of the Texas Forest Service."

54. Chapman, "Administrative History of the Texas Forest Service."

55. Maxwell, "One Man's Legacy"; Billings, *A Century of Forestry*.

56. Billings, *A Century of Forestry*.

57. Billings, *A Century of Forestry*.

58. Jones, "Texas Timber Resources," 48.

59. Jones, "Texas Timber Resources," 48–49.

60. Maxwell and Martin, *Short History of Forest Conservation*.

61. Bishop, "Texas National Forests."

62. Douglas Brinkley, *The Wilderness Warrior: Theodore Roosevelt and the Crusade for America* (New York: Harper Perennial, 2009).

63. Maxwell and Martin, *Short History of Forest Conservation*; Billings, *A Century of Forestry*.

64. W. Goodrich Jones, "Proceedings of the Conference of Governors," in Maxwell and Martin, *Short History of Forest Conservation*, 41–42.

65. Dan K. Utley, "With the Yalies in the Deep Woods, May 10–13, 1909," in *Eavesdropping on Texas History*, ed. M. L. Scheer (Denton: University of North Texas Press, 2017), 133–53.

66. Utley, "With the Yalies in the Deep Woods," 137.

67. Maxwell and Martin, *Short History of Forest Conservation*.

68. Maxwell, "One Man's Legacy."

69. Maxwell, "One Man's Legacy."

70. Maxwell and Baker, *Sawdust Empire*.

71. Billings, *A Century of Forestry*.

72. Letter from J. E. Ferguson to W. G. Jones, February 9, 1937.

73. Billings, *A Century of Forestry*.

74. Bizzell, quoted in Maxwell, "One Man's Legacy, 380.

75. Maxwell, "One Man's Legacy."

76. Mel Cooksey and Ron Weeks, *A Birder's Guide to the Texas Coast* (Asheville, NC: American Birding Association, 2006).

77. Chapman, "Administrative History of the Texas Forest Service."

78. Chapman, "Administrative History of the Texas Forest Service."

79. Chapman, "Administrative History of the Texas Forest Service."

80. Chapman, "Administrative History of the Texas Forest Service."

81. John H. Foster, *Grass and Woodland Fires in Texas*, Bulletin 1 (College Station: Agricultural and Mechanical College of Texas, Department of Forestry, 1916).

82. Chapman, "Administrative History of the Texas Forest Service."

83. Chapman, "Administrative History of the Texas Forest Service."

84. John H. Foster, H. B. Krausz, and George W. Johnson, *First Annual Report of the State Forester*, Bulletin 4 (College Station: Agricultural and Mechanical College of Texas, Department of Forestry, 1917).

85. E. O. Siecke, H. J. Eberly, W. E. Bond, C. B. Webster, H. F. Munson, J. M. Cravey, J. M. Turner, E. B. Long, B. D. Hawkins, V. V. Bean, and H. A. Budde, *Eleventh Annual Report of the State Forester*, Bulletin 18 (College Station: Agricultural and Mechanical College of Texas, Texas Forest Service, 1926); Walker, "Texas Forests: Their Fall and Rise."

86. Chapman, "Administrative History of the Texas Forest Service."

87. Chapman, "Administrative History of the Texas Forest Service."

88. Lenthall Wyman and L. Goodrich Jones, *Forest Fire Prevention in East Texas*, Bulletin 9 (College Station: Agricultural and Mechanical College of Texas, Department of Forestry, 1919), 10, 11.

89. J. H. Foster, H. B. Krausz, and G. W. Johnson, *Forest Resources of East Texas*, Bulletin 5 (College Station: Agricultural and Mechanical College of Texas, Department of Forestry, 1917), in McWilliams and Lord, *Forest Resources of East Texas*, 4.

90. Foster, Krausz, and Johnson, *Forest Resources of East Texas*, 4.

91. Foster, Krausz, and Johnson, *First Annual Report of the State Forester*.

92. Chapman, "Administrative History of the Texas Forest Service."

93. Chapman, "Administrative History of the Texas Forest Service."

94. "State Forester Hands Out a Needed Roast," *Southern Industrial and Lumber Review* 27 (1918): 16.

95. Chapman, "Administrative History of the Texas Forest Service."

96. Billings, *A Century of Forestry*; Maxwell and Baker, *Sawdust Empire*.

97. Maxwell and Martin, *Short History of Forest Conservation*.

98. Chapman, "Administrative History of the Texas Forest Service."

99. Wyman and Jones, *Forest Fire Prevention in East Texas*; Eric O. Siecke and Lenthall Wyman, *Forestry Questions and Answers*, Bulletin 12 (College Station: Agricultural and Mechanical College of Texas, Department of Forestry, 1920).

100. Billings, *A Century of Forestry*.

101. "Purpose of the 'News,'" *Texas Forest News* 8.3 (1926): 3.

102. Maxwell and Baker, *Sawdust Empire*; Maxwell and Martin, *Short History of Forest Conservation*.

103. Maxwell and Baker, *Sawdust Empire*.

104. Billings, *A Century of Forestry*, 53. The Department of Forestry was renamed the Texas Forest Service in 1926.

105. Siecke et al., *Eleventh Annual Report of the State Forester*, 30.

106. Siecke et al., *Eleventh Annual Report of the State Forester*.

107. Chapman, "Administrative History of the Texas Forest Service."

108. "Forest Fire Conviction Secured in Nacogdoches," *Tyler Journal* 11.50 (April 10, 1936): 1.

109. Chapman, "Administrative History of the Texas Forest Service."

110. "Congratulations," *Pineywoods Pickups* 3.2 (1942): 6.

111. Walker, "Texas Forests: Their Fall and Rise."

112. Maxwell and Martin, *Short History of Forest Conservation in Texas*, 31.

113. Chapman, "Administrative History of the Texas Forest Service."

114. The 120-foot Aermotor MC-39 standing on E. O. Siecke State Forest was originally constructed at Sheffield's Ferry in 1937 and moved to the site in 1948. It stands above the landing platform constructed in 1926 for the state's first fire tower.

115. Foster, *Grass and Woodland Fires in Texas*, 10.

116. Foster, *Grass and Woodland Fires in Texas*, 10.

117. Chapman, "Administrative History of the Texas Forest Service."

118. Wyman and Jones, *Forest Fire Prevention in East Texas*, 8.

119. Maxwell and Baker, *Sawdust Empire*.

120. John Ippolito, *A Cultural Resource Overview of the National Forests in Texas* (Lufkin: USDA Forest Service, 1983).

121. Wyman and Jones, *Forest Fire Prevention in East Texas*, 6.

122. "I Remember When . . . ," *Texas Forest News* 57 (1978): 12–13.

123. "Fire Towers Then and Now," *Rio Grande Herald* 35.26 (April 14, 1977): 6.

124. Billings, *A Century of Forestry*, 53, 68.

125. "CCC Performed Vital Role in the 1930's," *Polk County Enterprise* 105.78 (September 27, 1987): 75.

126. *A Manual of Practical Forestry for East Texas* (College Station: Texas Forest Service, 1938).

127. Texas Forest Service, *Seventeenth Annual Report*, Bulletin 23 (College Station: Agricultural and Mechanical College of Texas, 1932).

128. *A Manual of Practical Forestry for East Texas*.

129. E. A. Harris, *Management Plan, Davy Crockett National Forest*, unpublished USFS document, 1937.

130. "Forest Fires during 1927," *Texas Forest News* 10.7 (1928): 1.

131. Newton County reported 26.25 percent of all fires in 1937; Tyler, 15 percent; Polk, 14.3 percent; and Hardin, 8.8 percent. The fire statistics were reported in *A Manual of Practical Forestry for East Texas*.

132. Samuel J. Gerald, "52 Steel Towers Guard against Forest Fires in East Texas," *Claude News* 46.40 (June 7, 1935): 8.

Chapter 2. Neutral Ground: Geography and Climate

1. Hunters Randall, "Recreational Plan, Texas National Forests" (1936), in Robert D. Baker, "Timbered Again: The Story of the National Forests in Texas," unpublished manuscript, 1990.

2. Bishop, "Texas National Forests," 11.

3. Richard A. Anthes, John J. Cahir, Alistair B. Fraser, and Hans A. Panofsky, *The Atmosphere, Third Edition* (Columbus, OH: Charles E. Merrill, 1981).

4. E. R. Swanson, *Geo Texas: A Guide to the Earth Sciences* (College Station: Texas A&M University Press, 1995).

5. Three towers constructed in the Lost Pines Region during 1970 occupy

two additional 1:250,000 sheets (Austin and Seguin quadrangles). Because these towers do little to influence the historic focus of the Forest Service's protection program, they are treated separately.

6. Workers of the Writers' Program, Program of the Work Project Administration in the State of Texas, *Texas: A Guide to the Lone Star State* (New York: Hastings House Publishers, 1940).

7. Workers of the Writers' Program, *Texas: A Guide to the Lone Star State.*

8. James W. Cruikshank, *Forest Resources of Northeast Texas*, Forest Survey Release No. 40 (New Orleans: USDA Forest Service, 1938), 3.

9. National Geodetic Survey, US Coast and Geodetic Survey marker BL2191, accessed September 19, 2024, https://www.ngs.noaa.gov/cgi-bin/ds_mark.prl?PidBox=BL2191.

10. "Memorandum of Understanding between the Texas Forest Service and the United States Forest Service," unpublished document, rev. December 7, 1949.

11. "Forest Fires during 1927," 1.

12. Lawrence M. Whitfield, *Draft Environmental Impact Statement and Unit Plan, Sam Houston National Forest, Texas* (Lufkin: USDA Forest Service, 1977).

13. The Ancient and Honorable Order of Squirrels was a lookout-specific education program in which members received a wallet-sized certificate after climbing to the top of a tower and reciting a fire creed. The program was initiated in Minnesota in 1927. A 1949 article, "Honorable Order of Squirrels," by Elizabeth Bachmann provides a good historical summary of the program.

14. Croker, "The Longleaf Pine Story," 42.

15. Croker, "The Longleaf Pine Story," 40.

16. Truett and Lay, *Land of Bears and Honey.*

17. David A. Anderson, "History of the Texas Forest Service, a Part of the Texas A&M University System," unpublished report, Texas Forest Service, 1968; Richard T. Malouf, "Thematic Evaluation of Administrative and Fire Lookout Tower Sites on the Mark Twain National Forest, Missouri," unpublished report, 1991.

18. "News by the Grapevine," *Pineywoods Pickups* 3.2 (1942): 2.

19. Sherman L. Frost, "Operations Conroe," unpublished report, Texas A&M Forest Service collection, College Station, October 1947.

20. Anderson, "History of the Texas Forest Service."

21. "Forest Fire Danger Now More Easily Prevented through Aid of Instruments," *Lampasas Daily Leader* 34.149 (August 30, 1937): 3.

22. *Texas Forestry Progress, 1943–1944, 28th and 29th Annual Reports*, Bulletin 32 (College Station: Agricultural and Mechanical College of Texas, Texas Forest Service, 1944).

23. "Memorandum of Understanding between the Texas Forest Service and the United States Forest Service," December 7, 1949.

24. Richard G. Baumhoff, "Fighting Forest Fires: Protecting Timber in Government Owned Areas in Missouri Ozarks Requires Constant Vigilance," *Cassville Republican* (undated clipping in Mark Twain National Forest Historical File), cited in Malouf, "Thematic Evaluation," 39.

25. Texas Forest Service, staff meeting minutes, November 13, 1963, 3–4.

Chapter 3. Forest Protection and Fire Control

1. The Division of Forest Protection was the first department established within the Department of Forestry in 1922, demonstrating the significance of the mission and the objectives of the organization.

2. Towers were completed at Yellowpine, Ratcliff, Bird Mountain, Bath, Liberty Hill, Hortense, Conroe, and Etoile.

3. Gerald, "52 Steel Towers," 8.

4. Robert Sorrell, *Blue Ridge Fire Towers* (Charleston, SC: History Press, 2015).

5. Chapman, "An Administrative History."

6. "Private Owners Practice Fire Protection," *Texas Forest News* 9.3 (1927): 3.

7. Anderson, "History of the Texas Forest Service"; "Texas Timberland Owners Cooperate in Fire Protection," *Texas Forest News* 10.12 (1928): 1, 4.

8. "Texas Timberland Owners Cooperate in Fire Protection."

9. International Derrick and Equipment Company lookouts were constructed at the Bath, Yellowpine, Ratcliff, and Bird Mountain Duty Stations.

10. "Fire Detection Cab Constructed in Tree Top," *Texas Forest News* 11.4 (1929): 4; "Another Lookout Tower Being Erected," *Texas Forest News* 10.9 (1928): 1.

11. "Lookout Towers Valuable Aid in Fire Control Work," *Texas Forest News* 10.1 (1928): 1.

12. Billings, *A Century of Forestry*; "Division Patrolman J. M. Cravey Resigns," *Texas Forest News* 9.8 (1927): 2.

13. Isaac C. Burroughs, *Davids of Today Slay the Giant, Forest Fire*, Bulletin 2 (College Station: Agricultural and Mechanical College of Texas, 1929).

14. *A Manual of Practical Forestry for East Texas*.

15. *Texas Almanac and State Industrial Guide: The Encyclopedia of Texas* (Dallas: A. H. Belo, 1949), 198.

16. "Lookout Towers Valuable Aid in Fire Control Work."

17. Texas Forest Service, *Seventeenth Annual Report*.

18. Texas Forest Service, *Seventeenth Annual Report*, 25.

19. "Patrolmen Use Lookout Trees," *Texas Forest News* 11.12 (1929): 2.

20. Siecke et al., *Eleventh Annual Report*.

21. "Patrolmen Use Lookout Trees."

22. Faye Green, "Oral History Interview with Johnny Columbus and Columbus Lacey," unpublished interview, USDA US Forest Service Oral History no. 1239, August 19, 1996.

23. Texas Forest Service, *Seventeenth Annual Report*.

24. "Unique Fire Detection Structure Erected," *Texas Forest News* 14.7 (1932): 1–2.

25. Texas Forest Service, *Seventeenth Annual Report*.

26. Texas Forest Service, *Seventeenth Annual Report*, 24.

27. Larry J. Fisher was an accomplished photographer, musician, playwright, filmmaker, and aviator who was involved in the first efforts to preserve the Big Thicket.

28. Larry J. Fisher, "Ability in Acrobatics Saves Lookoutman Locked in Fire

Tower," condensed from *Beaumont Enterprise* article, appears in *Texas Forest News* 17.3 (1937): 2.

29. "Perricone Quadruplets Visit Fire Tower," *Texas Forest News* 19.7 (1939): 4.

30. Knox Ivie, "Life of a Smoke Chaser," *Have You Heard* 25 (1970): 4–5.

31. Lease between Alton M. and Mary Langston and the TFS, signed May 20, 1932, Texas A&M Forest Service Offices, College Station.

32. "Unique Fire Detection Structure Erected."

33. Ivie, "Life of a Smoke Chaser."

34. "Pictures Showing Texas Forest Service Activities," unpublished report, Texas Forest Service, July 1932.

35. National Geodetic Survey, US Coast and Geodetic Survey marker BL2049, accessed April 13, 2023, https://www.ngs.noaa.gov/cgi-bin/ds_mark.prl?PidBox=BL2049.

36. Pamela A. Conners, "Historic American Buildings Survey, North Mountain Lookout," HABS No. CA-2271 (San Francisco: National Park Service, 1988).

37. Peter L. Steer and Keith E. Miller, *Lookouts in the Southwestern Region*, Management Report No. 8 (Washington, DC: USDA Forest Service Cultural Resources, 1989).

38. Mark V. Thornton, "Fixed Point Fire Detection, the Lookouts," unpublished report prepared for the Forest Service, Region 5, 1986.

39. Correspondence between Aermotor president L. C. Walker and J. O. Burnside, August 26, 1944. The vast majority of the unpublished letters and memoranda pertaining to the Fire Control Department and referenced in this study are preserved in the archives of the Texas A&M Forest Service, College Station, Texas.

40. The MC-39 lookouts referred to in this letter relate to those purchased for expanding the network into North Texas.

41. Correspondence between Aermotor president D. R. Scholes and M. V. Dunmire, November 10, 1948.

42. Correspondence between Aermotor Company and J. O. Burnside, June 11, 1958.

43. Correspondence between J. O. Burnside and EMSCO Derrick and Equipment Company, October 11, 1944.

44. Correspondence between F. A. Scoval (EMSCO) and J. D. Young, October 1, 1953.

45. Memorandum between Director Folweiler and J. O. Burnside, May 27, 1958.

46. "Specifications for the Dismantlement and the Re-erection of Oil Derricks as Forest Fire Lookout Towers," unpublished Texas Forest Service document, September 9, 1958.

47. Correspondence between W. E. White and W. T. Murphey (National Creosoting Company), June 7, 1932.

48. Letter from the Service Bureau, American Wood Preservers Association, to R. S. Manley (Texas Creosoting Company), August 12, 1932.

49. "Wooden Observation Tower Completed by Texas Forest Service: First of Its Kind to Be Built in Texas," *Texas Forest News* 18.2 (1938): 1.

50. "Texas Forest Service: Fire Control Department. Record of Types of

Lookout Towers by Districts," Internal TFS Property Inventory, March 24, 1959, Texas A&M Forest Service Archives.

51. "Tallest Wooden Fire Tower Built in Texas Forest," *Deport Times* 34.52 (February 4, 1943): 8.

52. George E. Leigh, "Bilby Towers," NOAA 200th Feature Stories: Bilby Towers, July 1, 2021, https://celebrating200years.noaa.gov/magazine/bilby/welcome.html.

53. Correspondence between O. C. Braly and J. O. Burnside, March 29, 1961.

54. Malouf, "Thematic Evaluation."

55. Correspondence between W. E. White and H. H. Richter, July 7, 1938.

56. Winston O. Kirkland and Bob Erwin, "Texas Forest Service Tower Locations, from Jan. 1937 to –," Texas A&M Forest Service Archives.

57. Lease between the Angelina County Lumber Company and the TFS, signed December 18, 1937, Texas A&M Forest Service Archives.

58. "Forest Fire Lookout Towers Supervised by Texas Forest Service," unpublished inventory, Texas Forest Service, compiled May 28, 1944.

59. "Sets Town Erection Record," *Happy Days*, April 21, 1934, 2.

60. Joyce McKay and Mark Bruhy, "Fifield Fire Lookout Tower," National Register of Historic Places Registration Form, US Department of Interior, National Park Service, 1996.

61. "Fifield Fire Lookout Tower."

62. Conners, "Historic American Buildings Survey, North Mountain Lookout."

63. "Texas Forest Service Moves Two Fire Towers," *Silsbee Bee*, November 28, 1957, 6.

64. *Polk County Enterprise* 104.63 (August 7, 1986): 5A.

65. Memorandum between Jim Hull and Dolores S. Obee, April 20, 1993.

66. Correspondence between J. O. Burnside and O. C. Braly, April 10, 1956. The bottom portion of this letter includes an internal note indicating that Braly had worked with the TFS in the past.

67. "Plans for Lookout Towers." Memorandum between J. O. Burnside and A. D. Folweiler, discussing the "Report on Programs and Needed Facilities, 1960–1975," April 16, 1958.

68. Memorandum between J. O. Burnside and A. D. Folweiler, April 5, 1958.

69. "Plans for Lookout Towers."

70. Correspondence between O. C. Braly and J. O. Burnside, October 25, 1958.

71. Daingerfield, East Mountain, Gilmont, Latch, Leesburg, Pine Mills, Arp, Church Hill, Meadows, Salem, Starrville, Grayburg, Cairo Springs, Mauriceville, and Magnolia each had, at one time, cabs constructed by Braly.

72. Memorandum between A. D. Folweiler and District 1, June 27, 1962.

73. "Deaths," *Have You Heard* 57 (1975): 3.

74. Suiter, *Poets on the Peaks*.

75. David S. Rotenstein, "A Review of the State of Maryland's National Historic Preservation Act (Section 106) Compliance Efforts in Support of a Proposed Communications Facility at Lambs Knoll, Frederick County, Maryland," unpublished report, Federal Communications Commission File No. 0001601177, March 2, 2004.

76. Conners, *Fire Season*. 6.

77. William Oates, "Oral History Interview with Mervis Lowery," unpublished interview, Texas A&M Forest Service, January 20, 2015.

78. "Passing of an Era," *Texas Forest News* 60 (1981): 14–15.

79. "Fire Observation Towers Built," *Texas Forest News* 10.6 (1928): 1.

Chapter 4. Cussed and Discussed: The 1920s

1. "Another Mile Stone Passed," *Pineywoods Pickups* 2.4 (1941): 4.

2. "East Texas Forests Being Well Patrolled," *Timpson Weekly Times* 41.35 (August 27, 1926): 7.

3. "East Texas Forests Being Well Patrolled."

4. "Forest Fire Season—1926," *Texas Forest News* 9.3 (1927): 2.

5. "Forest Fire Season—1926," 2.

6. "Fire Situation Serious during Month of October," *Center Daily News* 1.147 (November 15, 1929): 2.

7. Billings, *A Century of Forestry*.

8. Billings, *A Century of Forestry*.

9. E. O. Siecke, H. J. Eberly, W. E. Bond, H. F. Munson, J. M. Cravey, and J. M. Turner, *Ninth Annual Report of the State Forester*, Bulletin 17 (College Station: Agricultural and Mechanical College of Texas, Department of Forestry, 1924).

10. Siecke et al., *Eleventh Annual Report of the State Forester*.

11. "East Texas Forest Fire Damage Is Decreased Last Year: Only One Per Cent of Protected Area Burned, Report Shows," *Corrigan Press* 44.9 (February 17, 1938): 1.

12. William T. Hartman, "The Texas Rangers," unpublished short story no. 48 (SS48), January 1998. This collection of short stories is in the archives at the Texas A&M Forest Service offices in College Station, Texas.

13. Billings, *A Century of Forestry*.

14. Untitled, *Texas Forest News*, May 1925, 2.

15. Untitled, *Texas Forest News*, May 1925, 2.

16. "Private Companies Erect Lookout Towers," *Texas Forest News* 8.8 (1926): 4.

17. "Lumber Firm Erects Fire Towers," *Texas Forest News* 8.9 (1926): 2.

18. "More Fire Lookout Towers for Alabama," *Texas Forest News* 9.3 (1927): 4.

19. "Patrolmen Use Lookout Trees," 2.

20. Untitled, *Texas Forest News* 6.4 (April 1924): 1.

21. Untitled, *Texas Forest News* 7.12 (December 1925): 2.

22. Siecke et al., *Eleventh Annual Report*.

23. "First Fire Tower for Texas," *Texas Forest News* 8.7 (1926): 4.

24. Untitled, *Texas Forest News*, December 1925, 2.

25. "Fire Fighting Scouts," *Texas Forest News* 9.11 (1927): 2.

26. "Hurrah for the Scouts," *Kirbyville Banner*, in "Fire Fighting Scouts," *Texas Forest News* 9.11 (1927): 2.

27. "First Fire Tower for Texas."

28. "Lookout Towers Valuable Aid in Fire Control Work," *Texas Forest News* 10.1 (1928): 1.

29. Billings, *A Century of Forestry*. State Forest Number 1 was renamed the Siecke State Forest in 1951, and State Forest Number 2 was renamed the Jones State Forest in 1949.

30. Untitled, *Texas Forest News* 7.11 (November 1925): 2.

31. "State Forest Put under Administration," *Texas Forest News* 9.4 (1927): 3.

32. "Research Work Being Conducted by Texas Forest Service," *Texas Forest News* 9.12 (1927): 2.

33. Untitled, *Texas Forest News* 1.5 (May 1925): 1.

34. Billings, *A Century of Forestry*.

35. "Kirby Donates Forest Land to A. and M," *Texas Forest News* 10.2 (1928): 1.

36. "With Apologies to Jeff O'Quinn," *Pineywoods Pickups* 2.5 (1941): 6.

37. J. M. Turner was the Kirbyville inspector; V. V. Bean was the superintendent, Kirbyville State Forest; H. J. Eberly was chief, Division of Forest Protection; and H. F. Munson was assistant state forester.

38. Siecke et al., *Eleventh Annual Report*.

Chapter 5. Come Up Sometime: The 1930s

1. "CCC Camp News," *Montgomery County News* 17 (October 4, 1934): 8. The author of the poem, Douglas Toland, worked with CCC Company 1804 in Conroe.

2. H. H. Buckles, "Moral Re-education of CCC Youth Fits Enrollees for Useful Futures," *Lufkin Daily News*, Centennial Edition, August 16, 1936, 2.

3. Cynthia Brandimarte and Angela Reed, *Texas State Parks and the CCC: The Legacy of the Civilian Conservation Corps* (College Station: Texas A&M University Press, 2013).

4. Buckles, "Moral Re-education of CCC Youth."

5. Maxwell and Baker, *Sawdust Empire*.

6. Roy Bedichek, *Adventures with a Texas Naturalist* (Austin: University of Texas Press, 1947), 67.

7. Randolph B. Campbell, *Gone to Texas: A History of the Lone Star State* (New York: Oxford University Press, 2003).

8. Francis E. Abernethy, "Foreword," in Truett and Lay, *Land of Bears and Honey*, xi–xiii.

9. M. A. Wellborn, "Texas and the CCC: A Case Study in the Successful Administration of a Confederated State and Federal Program" (master's thesis, University of North Texas, 1989), 18.

10. Wellborn, "Texas and the CCC."

11. James W. Steely, *Parks for Texas: Enduring Landscapes of the New Deal* (Austin: University of Texas Press, 1999), 16.

12. Steely, *Parks for Texas*.

13. Roosevelt proposed the idea to Congress on March 21, 1933. After negotiations, Congress returned the ECW bill to Roosevelt on March 31, 1933.

14. "Federal Government Purchases Land for Forest Purposes," *Texas Forest News* 15.3 (1933): 4.

15. Maranda Gilmore, "Civilian Conservation Corps: A Guide to Civilian Conservation Corps Camp and Enrollee Records, in the Holdings of the National Archives at Atlanta," unpublished finding guide. (Atlanta: National Archives and Records Administration, 2008).

16. Brandimarte and Reed, *Texas State Parks and the CCC*; Kenneth Hendrickson Jr., "Replenishing the Soil and the Soul of Texas: The Civilian Conservation Corps in the Lone Star State as an Example of State-Federal Work Relief during the Great Depression," *The Historian* 65.4 (2003): 801–16.

17. Wellborn, "Texas and the CCC," iv.

18. Steely, *Parks for Texas*.

19. Camps were administered through the army's "service corps area" regions. The Eighth Army Corps Area included the states of Texas, Arizona, Colorado, New Mexico, Oklahoma, and Wyoming. In Texas, the Eighth Corps Area headquarters were at Fort Sam Houston in San Antonio, and enrollees could be sent for conditioning there or at Fort Bliss, Texas; Fort Sill, Oklahoma; or Fort Bullis, Texas.

20. Hendrickson, "Replenishing the Soil and the Soul of Texas," 809.

21. Steer and Miller, *Lookouts in the Southwestern Region*. 9.

22. "CCC Money Sent Home," *Claude News* 45.39 (June 1, 1934): 8.

23. "CCC Camps Have Given Occupation to 76,804 Texans," *Mt. Pleasant Daily Times* 17.271 (October 26, 1936): 1.

24. "Federal Government Purchases Land for Forest Purposes," Internal TFS Property Inventory, Texas A&M Forest Service Archives, 3.

25. Wellborn, "Texas and the CCC."

26. *Official Annual for 1936, Lufkin District, 8th Corps Area* (Baton Rouge, Louisiana: Direct Advertising, 1936).

27. "Tyler Opens Station for Forest Service Recruits," *Tyler Journal* 8.52 (April 28, 1933): 1.

28. "Boys Leave for Forestry Camps," *Montgomery County News* 16.39 (June 1, 1933): 1.

29. *Eighteenth and Nineteenth Annual Reports*, Bulletin 25 (College Station: Agricultural and Mechanical College of Texas, Texas Forest Service, 1934), 5–6.

30. Steely, *Parks for Texas*.

31. Steely, *Parks for Texas*, 71–72.

32. *Official Annual for 1936, Lufkin District, 8th Corps Area*.

33. McKay and Bruhy, "Fifield Fire Lookout Tower."

34. Steely, *Parks for Texas*.

35. *Official Annual for 1936, Lufkin District, 8th Corps Area*, 7.

36. H. H. Buckles, "17 Camps over District Carry Out Projects: Administrative Unit Here a Major Lufkin Asset. Supervise Work: Staff of Officers and Civilians Guide Big Program," *Lufkin Daily News*, Centennial Edition, August 16, 1936, 1.

37. Wellborn, "Texas and the CCC," ii.

38. James W. Steely, "The Civilian Conservation Corps Invades East Texas, 1933–1942," *East Texas Historical Journal* 40.2 (2002): 34–39, quote on 36.

39. "Polk County Location for a Reforrestation [sic] Camp Recommended," reprinted from *Dallas News* in *Plain Dealer* 2.14 (May 11, 1933): 1.

40. "Headquarters for Forestry Army Are 312 Shepherd Ave," *Lufkin Daily News* 18.213 (July 11, 1933).

41. "Publication Texas Forest News Resumed," *Texas Forest News* 16.1 (1936): 3.

42. Between April 1933 and January 1936.

43. "Publication Texas Forest News Resumed."

44. Eric O. Siecke and R. Fechner, "Texas and National Reforestation Work," *Claude News* 44.44 (July 7, 1933): 6.

45. W. A. "Hank" Atkinson, "Camp Chatter," *Lufkin Daily News* 18.256 (August 30, 1933): 4.

46. "Tree Army Recruits Register Tuesday," *Plain Dealer* 2.35 (October 5, 1933): 3.

47. "Many Forestry Workers Plan to Re-enlist," *Orange Leader* 20.217 (September 14, 1933): 1.

48. Note the agreement with Aermotor's June 11, 1958, letter to J. O. Burnside.

49. "42 CCC Camps for Texas," *Claude News* 45.39 (June 1, 1934): 8.

50. *Official Annual for 1936, Lufkin District, 8th Corps Area*.

51. *Official Annual for 1936, Lufkin District, 8th Corps Area*, 23.

52. "To Enroll 7600: CCC Camps Will Take New Men to Fill Vacancies," *Lufkin Daily News* 19.287 (October 7, 1934).

53. "CCC to Move 13 New Camps to Texas Points," *Lufkin Daily News* 19.287 (October 7, 1934).

54. "Officials to Attend Meet: Texas Forest Chief Calls Session for Big Area in City," *Lufkin Daily News* 20.22 (November 26, 1934): 1.

55. "CCC Plans to Build in City: Expansion of Lufkin Headquarters Awaits Final Decision," *Lufkin Daily News* 19.6 (November 7, 1934).

56. "New Camp at Nancy Swells CCC Forces to over 5,000 in Area Serviced by Lufkin: Local Headquarters Now Has 22 Etex Camps under Command," *Lufkin Daily News* 20.21 (November 25, 1934).

57. "Grub for 5,000 CCC Boys Trucked across Town in Warehouse Change," *Lufkin Daily News* 20.50 (December 30, 1934).

58. "Tag $600,000,000 for CCC," *Happy Days*, March 9, 1935, 1; "Future of CCC Assured: Appropriations Bill Brought to Vote in Congress with Little Objection to Continuation and Extension of Emergency Conservation Work: Announcement of Plans for Future Expected Next Week," *Happy Days*, March 23, 1935: 1.

59. *Official Annual for 1936, Lufkin District, 8th Corps Area*, 76.

60. "CCC Wins Nation Wide Approval," *Texas Forest News* 16.8 (1936): 1.

61. Buckles, "17 Camps over District Carry Out Projects."

62. *Official Annual for 1936, Lufkin District, 8th Corps Area*.

63. "CCC Camp Set-Up Effected in Tyled [sic]," *Tyler Journal* 11.13 (July 26, 1935): 1; "4 Government Agency District Offices Are Now Located in Tyler: Citizens Are Requested to Familiarize Themselves with Them, to Help Others Having Business," *Tyler Journal* 11.17 (August 23, 1935): 1.

64. "4 Government Agency District Offices Are Now Located in Tyler"; "CCC Company to Occupy Location by September 1," *Mt. Pleasant Daily Times* 16.142 (August 19, 1935): 1.

65. These camps included SCS-25 Nacogdoches (C); F-25 Zavalla (C); SCS-27 Madisonville; SCS-20 Mt. Pleasant; F-17 Apple Springs; F-20 Kennard; F-19 Richards; F-21 Coldspring (C); F-22 Broaddus; SCS-8 Wolfe City; SCS-9 Bogata (C); SP-49 Daingerfield; P-90 Linden; SCS-23 Marshall; SCS-22 Winnsboro; SP-54 Tyler; SCS-19 Jacksonville; and F-18 Milam.

66. Sarah Jackson, "Interview with Annie M. Jimerson," unpublished interview, University of North Texas Oral History Collection no. 1139, October 2, 1996.

67. "85 CCC Camps Will Be Kept during Winter," *Mexia Weekly Herald* 37.45 (November 1, 1935): 4.

68. Buckles, "17 Camps over District Carry Out Projects."

69. "CCC Camp," *Huntsville Item*, April 2, 1936, 5.

70. Piney Woods (Weldon), Fails (Mossy Grove), Dodge, Four Notch, Bath (Possum Walk), and Farris (moved to Pool). Note that the Bath Tower was constructed before the ECW era, and the 1936 TFS annual bulletin does not credit any lookout tower construction to the Huntsville company.

71. "Maydelle CCC Camp Celebrates Third Anniversary," *Texas Forest News* 16.7 (1936): 3.

72. "Few Accidents in Texas Forest Service CCC Camps," *Texas Forest News* 16.7 (1936): 1.

73. "Texas CCC Boys Have Busy Season Ahead of Them in Pushing Forward Programs," *Tyler Journal* 12.30 (November 20, 1936): 6.

74. William P. Hardegree, "Moral Re-education of CCC Youth Fits Enrollees for Useful Futures," *Lufkin Daily News*, Centennial Edition, August 16, 1936, 4.

75. Buckles, "17 Camps over District Carry Out Projects."

76. "Lufkin CCC Motor Pool Makes Best Record in 8th Corps Area: Lauded by Officials for Efficiency and Economy," *Lufkin Daily News*, Centennial Edition, August 16, 1936, 4.

77. "Enrollees, like the Army, March on Their Stomachs: Feeding Hundreds of Hungry Camp Youths Is Giant Job but It's Done Tastily, Healthfully and Economically by Efficient Mess Officers," *Lufkin Daily News*, Centennial Edition, August 16, 1936, 2.

78. "CCC Wins Nation Wide Approval."

79. "Camps Guide Boys to Jobs: CCC Is Training Ground for Better Positions," *Lufkin Daily News*, Centennial Edition, August 16, 1936, 2.

80. Untitled, *Corrigan Press* 42.11 (March 6, 1936): 1; "Hagood Given Command of Chicago Area," *Cass County Sun* 61.17 (April 23, 1936): 2.

81. "CCC Camps Have Given Occupation to 76,804 Texans."

82. "Texas CCC Boys Have Busy Season Ahead of Them."

83. "CCC Progress Report," *Texas Forest News* 17.3 (1937): 1.

84. "Network of Telephone Lines Rapidly Spreads throughout N.E. Texas," *Rusk Cherokeean* 18.17 (November 13, 1936): 1.

85. "Texas CCC Camps Rank First in Safety First Program," *Rusk Cherokeean* 20.4 (August 19, 1938): 2.

86. L. Childs, "Coushatta Indians Help Texas Camp Celebration," *Happy Days*, June 5, 1937, 3.

87. *Twenty-Second and Twenty-Third Annual Reports of the Texas Forest Service*, Bulletin 27 (College Station: Agricultural and Mechanical College of Texas, Texas Forest Service, 1938).

88. "CCC Rock Fords Giving Good Service," *Texas Forest News* 17.4 (1937): 4.

89. *Twenty-Second and Twenty-Third Annual Reports of the Texas Forest Service*.

90. *Twenty-Second and Twenty-Third Annual Reports of the Texas Forest Service.*

91. "57 CCC Camps Will Be Operated in Texas between April 1–September 30: Camp F-13 in the Sabine National Forest in Shelby County Will Be Re-established, Reported," *Timpson Daily Times* 38.65 (April 1, 1939): 1.

92. The initials "DAA" appear on the map submitted to the regional forester. Likely, these are for David Adair Anderson, who began working with the TFS in 1936.

93. "CCC Camp Accomplishments Reviewed," *Texas Forest News* 19.9 (1939): 5.

94. "Four CCC Forestry Camps after October 1," *Texas Forest News* 19.9 (1939): 7.

95. Correspondence between W. E. White and H. M. Seaman, August 26, 1939.

96. Correspondence between W. E. White and H. M. Seaman, October 24, 1939.

97. "New Lookout Tower Erected in Newton County," *Texas Forest News* 20.3 (1940): 5.

98. "CCC Camp Will Not Move Here Until May," *Alto Herald* 39.48 (April 4, 1940): 1.

99. Billings, *A Century of Forestry*; "New Tree Nursery Established by Texas Forest Service near Alto: Annual Production of 10,000,000 Trees Planned," *Timpson Daily Times* 39.81 (April 22, 1940): 1; Walker, "Texas Forests: Their Fall and Rise."

100. R. M. Hayes, "Over Million Trees Growing on Tree Farm," *Alto Herald*, November 21, 1940, 1.

101. Workers of the Writers' Program, *Texas: A Guide to the Lone Star State.*

102. Maxwell and Baker, *Sawdust Empire*, 169.

103. Foster, Krausz, and Johnson, *First Annual Report of the State Forester.*

104. Foster, Krausz, and Johnson, *First Annual Report of the State Forester.*

105. *National Forests of the Southern Region: A Report of Progress, 1934–1954* (Atlanta: USDA Forest Service Bulletin, 1955).

106. Foster, Krausz, and Johnson, *First Annual Report of the State Forester.*

107. The bill was "an act to give the consent of the State of Texas to the acquisition by the United States Government of land in the state for national forests and parks; retaining to the state concurrent civil and criminal jurisdiction, and authorizing Congress to enact all such legislation as deemed necessary to protect and administer such national forests and parks."

108. Hyman, "A History of the Texas National Forests"; Campbell, *Gone to Texas.*

109. "A History of the Texas National Forests"; *Eighteenth and Nineteenth Annual Reports.*

110. According to Hyman, the acquisition crew consisted of C. E. Beaumont, chief of the party; Cary H. Bennet from the Bureau of the Biological Survey, chief of acquisition examination party; Russell Chipman; Sherman L. Frost; William F. Fisher; Albert Smith Jr.; John B. Fulton; Carter S. Sloan; L. B. Robinson; Ralph Morgan; and Gordon A. Hammon.

111. Hyman, "A History of the Texas National Forests."

112. Hyman, "A History of the Texas National Forests."

113. C. E. Beaumont, "Report on the Crockett Purchase Unit," unpublished report, dated November 15, 1933, and approved by the National Forest Reservation Commission March 26, 1934.

114. William Willard Ashe, "Proposed Neches (West) Purchase Unit No. 69, Texas," unpublished report, October 1923.

115. Joseph C. Kircher, "Supplement to Reports of the Houston, Crockett, Angelina, and Sabine Units, Texas," unpublished report, February 15, 1934.

116. The Tejas Unit was planned for 125,000 acres in portions of Newton, Jasper, and Orange Counties but was never realized.

117. Malouf, "Thematic Evaluation."

118. Peter L. Stark, *Names, Boundaries, and Maps: A Resource for the Historical Geography of the National Forest System of the United States*, January 30, 2021, https://forestservicemuseum.org/wp-content/uploads/2020/02/Northern-region-Feb2020.pdf.

119. Kircher, "Supplement to Reports of the Houston, Crockett, Angelina, and Sabine Units."

120. Stark, *Names, Boundaries, and Maps*.

121. Kircher, "Supplement to Reports of the Houston, Crockett, Angelina, and Sabine Units."

122. Stark, *Names, Boundaries, and Maps*.

123. Kircher, "Supplement to Reports of the Houston, Crockett, Angelina, and Sabine Units."

124. Correspondence between Joseph C. Kircher and E. O. Siecke, February 12, 1934.

125. Correspondence between E. O. Siecke and Joseph C. Kircher, April 6, 1934.

126. *National Forests of the Southern Region*.

127. Malouf, "Thematic Evaluation."

128. "Purchase of Big East Texas Timber Acreage Planned," *Montgomery County News*, March 29, 1934, 1; Hyman, "A History of the Texas National Forests"; "CCC Camps Have Given Occupation to 76,804 Texans."

129. Bishop, "Texas National Forests," 14.

130. "More Forest Land Bought in East Texas; Texas Additions to National Forests Cost Government $215,266," *Delta Courier* 54.10 (March 5, 1935): 4.

131. Bishop, "Texas National Forests," 14.

132. "Texas Purchase Units Given Permanent Status under U.S. Forest Service," *Texas Forest News* 16.12 (1936): 2.

133. Houston County Timber Company, Pickering Lumber Company, Temple Lumber Company, Delta Land and Timber Company, Long Bell Lumber Company, Trinity County Lumber Company, and Kirby Lumber Company.

134. Paul W. Schoen, "Sale Price versus Assessed Value of Timberland Purchased for Texas National Forests," memorandum, June 24, 1937.

135. "Vast Areas of Bare, Gaunt East Texas Cut-Over Lands Being Restored to Forest," *Tyler Journal* 11.9 (June 28, 1935): 7.

136. "National Parks May Be Located in East Texas," *Rusk Cherokeean* 16.12 (September 29, 1933): 6.

137. Steely, *Parks for Texas*, 39.

138. Wellborn, "Texas and the CCC," 115.

139. Wellborn, "Texas and the CCC," 115.

140. Connie Ford McCann, unpublished diary, University of North Texas Libraries, Portal to Texas History Collection, April 14, 2023, https://texashistory.unt.edu/ark:/67531/metapth121794.

141. Wellborn, "Texas and the CCC," 116.

142. Steely, *Parks for Texas*, 95.

143. *Official Annual for 1936, Lufkin District, 8th Corps Area*, 76.

144. The poem was written by an anonymous Lufkin enrollee and appeared in an untitled article in the *Lufkin Daily News* 18.289 (October 9, 1933).

145. W. A. "Hank" Atkinson, "Gorgeous, Beautiful, Plentiful," *Happy Days*, July 29, 1933, 8.

146. Atkinson, "Gorgeous, Beautiful, Plentiful."

147. "Headquarters for Forestry Army Are 312 Shepherd Ave."

148. Untitled, *Lufkin Daily News* 18.250 (August 23, 1933).

149. W. A. "Hank" Atkinson, "Camp Chatter," *Lufkin Daily News* 18.235 (August 5, 1933).

150. "Open House Held at Forestry Camp," *Lufkin Daily News* 18.239 (August 10, 1933).

151. W. A. "Hank" Atkinson, "Camp Chatter," *Lufkin Daily News* 18.240 (August 11, 1933).

152. W. A. "Hank" Atkinson, "Camp Chatter," *Lufkin Daily News* 19.34 (December 12, 1933).

153. W. A. "Hank" Atkinson, "To the Editor," *Happy Days*, October 28, 1933, 4.

154. Thomas Hool, letter to the author, November 23, 2016.

155. "Has Unique Fire Tower Record," *Happy Days*, July 21, 1934, 2.

156. "Tower-Building Outfit Records Another One," *Happy Days*, August 4, 1934, 2.

157. *Official Annual for 1936, Lufkin District, 8th Corps Area*.

158. "Four CCC Forestry Camps after October 1."

159. Much of this camp remains as the East Texas Baptist Encampment.

160. William T. Hartman, "The Charge," unpublished short story (SS43), January 1998.

161. Historical marker: Atlas Number 5507013709.

162. "National Forest Service Camp Now Located Ratcliff," *Crockett Courier*, August 30, 1934, 6.

163. *Official Annual for 1936, Lufkin District, 8th Corps Area*; "Second Tree Army Camp at Ratcliff," *Crockett Courier*, October 25, 1934, 6.

164. "CCC Enrollees Work on Development of Parks," *The J-TAC* 17.21 (March 2, 1937): 4.

165. Bobby Johnson, "Oral History Interview with Hubert F. Doss," unpublished interview, USDA US Forest Service Oral History no. 1281, October 12, 1991.

166. Bobby Johnson, "Oral History Interview with Fount Kelley," unpublished interview, USDA US Forest Service Oral History no. 1287, October 12, 1991.

167. "Jailer Shot as Felons Escape Crockett Jail: Break Occurs as Jail Keeper Enters Cell with Pair," *Mexia Weekly Herald* 39.18 (May 7, 1937): 1.

168. Bobby Johnson, "Oral History Interview with A. Coy Allen,"

unpublished interview, USDA US Forest Service Oral History no. 1286, October 12, 1991.

169. Texas Society of American Foresters, *Highlights of Texas Forestry*, undated pamphlet.

170. Walker, "Texas Forests: Their Fall and Rise."

171. Bobby Young, "The Good Old Days: Recollections of CCC," *Texas Forest News* 56.7 (1977): 8–9.

172. Walker, "Texas Forests: Their Fall and Rise," 14.

173. "Forest Service Fire Observation," *Crockett Courier* 39.12 (April 12, 1928): 1.

174. "Location and Field Notes," field notebook, Texas A&M Forest Service Archives, 13.

175. Correspondence between W. E. White and E. O. Siecke, April 23, 1938.

176. Correspondence between E. O. Siecke and W. E. White, April 28, 1938.

177. Correspondence between B. B. Whipple and W. E. White, June 8, 1938.

178. *Twentieth and Twenty-First Annual Reports*, Bulletin 26 (College Station: Agricultural and Mechanical College of Texas, Texas Forest Service, 1936), 12.

179. *Official Annual for 1936, Lufkin District, 8th Corps Area*, 25.

180. "Memorandum of Understanding between the Texas Forest Service and the United States Forest Service," December 7, 1949.

181. "Ratcliff Lookout Tower Saves Place," *Crockett Courier* 58.41 (October 9, 1947): 1.

182. "Aerial Detection Feasibility Study—Texas," memorandum from John Olson, assistant forest supervisor, to district rangers, June 16, 1971.

183. "CCC Camp News," *Montgomery County News*, August 22, 1935, 4.

184. "Tree Army for the County," *Montgomery County News*, June 1, 1933, 1.

185. "Tree Army Unloaded Today," *Montgomery County News*, June 15, 1933, 1.

186. "Tree Army Getting Settled," *Montgomery County News*, June 22, 1933, 1.

187. "CCC Camp News," *Montgomery County News*, October 4, 1934, 2.

188. "Road Project Is Educational Aid," *Happy Days*, June 2, 1934, 2.

189. "Texas Forest Service Supervision of CCC Netted Montgomery County Large Gains: County Map Dotted Today with Improvements of State Objectives," *Montgomery County News*, February 14, 1935, 1.

190. "Conroe CCC Has Fine Fire Fighting Record," *Montgomery County News*, January 11, 1934, 1.

191. "CCC Camp News," *Montgomery County News*, January 17, 1935, 8.

192. "CCC Camp News," *Montgomery County News*, January 24, 1935, 5.

193. "Educational Aim at Camp Proves Benefit to All: Brief Outline Shows That Enrollees Are Offered Best Facilities," *Montgomery County News*, January 31, 1935, 5.

194. *Official Annual for 1936, Lufkin District, 8th Corps Area*.

195. "Texas Forest Service Supervision of CCC Netted Montgomery County Large Gains." The article indicates four 100-foot steel lookout towers were constructed by Conroe enrollees. While the towers are not identified, Splendora, Keenan, Willis, and Farris can be assumed based on distance from camp and the relative construction dates for these four towers.

196. "Work Carries Men Far Afield," *Happy Days*, April 14, 1934, 2.

197. "Maydelle CCC Camp Closed," *Texas Forest News* 18.2 (1938): 2.

198. "The Maydelle CCC Camp," *Rusk Cherokeean* 17.50 (July 3, 1936): 2.

199. Correspondence between D. J. Fox and P. Ebarb, February 17, 1989.

200. "New Jacksonville Tower Attracts Many Visitors," *Texas Forest News* 16.10 (1936): 1.

201. Workers of the Writers' Program, *Texas: A Guide to the Lone Star State*.

202. "News about Forestry Personnel," *Texas Forest News* 38.4 (1959): 7.

203. Tim Parker, "Forester Firewatcher Is Mistaken for St. Nick," *Sweetwater Reporter* 58.280 (November 27, 1955): 7.

204. "Tower Repair," memorandum from District 1 forester to head, Fire Control Department, October 18, 1957.

205. "Forest Fire Fighters Hold Maydelle Meeting," *Rusk Cherokeean* 97.32 (September 16, 1943): 5.

206. "Rusk-Maydelle Chapters Visit Forest Station," *Rusk Cherokeean* 105.42 (April 16, 1953): 1.

207. W. A. "Hank" Atkinson, "What Is Success?," *Happy Days*, March 24, 1934, 5. Atkinson worked in Company 838 in Lufkin.

208. "CCC Workers Seek Education," *Tyler Journal* 10.9 (June 29, 1934): 15.

209. "Educational Program at CCC Camps," *Timpson Daily Times* 34.77 (April 17, 1935): 1.

210. "Ten Livingston CCC Boys Attend Local High School," *Tyler Journal* 11.24 (October 11, 193): 5.

211. "Brainstorming in the CCC," *Happy Days*, October 13, 1934, 7.

212. "Forestry Personnel Aid CCC Educational Program," *Texas Forest News* 16.8 (1936): 1.

213. "Obituary—Dwight Thurman Smith, Sr.," *Lufkin Daily News*, September 8, 1998, https://www.findagrave.com/memorial/88585389/dwight-thurmond-smith.

214. H. H. Buckles, "CCC Educational Program One of Major Phases of Camp Life," *Lufkin Daily News*, Centennial Edition, August 16, 1936, 1.

215. With two exceptions, each camp has a full-time adviser. One "white" camp had the post temporarily vacant through resignation of the adviser to accept a better position, while the two "colored" camps at Lufkin and Nacogdoches were served by the same adviser.

216. "Camp Chatter: News and Gossip from the Lufkin CCC Encampment," *Lufkin Daily News* 19.102 (March 2, 1934). Though the tower is not named in the article, the location is described as "near Nacogdoches." Given location and date, Fern Lake is assumed.

217. Thomas Hool, letter to the author, November 23, 2016.

218. R. F. Balthis, *Forestry Lessons for Students in Vocational Agriculture in Texas: A Forestry Text for Instructors and Students*, Bulletin 24 (College Station: Agricultural and Mechanical College of Texas, Texas Forest Service, 1934).

219. *Eighteenth and Nineteenth Annual Reports*, 19.

220. Utley, "With the Yalies in the Deep Woods," 133–51.

221. Correspondence between W. E. White and Allen M. Riley, August 19, 1937.

222. "Texas Forest Service to Get New Radio Tower," *Silsbee Bee* 40.44 (January 8, 1959): 7.

223. Correspondence between A. D. Folweiler and F. J. Kanatzar, November 8, 1955.

224. Memorandum between head, Fire Control Department, and district foresters, May 9, 1962.

225. Gerald, "52 Steel Towers," 8.

226. "Check 'Blind Spots' on National Forests," *Hereford Brand* 36.38 (September 24, 1936): 4.

227. *Twentieth and Twenty-First Annual Reports*.

228. These were towers at Yellowpine and Ratcliff. Both were erected in 1928. The area around these facilities was absorbed by the Sabine and Davy Crockett National Forests.

229. Correspondence between B. B. Whipple and W. E. White, June 8, 1938.

230. Apple Springs Lookout was moved to Mt. Zion in 1939 during a reorganization of the Davy Crockett National Forest network.

231. Correspondence between B. B. Whipple and W. E. White on June 8, 1938, indicates that the Willow Springs Tower was still on privately owned land. In this letter, Whipple requested that the TFS maintain its lease for the lookout until a new lease was negotiated. White communicated with the TNFs on March 16, 1939, inquiring into the status of those negotiations. A week later Forest Supervisor P. F. W. Prater wrote to White indicating that there was "some question to the actual ownership of the land" on which the tower sat and advised that the USFS would provide an update to TFS as soon as possible. It was not until December 13, 1939, that Whipple wrote to White to inform him that a new lease existed between the government and landowner and recommended that the TFS could cancel its property lease. A July 22, 1943, letter between E. E. Wagner and J. O. Burnside summarizes the conditions of the government lease, which was subject to yearly renewal. At the time of letter, Wagner informed Burnside that the lease was renewed as late as June 30, 1943, "but when we recently sent a voucher to the Carey Land & Development Company to cover the rental from July 1, 1942 to June 30, 1943, we were advised that the surface of this land had been sold to the Sabine Lumber & Refining Co. on July 1, 1942, and that the latter company had transferred their title to the Sabine Land & Cattle Company." At about the same time, the USFS returned the tower to the TFS. J. O. Burnside then reached out to the Sabine Land & Cattle Company on July 24, 1943, and was able to secure a lease by August 2, 1943.

232. Lease between J. B. Stanton and the TFS, signed September 2, 1933, Texas A&M Forest Service Archives.

233. Correspondence between Eunice M. and C. A. Bryan to W. A. Norman for A. D. Folweiler, January 7, 1961.

234. "Twenty–Forty Years Ago: From the Files of the Clarksville Times," *Clarksville Times* 85.7 (March 1, 1957): 7.

235. Red River, Bowie, Morris, Titus, Camp, Upshur, Cass, Marion, Harrison, Gregg, Rusk, and Panola Counties.

236. "Organized Fire Protection Comes to Northeast Texas," *Texas Forest News* 16.4 (1936): 1.

237. "Locations and Field Notes: Texas Forest Service Division of Forest Protection," unpublished field notebook, Texas A&M Forest Service Archives.

238. Correspondence between R. C. Lane and W. E. White, March 27, 1936.

239. Correspondence between W. E. White and R. C. Lane, March 30, 1936.

240. Locations and Field Notes: Texas Forest Service Division of Forest Protection."

241. "Organized Fire Protection Comes to Northeast Texas."

242. "A Summary of Local News Happenings: Twenty–Forty Years Ago from the Files of the Clarksville Times," *Clarksville Times* 88.31 (August 20, 1954): 9.

243. "A Summary of Local News Happenings: Twenty–Forty Years Ago from the Files of the Clarksville Times," *Clarksville Times* 83.13 (April 6, 1956): 9.

244. "A Summary of Local News Happenings: Twenty–Forty Years Ago from the Files of the Clarksville Times," *Clarksville Times* 86.34 (September 12, 1958): 7.

245. "Network of Telephone Lines Rapidly Spreads throughout Northeast Texas," *Texas Forest News* 16.11 (1936): 3.

246. Correspondence between J. O. Burnside and W. E. White, December 23, 1942.

247. Correspondence between J. O. Burnside and Don Young, January 4, 1942.

248. Correspondence between Don Young and J. O. Burnside, January 6, 1943.

249. "Fire Fighting Instructions," *The Torch*, March 1939, 4.

250. McCann, unpublished diary.

251. *Eighteenth and Nineteenth Annual Reports*, 17.

252. W. A. "Hank" Atkinson, "Camp Chatter," *Lufkin Daily News*, September 8, 1933.

253. "Over Million Trees Growing on Tree Farm."

254. Billings, *A Century of Forestry*.

255. "Planted Pine Trees at Conroe," *Rockdale Reporter and Messenger* 58.25 (August 7, 1930): 14.

256. Billings, *A Century of Forestry*. Billings explains the history of nurseries at Siecke and Jones State Forests and provides the information on capacity.

257. "South's Largest Tree Planting Program Has Been Completed by CCC," *Tyler Journal* 12.14 (July 31, 1936): 3.

258. "Federal Reforestation," *Bonham Herald* 10.77 (May 24, 1937): 3.

259. Baker, "Timbered Again."

260. "CCC Finishes Planing [sic] of Six Million Trees," *Rusk Cherokeean* 20.30 (February 17, 1939): 1.

261. "Over Million Trees Growing on Tree Farm."

262. "Over Million Trees Growing on Tree Farm."

263. Texas Forest Service, *Seventeenth Annual Report*, 14.

264. "CCC Camp News," *Montgomery County News*, October 11, 1934, 8.

265. "Forestry Service to Erect Watch Tower Southeast of Timpson," *Timpson Daily Times* 34.241 (December 4, 1935): 1.

266. "Forestry Service to Erect Watch Tower Southeast of Timpson."

267. Texas Forest Service Supervision of CCC Netted Montgomery County Large Gains."

268. "Engineering Class Being Taught Now at J'Ville CCC Camp," *Rusk Cherokeean* 18.10 (September 25, 1936): 3.

269. William T. Hartman, "The CCC Camp," unpublished short story (SS40), January 1998.

270. "Forests of East Texas Being Mapped," *Texas Forest News* 20.5 (1940): 5.

271. Green, "Oral History Interview with Johnny Columbus and Columbus Lacey."

272. Locations and Field Notes: Texas Forest Service Division of Forest Protection."

273. "Type Mapping of East Texas Forests Progressing Rapidly," *Texas Forest News* 17.10 (1937): 2.

274. "Timber Type Survey Progressing," *Texas Forest News* 18.1 (1938): 1.

275. Cruikshank, *Forest Resources of Northeast Texas*; Cruikshank and Eldridge, *Forest Resources of Southeastern Texas*.

276. "Forests of East Texas Being Mapped."

277. "Timber Mapping Covers Two-Thirds of East Texas," *Texas Forest News* 21.1 (1941): 3.

278. Harvey H. Kaiser, *Great Camps of the Adirondacks*, 7th ptg. (Hong Kong: South China Printing, 2003); Malouf, "Thematic Evaluation."

279. Steely, *Parks for Texas*.

280. *Eighteenth and Nineteenth Annual Reports*, 25.

281. Ippolito, *Cultural Resource Overview* (emphasis added).

282. Mark F. DeLeon, *Cultural Resources Overview for the National Forests in Mississippi* (Jacksonville, FL: USDA Forest Service, 1983).

283. Jeffrey A. Owens, "Placelessness and the Rationale for Historic Preservation: National Contexts and East Texas Examples," *East Texas Historical Journal* 43.2 (2005): 3–31.

284. Mark D. Spence, "Rankin Ridge Lookout Tower, Wind Cave National Park: Assessment of Qualification for Listing on National Register of Historic Places," unpublished report, 2010.

285. Malouf, "Thematic Evaluation."

286. Malouf, "Thematic Evaluation." The Mark Twain National Forest is in USDA Forest Service Region 9.

287. Steer and Miller, *Lookouts in the Southwestern Region*.

288. The Four Notch dwelling appears to stand in private ownership across the road from the original guard compound.

289. Alison T. Otis, William D. Honey, Thomas C. Hogg, and Kimberly K. Lakin, *The Forest Service and the Civilian Conservation Corps, 1933–42*, USDA Forest Service Report 395 (USDA Forest Service, 1986), https://www.nps.gov/parkhistory/online_books/ccc/ccc/index.htm.

290. Maxwell and Baker, *Sawdust Empire*, 98.

291. Correspondence between J. E. McPherson and Texas National Forest supervisor, December 15, 1958.

292. Groben, *Acceptable Plans*.

293. John R. Grosvenor, *A History of the Architecture of the USDA Forest Service*. USDA Forest Service Report EM-7310-8 (Washington, DC: USDA Forest Service, 1999).

294. Groben, *Acceptable Plans*.

295. Groben, *Acceptable Plans*.

296. Groben, *Acceptable Plans*.

297. Knox B. Ivie, "Forestry News of Richards CCC," *Navasota Daily Examiner* 39.20 (March 17, 1937): 1.

298. *Official Annual for 1936, Lufkin District, 8th Corps Area*.

299. "Forestry News of Richards CCC."

300. Ivie, "Forestry News of Richards CCC."

301. Memorandum between Texas National Forests supervisor and assistant regional forester, May 21, 1962.

302. Memorandum between Raven District ranger and forest supervisor, October 29, 1964.

303. "Survived Two Disasters: Finest Timber in East Texas Is Not for Sale," *Pine Needle* 4.29 (July 20, 1967): 5.

Chapter 6. Chestley Dickens and Betty Huffman: The 1940s

1. Naomi Cromeens, "Alsey's Fire—as told by Cromeens," Texas Forest Service employee newsletter, March–April 1940, 6.

2. *Twenty-Fourth and Twenty-Fifth Annual Reports*, Bulletin 30. (College Station: Agricultural and Mechanical College of Texas, Texas Forest Service, 1940).

3. "New Lookout Tower Erected in Newton County," 5.

4. "New Protection Unit in Tyler County: Landowners Are Contributing Half of Fire Protection Costs," *Texas Forest News* 21.9 (1941): 4.

5. *Texas Forest Progress, 1941–1942: 26th and 27th Annual Reports*, Bulletin 31 (College Station: Agricultural and Mechanical College of Texas, Texas Forest Service, 1942).

6. *Texas Forest Progress, 1941–1942: 26th and 27th Annual Reports*.

7. Correspondence between W. E. White and William H. Parker, December 15, 1945.

8. "May Use Emergency Fire Fighting Crews: Lumber Company and Volunteer Assistance Studied," *Texas Forest News* 22.7 (1942): 2.

9. J. O. Moosberg, "Forest Fire Fighter Service to Be Organized," *Timpson Daily Times* 41.230 (November 21, 1942): 4.

10. "Forest Fire Fighter Service to Be Organized."

11. "Toyko Loves an American Forest Fire!," advertisement, *Rusk Cherokeean* 96.43 (November 26, 1942): 3.

12. "Shelby Gets Fire Tools to Help Fight Forest Fires," *Timpson Daily Times* 42.240 (November 29, 1943): 1.

13. Anderson, "History of the Texas Forest Service."

14. Anderson, "History of the Texas Forest Service."

15. William T. Hartman, "World War II and the Texas Forest Service," unpublished short story (SS55), January 1998.

16. Anderson, "History of the Texas Forest Service."

17. "Forest Areas to Get Air Patrols," *Clarksville Times* 72.7 (March 3, 1944): 4.

18. William T. Hartman and P. R. Wilson, "Use of Aircraft in Forest Protection in the South," unpublished report, Texas Forest Service, May 1948.

19. Joseph O. Burnside and Nort Baser, "Fire! Texas Forest Threat," *Texas Forest News* 27.3 (1948): 3.

20. "Plans Made at Woodville Meeting for Concentrated Aerial Patrol," *Silsbee Bee*, October 11, 1945, 6.

21. Correspondence between Don Young and W. E. White, September 23, 1942.

22. Correspondence between W. E. White and Frank Quinn, September 16, 1942.

23. Correspondence between Frank Quinn and W. E. White, November 16, 1942.

24. "Texas Forest Service, Fire Control Department Forest Fire Lookout Towers," December 21, 1951.

25. Correspondence between Don Young and B. Winston, September 25, 1941.

26. Correspondence between TFS and J. C. Roak, June 16, 1941.

27. Correspondence between Don Young and W. E. White, August 8, 1941.

28. Correspondence between Don Young and W. E. White, August 17, 1941.

29. Correspondence between the TFS and chief of ordnance, War Department, September 18, 1941.

30. Correspondence between E. O. Siecke and W. E. White, September 25, 1941.

31. Correspondence between TFS and chief of ordnance, September 18, 1941.

32. Correspondence between E. O. Siecke and W. D. McFarlane, September 18, 1941.

33. Correspondence between E. O. Siecke and W. E. White, September 29, 1941.

34. Correspondence between Don Young and W. E. White, October 21, 1941.

35. Correspondence between W. E. White and E. O. Siecke, November 5, 1941.

36. "Changes of Personnel," *Pineywoods Pickups* 3.1 (1942): 2.

37. Correspondence between J. O. Burnside and Chester Johnson, March 7, 1944.

38. "Fire Lookout Towers in East Texas to Be Reorganized," *Sweetwater Reporter* 51.179 (July 28, 1948): 7.

39. National Geodetic Survey, National Coast and Geodetic Survey Data Sheet DM1239, accessed September 22, 2024, https://www.ngs.noaa.gov/cgi-bin/ds_mark.prl?PidBox=DM1239.

40. Hartman, "World War II and the Texas Forest Service."

41. Anderson, "History of the Texas Forest Service."

42. "Citizenship-Loyalty Affidavit," signed by John W. Huffman, May 7, 1942.

43. Correspondence between E. O. Siecke and John W. Huffman, June 16, 1942.

44. Correspondence between E. O. Siecke and John W. Huffman, June 16, 1942.

45. Memorandum between J. O. Burnside and Patrolmen Huffman, Johnson, and Pitts, August 28, 1942.

46. *Pineywoods Pickups* 2.5 (1941): 2.

47. "General Circular Letter—1943: AWS Work Changes," written by J. O. Burnside, January 18, 1943.

48. Suiter, *Poets on the Peaks*, 50.

49. "Appointment of Mrs. Mabel Pitts—Emergency Lookout Hyatt State Forest," signed by Mabel Pitts, February 2, 1943.

50. "Appointment Form for Forest Patrolman State Forest & Blanket Patrol," signed by Enoch M. Pitts, August 12, 1937.

51. Correspondence between J. O. Burnside and W. E. White, November 6, 1942.

52. "Appointment of Mrs. Mabel Pitts."

53. Correspondence between Enoch M. Pitts and W. E. White, May 4, 1943.

54. "Application for Position in Texas Forest Service, Division of Forest Protection, Lufkin, Texas," completed by Lettie Hughes, November 18, 1943.

55. Correspondence between Lettie Hughes and J. O. Burnside, November 27, 1943.

56. "Appointment of Lettie Hughes—Emergency Lookout—Shortleaf Type," signed by Lettie Hughes, November 27, 1943.

57. "Safety at Ladder Towers," Texas Forest Service, staff meeting minutes, May 12, 1959.

58. "Appointment of Mrs. Betty A. Huffman—Lookout Duty State Forest Tower," signed by Betty A. Huffman, February 9, 1943.

59. "Emergency Forest Patrolman's Circular Letter—Notice of Appointment and Instructions," addressed to J. W. Huffman, February 1930.

60. "Appointment of Mrs. Betty A. Huffman."

61. *Texas Forestry Progress 1943–1944: 28th and 29th Annual Reports*; "Service Selling Telephone Lines," *Texas Forest News* 29.2 (1950): 2.

62. "Fire Lookout Towers in East Texas to Be Reorganized."

63. Billings, *A Century of Forestry*.

64. Chapman, "An Administrative History."

65. Correspondence between J. O. Burnside and W. H. Parker, July 20, 1945.

66. Correspondence between W. E. White and W. H. Parker, December 15, 1945.

67. Billings, *A Century of Forestry*.

68. "Burnside to Retire," *Texas Forest News* 50 (Spring 1971): 6.

69. "Retired," *Have You Heard* 30 (1971): 12; Young, "The Good Old Days," 8–9.

70. Young, "The Good Old Days," 8–9.

71. "Retired."

72. Correspondence between J. O. Burnside and F. F. Neely, July 28, 1943.

73. "Burnside to Retire."

74. "Appointment Form for Lookoutman for Patrol Districts Organized within the Ice Damage Area," addressed to Neal W. Kincel, effective April 1, 1944.

75. "Appointment of Barney S. Parrish—Lookoutman Hortense Lookout Tower PU#6," signed by Barney Parrish, March 13, 1944.

76. "Appointment of Cleo C. Horn—Forest Patrolman—Longleaf Type," signed by Cleo C. Horn, August 16, 1943.

77. Frost, "Operations Conroe."

78. "Fire Loss at Conroe Worst in TFS Annals," *Texas Forest News* 27.1 (1948): 9.

79. "Forest Fires Rank No. 7 Story in '48," *Texas Forest News* 27.1 (1948): 8.

80. "Fire! Texas Forest Threat."

81. "Fire! Texas Forest Threat."

82. Chapman, "An Administrative History."

83. "W. E. White Leaves T.F.S.; S. L. Frost Acting Director." *Texas Forest News* 27.3 (1948): 2.

84. "Fire! Texas Forest Threat."

85. "Fire Rages in Texas Forests," *Rusk Cherokeean* 101.14 (October 7, 1948): 7.

86. "Less Forest Fires during Last Year," *Rusk Cherokeean* 102.33 (February 16, 1950): 7.

87. "Fire Jeep," *Buzz Saw*, November 30, 1948.

88. "More Funds for Forest Protection," *Rusk Cherokeean* 101.7 (August 19, 1948): 5.

89. "More Funds for Forest Protection."

90. Correspondence between C. H. Hebert and W. E. White, April 1, 1937.

91. Correspondence between J. O. Burnside and George C. Reed, July 26, 1948.

92. Anderson, "History of the Texas Forest Service."

93. Memorandum from Angelina district ranger to TNF supervisor, April 10, 1959.

94. Memorandum from Yellowpine district ranger to TNF forest supervisor, April 9, 1959.

95. "New Figures Out on Telephone Line Mileage," *Pineywoods Pickups* 2.4 (1941): 4.

96. "Maggie's Gossip Column," *Pineywoods Pickups* 2.4 (1941): 2.

97. *Twenty-Second and Twenty-Third Annual Reports of the Texas Forest Service*.

98. "Forest Service Telephone Lines Hit by Hurricane," *Texas Forest News* 21.9 (1941): 6.

99. "Supplemental Duties and Responsibilities for Texas Forest Service Patrolmen within the Ice Damage Area," appointment letter signed by William P. James, May 12, 1944.

100. "Fire! Texas Forest Threat."

101. Anderson, "History of the Texas Forest Service."

102. "Service Selling Telephone Lines."

103. "Radio Helps Cut Forest Fire Loss," *Rusk Cherokeean* 101.49 (June 23, 1949): 7.

104. *Texas Almanac and State Industrial Guide*.

105. "Radio Will Help to Prevent Forest Fires," *Palacios Beacon* 42.13 (March 31, 1949): 6.

106. "Radio Helps Cut Forest Fire Loss."

107. "Memorandum of Understanding between the Texas Forest Service and the United States Forest Service," December 7, 1949.

108. "Memorandum of Understanding between the Texas Forest Service and the United States Forest Service," December 7, 1949.

109. Billings, *A Century of Forestry*.

110. A. D. Folweiler and A. A. Brown, *Fire in the Forests of the United States* (St. Louis: John H. Swift, 1946).

111. Patrick Ebarb, letter to the author, July 26, 2023.

112. "Organizational Plan, Texas Forest Service: Effective Date 14 February 1949," Texas Forest Service, staff meeting minutes, February 8, 1949.

113. "Organizational Plan, Texas Forest Service."

114. Billings, *A Century of Forestry*.

115. Texas Forest Service, staff meeting minutes, July 27, 1950.

116. "U.S. Forest Service Surplus Tower," memorandum between A. D. Folweiler and J. O. Burnside, September 3, 1957.

117. Memorandum between J. O. Burnside and A. D. Folweiler, January 17, 1963.

118. Memorandum between A. D. Folweiler and J. O. Burnside, January 18, 1963.

119. Memorandum between A. D. Folweiler and J. O. Burnside, January 21, 1963.

120. Texas Forest Service, staff meeting minutes, November 16–17, 1965.

121. Ebarb, letter to the author, July 26, 2003.

Chapter 7. Guardians of the Pineywoods: The 1950s

1. Herbert F. Hare, "A Man Looks over His Kingdom," *Have You Heard* 6 (1964): 3. Hare served at the Piney Woods (Weldon) Lookout.

2. Chester O'Donnell, author and ed., *Bob Jackson, Guardian of the Pineywoods*, photography by Jim Felts. This undated film was produced by the TFS sometime during the early 1950s. The earliest reference to the motion picture appears on page 12 of the December 8, 1953, TFS meeting minutes. In them, the director requested the name and effectiveness of all films available to the TFS for education. The title *Bob Jackson* is listed with a comment that the film is "good."

3. "Guardians of the Pineywoods," *Alto Herald*, no. 10 (August 15, 1957): 1.

4. "The Fire Control Department," *Texas Forest Service Spotlight* 8 (1955): 1.

5. Patrick Ebarb, "Aerial Detection of Forest Fires in Texas," *Texas Forest News* 59 (1980): 16–17.

6. "Help Wanted," *Timpson Times* 79.44 (October 30, 1964): 4.

7. "The Fire Control Department."

8. Ebarb, "Aerial Detection of Forest Fires in Texas."

9. Ebarb, "Aerial Detection of Forest Fires in Texas."

10. Texas Forest Service, staff meeting minutes, May 16–17, 1962.

11. "Proposed Outline of Positions, Duties, and Salaries: Texas Forest Service Personnel, January 1950," Texas Forest Service, staff meeting minutes, December 6, 1949.

12. Texas Forest Service, staff meeting minutes, February 14–15, 1957.

13. Hyman, "A History of the Texas National Forests."

14. "Proposed Outline of Positions, Duties, and Salaries."

15. Hyman, "A History of the Texas National Forests."

16. Texas Forest Service, staff meeting minutes, May 16–17, 1962.

17. "Forestry Man in This District Is Recognized," *Rusk Cherokeean* 109.2 (July 5, 1956): 8.

18. "Recognition to Crewmen," Texas Forest Service, staff meeting minutes, August 13, 1957.

19. "Highway Fires Proving Smokers Carelessness," *Alto Herald*, no. 13 (September 9, 1954): 8.

20. "Floy M. Creel Tells Pct. 2 Fire Safety," *Rusk Cherokeean* 110.46 (May 8, 1958): 2.

21. "Tid Bits," *Have You Heard* 1 (April 1963): 3.

22. "Forestry Man in This District Is Recognized," 8.

23. Correspondence between A. D. Folweiler and H. F. Peck, May 16, 1963.

24. "Nine Texas Forest Service Men Retire," *Texas Forest News* 36.2 (1957): 3.

25. "Locations and Field Notes: Texas Forest Service Division of Forest Protection."

26. "Nine Texas Forest Service Men Retire."

27. "Crewleader Kelley with 28 Years Service Witnesses Many Changes in Equipment," *Texas Forest News* 34.4 (1955): 4.

28. Correspondence between J. O. Burnside and Claud C. Purvis, August 8, 1957.

29. "Falls from Tower," *Sweetwater Reporter* 53.71 (March 24, 1950): 8.

30. "Forest Tower Fall Is Fatal," *Timpson Weekly Times* 65.13 (March 31, 1950): 13.

31. "News about Forestry Personnel," *Texas Forest News* 37.4 (1958): 3.

32. "Evadale Man Dies of Sudden Heart Attack Wednesday," *Silsbee Bee* 40.25 (August 28, 1958): 1.

33. "Fox Trees Towerman," *Texas Forest News* 34.5 (1955): 7.

34. "Tid Bits," *Have You Heard* 5 (April 1964): 3.

35. "Employment of Crewleaders," Texas Forest Service, staff meeting minutes, November 20–21, 1956.

36. "Employment of Very Young Men as Crewmen," Texas Forest Service, staff meeting minutes, August 31–September 1, 1961.

37. Memorandum between J. O. Burnside and District 5 forester, April 15, 1966.

38. Memorandum between District 5 forester and J. O. Burnside, April 18, 1966.

39. "Employment of Women as Lookouts," Texas Forest Service, staff meeting minutes, October 31, 1950.

40. "Texas Woman Says Her Job Lonely," *Timpson Weekly Times* 69.49 (December 3, 1954): 3.

41. Memorandum between Edwin Dale and head, Fire Control Department, December 20, 1951.

42. Memorandum between District 1 forester and head, Fire Control Department, October 30, 1958.

43. Anderson, "History of the Texas Forest Service."

44. "Forest Fire Losses in East Texas Show Decline in Last Year," *Wood County Record* 23.29 (October 20, 1953): 3.

45. "Ways to Reduce the Time Spent by Crewleaders on Towers," Texas Forest Service, staff meeting minutes, December 8–10, 1953.

46. "Use of Airplanes and Lookout Towers, When to Use Which," Texas Forest Service, staff meeting minutes, December 7–9, 1954.

47. "Memorandum of Understanding between the Texas Forest Service and the United States Forest Service," December 9, 1953.

48. "Buildings, Water and Sanitation," memorandum between acting forest supervisor TNFs and Tenaha district ranger, November 25, 1959.

49. Texas Forest Service, *Seventeenth Annual Report*.

50. Memorandum between J. O. Burnside and Director Folweiler, March 7, 1958.

51. Ebarb, "Aerial Detection of Forest Fires in Texas."

52. "1951–53 Appropriations Are Announced for Texas Forest Service," *Texas Forest News* 30.4 (1951): 5.

53. "Locations and Field Notes: Texas Forest Service Division of Forest Protection."

54. National Geodetic Survey, National Coast and Geodetic Survey Data Sheet BL2109, September 26, 2024, https://www.ngs.noaa.gov/cgi-bin/ds_mark.prl?PidBox=BL2109.

55. Correspondence between J. O. Burnside and O. C. Braly, March 20, 1961.

56. Correspondence between O. C. Braly and J. O. Burnside, March 29, 1961.

57. The new towers were erected at Salem, Mauriceville, Magnolia, Daingerfield, Cairo Springs, and Gilmer. They varied in height from 87 to 114 feet.

58. "Towers Are Built from Oil Derricks," *Texas Forest News* 31.5 (1952): 2.

59. Memorandum between District 3 forester and J. O. Burnside, November 25, 1957.

60. Memorandum between District 3 forester and J. O. Burnside, November 25, 1957.

61. Memorandum between Director Folweiler and J. O. Burnside, May 27, 1958.

62. "Candy Hill Tower Lease," note to director regarding the Candy Hill Lookout, April 3, 1964.

63. Memorandum between J. O. Burnside and Director Folweiler, May 31, 1961.

64. "Contract No. 2172 with Moon Builders Supply of Overton, Texas, for Relocating Candy Hill Lookout Tower to New Site Near Grapeland, Texas," Internal TFS document, College Station, Texas.

65. "Proposed Sale of Texas National Forests," Texas Forest Service, staff meeting minutes, December 8–10, 1953.

66. "Proposed Sale of Texas National Forests."

67. Sherman L. Frost, *Texas National Forest Study* (Longview: East Texas Chamber of Commerce, 1954).

68. Frost, *Texas National Forest Study*.

69. Stark, *Names, Boundaries, and Maps*.

70. An act for the relief of unemployment through the performance of useful public work and for other purposes.

71. Stark, *Names, Boundaries, and Maps*.

72. Stark, *Names, Boundaries, and Maps*, 20.

73. Frost, *Texas National Forest Study*, 5.

74. Frost, *Texas National Forest Study*.

75. "Agreeing and Consenting for Federal Government to Purchase Lands

in Texas for National Parks or National Forests, S.C.R. No. 73," in Frost, *Texas National Forest Study*, "Exhibit C," 43.

76. Suiter, *Poets on the Peaks*, 38.

77. Jack Kerouac, *Desolation Angels*, introduction by Joyce Johnson (New York: Riverhead Books, 1995), 48.

78. Frost, *Texas National Forest Study*, 1.

79. Frost, *Texas National Forest Study*, 2.

80. Frost, *Texas National Forest Study*, IX.

81. USDA Forest Service, *Texas National Forests: Facts about the Forest* (Atlanta: USDA Forest Service, ca. 1955).

82. Johnie Page patrolman appointment, signed September 1, 1945.

83. Stark, *Names, Boundaries, and Maps*.

84. Frost, *Texas National Forest Study*, 11.

Chapter 8. Unit 308: The 1960s

1. The earliest reference for crewleader diaries is in the September 1954 staff meeting minutes. Green felt "it desirable if his crewleaders were supplied with diaries of the kind he is required to use." The director agreed to purchase a number of 1955 diaries.

2. E. O. Lowery, crewleader diaries for 1960 and 1964, Texas A&M Forest Service Archives.

3. "Tid Bits," *Have You Heard* 14 (1966): 6.

4. Oates, "Oral History Interview with Mervis Lowery."

5. National Geodetic Survey, National Coast and Geodetic Survey Data Sheet BY0469, April 17, 2023, https://www.ngs.noaa.gov/cgi-bin/ds_mark.prl?PidBox=BY0469.

6. "Request for Sale of Surplus Lookout Towers," memorandum between Bruce Miles and Patrick Ebarb, January 12, 1981.

7. Correspondence between S. L. Lindsey and C. Wells, July 31, 1984.

8. "Disposition of Diaries," Texas Forest Service, staff meeting minutes, November 15–16, 1960.

9. "List of Items Supplied to Crewleaders," Texas Forest Service, staff meeting minutes, August 13, 1957.

10. Lowery, crewleader diaries for 1960 and 1964.

11. Lowery, crewleader diaries for 1960 and 1964.

12. "The Place of Lookout Towers in Forest Fire Detection," Texas Forest Service, staff meeting minutes, February 13–14, 1962.

13. "The Place of Lookout Towers in Forest Fire Detection."

14. "A Report on a Visit to Texas Forest Service Watchtower at District No. 6 Headquarters South of Conroe," unpublished report, Texas Forest Service, August 1961.

15. The Jones Tower was inspected by R. H. Bloom, Tallie H. Cotton, D. W. Fate, and Zeb Rabon on August 29, 1961.

16. "Fire Lookout Tower Completed at State Forest near Conroe," *Rusk Cherokeean* 13.20 (November 20, 1931): 1.

17. Texas Forest Service, *Seventeenth Annual Report*, 32.

18. "A Report on a Visit to Texas Forest Service Watchtower."

19. "A Report on a Visit to Texas Forest Service Watchtower."

20. "Tower Joint Data, District #1 thru District #4" and untitled notebook with field observation between 1962 and 1965, Texas A&M Forest Service Archives.

21. Memorandum between J. O. Burnside and district foresters, May 9, 1962.

22. "New Towers Manned," *Texas Forest News* 41.2 (1962): 7.

23. Memorandum between J. O. Burnside and Director Folweiler, September 3, 1957.

24. Memorandum between J. O. Burnside and K. L. Burton, August 28, 1956.

25. Memorandum between J. O. Burnside and K. L. Burton, December 6, 1960.

26. Memorandum between J. O. Burnside and District 5 forester, August 30, 1962.

27. Memorandum between J. O. Burnside and Director Folweiler, March 6, 1957.

28. "Isolation of Lookout Towers," memorandum from J. O. Burnside to district foresters, March 11, 1957.

29. Memorandum between J. O. Burnside and Arthur Green, March 25, 1957.

30. "Highway Lookout Tower Signs," Texas Forest Service, staff meeting minutes, November 14–15, 1961.

31. "Forest Service Workers Essential to Woodland Areas," *Polk County Enterprise* 75.37 (May 30, 1957): 2.

32. "T.F.S. Crewleader and Tree Farmer Profitably Improving His Pine Woodlands," *Texas Forest News* 37.1 (1958): 5.

33. "Death Takes Two Fire Fighters," *Pineywoods Pickups* 3.1 (1942): 4.

34. Nora Holder, "Retirement for Two," *Polk County Enterprise* 83.4 (September 30, 1965): 20.

35. "Forest Fire Lookout Towers Supervised by Texas Forest Service," Internal TFS Property Inventory, May 26, 1944, Texas A&M Forest Service Archives.

36. Correspondence between W. E. White and A. L. Carter, February 1, 1937.

37. Correspondence between W. T. Carter and W. E. White, February 6, 1937.

38. Correspondence between W. E. White and A. L. Carter, February 8, 1937.

39. "I Remember When . . . ," 12–13.

40. "Another Protection Unit Placed under Administration," *Texas Forest News* 12.9 (1930): 1.

41. "Scouts Camp at Indian Village, Polk Co.," *Corrigan Tribune* 1.9 (August 29, 1931): 5.

42. "I Remember When . . . ," 12–13.

43. "I Remember When . . . ," 12–13.

44. "Illness," *Have You Heard* 45.9 (1974): 7.

45. "Personnel Changes," *Have You Heard* 1 (1963): 2.

46. "E. E. Covington Retires from Forest Service," *Timpson Times* 83.12 (March 22, 1968): 1.

47. "Texas Forest Service Erecting Large Steel Tower near Timpson," *Timpson Daily Times* 35.58 (March 21, 1936): 1.

48. "Assistant Forester Visits Timpson," *Timpson Daily Times* 35.123 (June 20, 1936): 1.

49. "E. E. Covington Retires from Forest Service."

50. Joseph O. Burnside, "Looking Ahead in Fire Control," Texas Forest Service, staff meeting minutes, August 23–24, 1966, 9–24.

51. Burnside, "Looking Ahead in Fire Control," 9–24 (emphasis added).

52. Ebarb, "Aerial Detection of Forest Fires in Texas."

Chapter 9. A Lot of Smoke but No Fire: The 1970s

1. Mike Browning, "Gallant Men of 213," *Have You Heard* 65 (1976): 8.

2. Memorandum between J. O. Burnside and District 3 forester, June 11, 1969.

3. Memorandum between J. O. Burnside and Paul Kramer, June 24, 1969.

4. Memorandum between J. O. Burnside and District 3 forester, July 10, 1969.

5. "Tid Bits," *Have You Heard* 30 (June 1971): 9.

6. Billings, *A Century of Forestry*.

7. System Form 201, completed by Paul R. Kramer, January 31, 1949.

8. Biographical sketch, Paul R. Kramer, October 1980.

9. Patrick Ebarb, letter to the author, July 26, 2023.

10. Joseph O. Burnside, "Lost Pines Protection Project," unpublished report, Texas Forest Service, September 1967; "Lost Pines Get Forest Fire Protection," *Texas Forest News* 48.4 (1969): 4.

11. "Lost Pines Get Forest Fire Protection."

12. Correspondence between J. O. Burnside and O. C. Braly, July 22, 1969.

13. Correspondence between J. O. Burnside and John Courtenay, October 4, 1967.

14. Memorandum between Paul Kramer and Fire Control Department, August 2, 1968.

15. R. B. Emery, "Rural Job Corps Center in Operation near New Waverly," *Texas Forest News* 44.4 (1965): 2.

16. William S. Clayson, *Freedom Is Not Enough: The War on Poverty and the Civil Rights Movement in Texas* (Austin: University of Texas Press, 2010).

17. Draft Environmental Impact Statement and Unit Plan, Sam Houston National Forest, unpublished document USDA-FS-R8 DES ADM 78-01, September 1977.

18. Memorandum between Director Kramer and J. O. Burnside, April 22, 1968.

19. Memorandum between J. O. Burnside and Director Kramer, May 1, 1968.

20. Correspondence between Ivan Blauser and George Chase, August 16, 1968.

21. "To Conduct Tour from Smithville," *Bastrop Advertiser and Bastrop County News* 117.17 (June 25, 1970): 1.

22. "Tid Bits," *Have You Heard* 31 (April 1972): 10.

23. Memorandum between District 7 forester and Paul Kramer, March 15, 1972.

24. Correspondence between P. Kramer and Clayton Garrison, April 3, 1973.

25. "Forest Service to Remain Active in Bastrop County," *Bastrop Advertiser and Bastrop County News* 122.12 (May 22, 1975): 1.

26. "Conveyance of One (1) Acre Tract, Bastrop County, Texas," agenda item submitted by Bruce Miles, director TFS, to Chancellor Dr. Frank W. R. Hubert, February 3, 1982.

27. "Lost Pines Get Forest Fire Protection."

28. "New Lookout Towers Loom on the Horizon," *Texas Forest News* 49 (Summer 1970): 2.

29. "Tid Bits," *Have You Heard* 29 (1971): 9.

30. "Atma Stanford," *Polk County Enterprise* 106.100 (December 15, 1988): 8.

31. Bill Dove, "Polk Paragraphs," *Polk County Enterprise* 82.1 (September 12, 1963): 1.

32. "Tid Bits," *Have You Heard* 29 (February 1971): 9.

33. "Lookout Tower Provides Detection," *Bastrop Advertiser and Bastrop County News* 117.9 (April 30, 1970): 6.

34. Correspondence between J. O. Burnside and O. C. Braly, August 4, 1970.

35. Memorandum between Paul Kramer and District 2 forester, September 23, 1970.

36. Memorandum between Paul Kramer and J. O. Burnside, February 11, 1971.

37. This discounts the erection of the East River Lookout at the Texas Forestry Museum in 1974.

38. Chandler Tower memo written by J. O. Burnside, September 25, 1970.

39. "Tid Bits," *Have You Heard* 29 (February 1971): 8.

40. "Request for Sale of Surplus Lookout Towers," memorandum between Pat Ebarb and Bruce Miles, January 12, 1981.

41. Memorandum between director and head, Fire Control Department, April 28, 1972.

42. Phone message: Wilbur Strong regarding wife's Weches Lookout lease, July 1, 1974.

43. "Lookout Towers, Maintenance and Use," memorandum to district and sub-district foresters from Pat Ebarb, June 30, 1975.

44. Memorandum between head, Fire Control Department and District 3 forester, February 11, 1975.

45. "Fire in Votaw Lookout Tower," memorandum between Area 4 forester and director, January 16, 1978.

46. "Disposal of Fire Lookout Tower," memorandum between District 2 forester and Area 1 forester, April 20, 1979.

47. Memorandum between District 3 forester and Area 1 forester, April 17, 1979.

48. Memorandum between Stephen Adams and head, Fire Control Department, December 12, 1980.

49. Don Staples, "Airmen Pilot Way through Danger, Blazes," *Polk County Enterprise* 95.49 (June 23, 1977): 13.

50. "Forest Service to Remain Active in Bastrop County."

51. Memorandum between head, Fire Control Department and District 6 forester, February 15, 1972.

52. Memorandum between District 6 forester and head, Fire Control Department, February 17, 1972.

53. Memorandum between director and head, Fire Control Department, August 19, 1975.

54. Memorandum between B. Miles and K. Burton, October 5, 1982.

55. Correspondence between J. Blott and M. I. Waller, June 7, 1977.

56. Patrick Ebarb, letter to the author, July 26, 2023.

57. Ebarb, letter to the author.

58. Ebarb, letter to the author.

59. Ebarb, letter to the author.

60. Correspondence between E. R. Wagoner and P. Kramer, July 7, 1972.

61. Correspondence between P. Kramer and E. R. Wagoner, July 10, 1972.

62. Ebarb, letter to the author.

63. "Aerial Detection Feasibility Study—Texas," memorandum from Assistant Forest Supervisor John Olson to district rangers, national forests in Texas, June 16, 1971.

64. Tommy Read, "Red Hills Lake Recreation Area Is Real Family Camper's Delight," *Silsbee Bee* 50.10 (May 4, 1967): 6.

65. "Lookout Towers, Maintenance and Use."

66. "1976 Getting Hotter," *Have You Heard* 60 (1976): 2.

67. "1976 Getting Hotter."

68. "Fire Tower Vandalism Is a Danger, Expense," *Rusk Cherokeean* 127.9 (April 22, 1976): 2.

69. Correspondence between B. Ross and J. Fox, August 14, 1978.

70. Memorandum between J. Fox and associate director, August 16, 1978.

71. Correspondence between W. B. Flynn and G. West, May 24, 1979.

72. "Spotlight," *Have You Heard* 32 (1972): 5.

73. Correspondence between W. E. White and E. O. Siecke, November 5, 1941.

74. "Retired," *Have You Heard* 49 (1975): 3.

75. "Spotlight."

76. Correspondence between Stephen Adams and Mrs. Dell McCann Jacobs, August 1, 1996.

Chapter 10. Potshots or Pearls: The 1980s

1. Mark V. Thornton, "Fixed Point Fire Detection, the Lookouts," unpublished report prepared for the Forest Service, Region 5, 1986.

2. Correspondence between Bruce Miles and Jimmie Breeding, November 4, 1980.

3. "Disposistion [sic] of District Fire Towers," memorandum between Area 1 forester and District 1 forester, June 25, 1979.

4. "Strategic Fire Tower Network," memorandum between director and Pat Ebarb, October 4, 1979.

5. "Disposition of Lookout Towers," memorandum between Pat Ebarb and Paul Kramer, April 25, 1980.

6. Memorandum between Jimmy Hull and Ken Burton, March 17, 1981.

7. "Invitation to Bid," Texas Forest Service announcement, November 3, 1980.

8. "Invitation to Bid."

9. "Kenefick Fire Tower Sales Agreement," signed by P. E. Perlitz, October 17, 1983.

10. Correspondence between J. Hull and B. Miles, January 16, 1981.

11. "Removal of Fire Tower, Shelby County, Texas," memorandum between Jim Hull and Genevieve Graffeo, January 16, 1981.

12. "Request for Sale of Surplus Lookout Towers," memorandum between Bruce Miles and Pat Ebarb, January 12, 1981.

13. "Authority to Sell Lookout Towers," memorandum between Pat Ebarb and Bruce Miles, January 14, 1981.

14. David Crisp, "County May Purchase Fire Tower," *Palestine Herald-Press*, June 10, 1981.

15. "Passing of an Era," 14–15.

16. Memorandum between area foresters and Ken Burton, April 20, 1981.

17. Memorandum between area foresters and Ken Burton, August 6, 1982.

18. Memorandum between area foresters and Ken Burton, October 4, 1982.

19. Standard bid form signed by E. W. Nerren, July 22, 1983.

20. Memorandum between B. Young, G. Lacox, and J. Blott from K. Burton and P. Ebarb, September 23, 1983.

21. "Cancellation of Lease and Reconveyance of Title for Keenen Tower," signed by B. Miles, August 10, 1983.

Chapter 11. Epilogue

1. Steer and Miller, *Lookouts in the Southwestern Region* (emphasis added), 155.

2. Steer and Miller, *Lookouts in the Southwestern Region*, 130.

3. Spence, "Rankin Ridge Lookout Tower, Wind Cave National Park."

4. John P. Freeman, with Wesley H. Haynes, *Views from on High: Fire Tower Trails in the Adirondacks and Catskills* (Lake Placid, NY: Adirondack Mountain Club, 2001).

5. Steer and Miller, *Lookouts in the Southwestern Region*, 130.

Appendix 2. Palestine Quadrangle

1. There are indications this tower was never erected, but a garage and dwelling were built.

2. Piers in place; one oral history suggests the tower was never built.

Index

Page numbers in italics refer to illustrations.

Abbey, Edward, 59
Adams Hill Lookout, 188, 213, 303
Adams, Steve, 210
aerial patrols, fire, 139–41, 148, 152–55, 162, 170–175, 183–84, 188–89, 196, 198, 200–201, 205–207, 211
Aermotor Company, 41–42, 44, 53–54, 183, 200, 311; LS-40, *31*, 53, 67, 117, 173, 183, 187, 194, 203, 311; LX-24, 311; MC-39, 41, 44, 46, 50, 67, 95, 97, 101, 116, 117, 135, 166, 181, 187, 192, *193*, 195, 210, 220, 311, 320n40; MC-40, 41–42, 44, 46, 102, 117, *128*, 129, 145, *146*, 164, 173, 189–90, 220, 311; MJ-66, 42, 48
Aircraft Warning Service (AWS), 66–67; 144–45, 149
Ais people, 28
Alabama Commission of Forestry, 63
Aldridge Lumber Company, 10; Aldridge Mill, 10, 28, 126
alidade, 32, 36, 40, 76
Allred, James, 91
Alto Lookout, 108, 151, 157, 164, 213, 217, 255
American Wood Preservers Association, 42
Ancient and Honorable Order of Squirrels, 28, 187, 318n13
Anderson, David A. (Andy), 10, 18, 96, 125, 135, 157, 327n92
Anderson, Mrs. Gene, 166
Angelina County Lumber Company, 9, 121
Angelina National Forest, 10, 24, 28, 125, 136, 158, 180, 199. *See also* US Department of Agriculture (USDA) Forest Service
Apple Springs Lookout, 108, 113, 118, 152, 256

Appleby Lookout, 108, 204, 213, 256
Arab Mountain (New York), 220–21
Arbor Day, 11
Ariola Lookout, 108, 172–73, 186, 213, 223
Arp Lookout, 47, 188, 213, 288, 321n71
Ashe Nursery (Mississippi), 122
Ashe, William W., 85
Atkinson, W. A. (Hank), 92–93
Ayish Bayou, 28, 200

B. R. Parkersburg, 41, 49–50
Bagwell, potential lookout site, 115
Ballard, Calvin F., 113, 115–16, 123–25
Balhis, F. R., *184*
Bannister, 37, 124
Barker, Raymond, 198
Barnes, Joe E., 107
Barnum Lookout, 47, 109, 138, 149, 188–89, 213, 224
Baser, Nor, 153
Bass, Bill, 203
Bastrop Lookout, 47, 201, 202–203, 213, 216, 309
Bastrop State Park, 125
Bath Lookout, 26, 47, 64, 129, 135, 158, 173, 225, 319n2, 319n9, 326n70
Battle of Shreveport, 145
Bean, V. V., 66, 323n37
Beat generation poets, 59
Beaumont, C. E., 84–87, 327n110
Beckville Lookout, 108, 116, 202, 209, 213, 217, 289
Bennet, Cary H., 327n110
Big Thicket, 21
Big Thicket Work Center. *See* Sam Houston National Forest
Bilby surveying tower, 43–44, 124, 171–72

349

Bilby, Jasper S., 44
Bird Mountain Lookout, 47, 64, 97, 101, 164–65, 186, 213, 257, 319n2, 319n9
Bishop, Loren L., 21, 88–89, 113
Bizzell, William, 16
Blackland Prairie, 21
Blott, James, 206
Bob Jackson, Guardian of the Pineywoods, 161–62
Bon Wier Lookout, 43, 82, 138, 159, 173, 213, 225
Bond, Walter, 33, 97
Boykin Springs, 28, 126
Bradford, Floyd, 199–200
Braly, Oley Cecil (O.C.), 44, 46–50, 141, 169, 172, 174–75, 184, 194, 199, 201, 203
Bramlett, Floyd A., 155
Brashears, Murray E., 116
Bray, William L., 6, 8, 21
Bronson Lookout, 108, 158, 179, 213, 220, 257
Bronson Tree Cab, 32, 37
Brown Paper Company, 63
Bryan, Eunice, 114
Buck Tree Cab, 46, 149
Buescher State Park, 125, 201
Buildup Index (BI), 208
Buna Lookout, 138, 213, 217, 226
Burkeville Lookout, 82, 137–38, 213, 226
Burnside, John Oliver (Joseph), 47, 150–51; aerial patrols, 140, 184; as head, Fire Control Department, 150, 152–55, 157, 159–60, 164–65, 184, 186 192–93, 196–98, 203; during CCC program, 96; during World War II, 141–42; employment opportunities, 167; lookout tower construction, correspondence and maintenance, 116–17, 189–90, 198, 200, 203
Burroughs, Isaac C., 33–34, 75, 93, 97, 113
Burton, Kenneth, 189–90, 216
Busselle, Theodore, 30, 140, 146
Byron, H. K. (Dude), 167

Caddo Lake State Park, 125, 141, 189; Caddo Lake State Park Lookout. *See* Karnack Lookout
Caddo people, 82
Cain, Charles, 194
Cairo Springs Lookout, 46–47, 173, 213, 227, 321n71, 341n57
Camden Lookout, 108, 138, 149, 193, 227
Camp Pershing Lookout. *See* Etoile Lookout
Campbell, Thomas, 12
Candy Hill Lookout, 47, 173–74, 188, 258
Carroll, F. L., 101
Carter Brothers Lumber Company, 9
Central Coal and Coke Company (4-Cs), 9–10, 95
Central School Lookout, 40, 94, 108, 154–55, 213, 258
Chalk, A. T., 125
Chambers Hill Lookout, 108, 118, 158, 207–208, 259
Chandler Lookout, 47, 201, 202–203, 213, 217, 290
Chapman, Herman H., 7, 28, 110
Chase, George, 199–200
Chatham, Alfred L., 113, 129–30
Childs, Lillie, 81
Chipman, Russell, 327n110
Chireno Lookout, 108, 124–25, 183, 188–89, 259
Chita Lookout, 47, 108, 111, 201–202, 213, 228, 260
Church Hill Lookout, 47, 188, 213, 217, 290, 321n71
Civil Air Patrol (CAP). *See* aerial patrol, fire
Civilian Conservation Corps (CCC): 1, 69, 165, 176; administrative structure 70–71, 74; camp segregation, 77, 89–91, 94, 100; cultural significance, 126–27; enrollee accidents, 78, 81; enrollee discharges and dissatisfaction, 79, 96; enrollee education, 93–94, 100, 105–107, 123; enrollees and enrollee stories, 19;

enrollee firefighting, 100, 118–20; enrollee health and nutrition, 71; enrollee mapping, 101, 122–25, 138; enrollee projects, 81–82, 88, 99–101, 107, 138; enrollee recreation, 92, 100; enrollee recruitment, 71, 76; enrollee wages, 70; public opinion, 79; reforestation and planting, 79, 82, 121–22

Civilian Conservation Corps (CCC), companies: Company 827 (F-6-T), Nancy, 73, 121–22, 150; Company 833 (S-54-T), Maydelle, 72, 78, 101–105, 125; Company 838 (P-57-T), Lufkin, 43, 72, 74, 81, 91–95, 107, 113, 120, 123, 147, 154, 166; Company 839 (P-59-T), Trinity 72, 76, 94, 106, 125, 157; Company 839 (P-92-T), Newton, 72, 82, 94, 329n159; Company 840 (P-61-T), Livingston, 57, 72, 74, 81–82, 106, 119, 125, 149, 166; Company 840 (P-91-T), Humble, 72, 82, 125, 157; Company 845 (S-53-T), Kirbyville, 66, 72, 74; Company 865 and 886 (P-73-T), Cleveland, 72; Company 880 (P-51-T), Center, 72; Company 880 (F-13-T), Patroon, 73; Company 880 and 2878 (C) (P-51-T), San Augustine, 72; Company 888 (P-58-T), Weches, 72, 90; Company 890 (P-55-T), Honey Island, 45, 72; Company 891 (P-56-T), Woodville, 72, 81, 110, 125, 129, 138; Company 893 (P-52-T), Pineland, 72, 89–90, 121; Company 897 (P-60-T), Oakhurst, 72, 91; Company 899 and 873 (P-74-T), Huntsville 72, 78, 125, 326n70; Company 1803 (F-4-T), Ratcliff, 10, 73, 76, 95–97, 121, 150; Company 1804 (S-62-T), Conroe, 72, 98–100, 123; Company 1806 and 2877 (C) (F-1-T), New Waverly 73, 76; Company 1814 (P-90-T), Linden, 72; Company 1814 (P-71-T), Groveton, 72, 74; Company 1816 and 899 (P-70-T), Woden, 72; Company 1820 (P-72-T), Jasper, 72, 121, 125; Company 2879 (F-20-T), Kennard, 73, 325n65; Company 2880 (SCS-27-T), Madisonville, 325n65; Company 2885 (SCS-19-T), Jacksonville, 123, 325n65; Company 2887 (F-18-T), Milam, 73, 121, 325n65; Company 2877 (C) (F-21-T), Coldspring, 73, 150, 325n65; Company 2882 (SCS-23-T), Marshall 325n65; Company 2884 (SCS-22-T), Winnsboro, 325n65; Company 2886 (SCS-20-T), Mt. Pleasant, 325n65; Company 2888 (SP-54-T), Tyler, 325n65; Company 2889 (C) (SCS-9-T), Bogata, 116, 325n65; Company 2891 and 1801 (SP-49-T), Daingerfield, 325n65; Company 2892 (F-19-T), Richards, 1, 73, 132–33, 150, 325n65; Company 2893 (F-17-T), Apple Springs, 73, 106, 121, 325n65; Company 2890 (C) (SCS-25-T), Nacogdoches, 325n65; Company 2899 (SCS-8-T), Wolfe City, 325n65

Clarke-McNary Law, 17, 31, 80, 81, 84, 177

Cleveland Lookout, 108, 114, 145, 188, 228

Clifton, L. D., 46

Coats, S. W., 57, 58

Columbus, Johnny, 37, 124

commercial logging: during the Great Depression, 69, 85; early practices, 8, 19; sawmills, 12; trams, 10. See also Piney Woods

Conners, Philip, 2, 59

Conroe Lookout (Jones State Forest), 47, 108, 184–86, 213, 229, 319n2

Conservation Association of Texas, 13

controlled burn. See prescribed fire

cooperation, Texas Forest Service and US Forest Service, 26, 30, 97–98, 135, 147, 152, 158, 168, 179, 199–200, 205, 208; aerial patrols, fire, 170–71, 207, 216; fire action boundary, 179, 216

cooperation, Texas Forest Service and US military, 111, 139–40, 143
Coushatta people, 81
Covington, Emory, 35, 194–96, 214
Creel, Floy, 164
creosoted wood lookout towers. *See* lookout towers
crewleaders and crewmen, 192; African American, 168–69; 162–65; diaries, 182–83, 342n1; educational efforts, 164–65; employment opportunities, 167–69; nonproductive time, 59, 206; Outstanding Crewleader Awards, 163–65; retention, 167–68. *See also* Forest Patrolmen
Cromeens Tree Cab, 40, 165
Cromeens, Alsey, 165, 194
Crossett Lumber Company, 63
cross-shots. *See* triangulate
Cudlupp Forestry Center, 199
Cushing Lookout, 47, 64, 97, 108, 186, 188, 213, 260–61
Cutler Tree Cab, 40
Cutler, Henry, 36
Cyclone Hill Lookout, 48, 50, 108, 113, 118, 133, 136, 173, 180, 261
Cypress Lake Lookout, 108, 213, 262

Daingerfield Lookout, 47, 168–69, 173, 213, 303, 321n71, 341n57
Damuth, Bannon, 172–73, 174
Danger Rating Station. *See* fire weather
Davy Crockett National Forest, 10, 20, 24, 98, 133, 152, 158, 178, 180; Ratcliff Lake Recreation Area, 10, 95, 126; Trinity District, 180. *See also* US Department of Agriculture (USDA) Forest Service
Deadwood Lookout, 108, 116, 213, 217, 220, 291
Deiterman, Betsy, vii
DeKalb Lookout, 108, 115, 186, 209, 213, 304
DeLeon, Mark, 126
Delta Land and Timber Company, 86, 89, 121, 328n133

Denning Lookout, 58, 108, 138, 262
Department of Agriculture. *See* US Department of Agriculture (USDA) Forest Service
Department of Forest Fire Control (Texas). *See* Fire Control Department
DePriest, Oscar Stanton, 90
Dern, George H., 176
Devils Lookout, 47, 108, 118, 199–200, 202, 208, 263
Dickens, Chestley, 145, *146*
Division of Forest Protection (1922–49). *See* Fire Control Department
Dodge Lookout, 108, 205, 213, 229, 326n70
Doss, Hubert, 95–96
Double Lake Recreation Area. *See* Sam Houston National Forest
Douglassville Lookout, 108, 210, 213, 304
Dreka Lookout, 36, 108, 113, 118, 147, 158, 173, 180–81, 207, 263
Dresser Industries, 200
Dunlap Manufacturing Company, 41, 203
Dunmire, Maurice V., 19, 155
Durham, William O., 33, 97, 113, 115
duty station, 162–63

East Mountain Lookout, 47, 188, 204–205, 213, 291, 321n71
East River Lookout, 47, 50, 108, 138, 149, 202–204, 213, 230
East Texas Chamber of Commerce, 175
Ebarb, Patrick, vii, 50, 104, 204-208, 211, 215–16
Eighth Corps Area (US Army), 24, 70, 74, 77, 79, 106, 324n19; during World War II, 141–42; sub-districts, 74, 77
Eliot, T.S., 19
Elkhart Lookout, 47, 64, 173, 186, 213, 264
Elysian Fields Lookout, 47, 64, 173, 180, 186–87, 213, 292
Emergency Conservation Act, 69–71

Emergency Conservation Work (ECW), 69–70, 77, 125
Emergency lookout tower. *See* lookout towers
Emergency Relief Act, 175
Emilee Lookout, 43, 138, 230
EMSCO Derrick and Equipment Company, 1, 41
Erwin, R. E. (Bob), 93
Etoile (Camp Pershing) Lookout, 33, 47, 59, 108, 124, 158, 182–83, 186, 213, 264–65, 319n2
Evadale Lookout, 47, 64, 67, 138, 166–67, 186, 213, 231
Ewing Tree Cab, 40

Fails Lookout, 40, 108, 165, 213, 217, 231, 326n70
Fairchild State Forest (I. D. Fairchild State Forest). *See* Texas State Forests
Fairchild, I. D., 65
Farris Lookout, 100, 108, 113, 118, 129, 132–33, 232, 326n70
Fechner, Robert, 70, 74–75, 91
Federal Forest Reserve Commission, 63, 86
Ferguson, James E., 13, 16, 69
Ferguson, Miriam, 69, 76
Fern Lake Lookout, 107–108, 197–98, 213, 265
Fernow, Bernard, 11–12
fire climax forests, 7
Fire Control Department (FCD), 29; administering protection units, 31–32; headquarters, 43, 75; inauguration as Division of Forest Protection, 14–15, 17, 61, 75, 319n1; rebranding to FCD, 159
fire danger. *See* fire weather
fire fighting tools: comparison of mechanized technology, TFS and USFS, 156; fire fighting Jeep, 2, 50, 153–57, 162–63, 168; hand tools, 34–36, 139, 152, 194; mechanized, 139, 151, 154–56, 163, 180
fire law. *See* law enforcement

fire weather, 29; danger rating station 29–30, 136, 179
Fisher, Larry J., 38, 319n27
Fisher, William F., 327n110
fixed point detection, 36–40, 64, 76, 112, 122–23, 140, 184, 199, 206, 218–19. *See also* lookout towers
Flash message. *See* Aircraft Warning Service (AWS)
Folweiler, Alfred D.: as director, 48, 50, 59, 158–60, 162–65, 182, 198; employment opportunities, 167–68; erecting lookouts, 173–74, 186, 184, 188; lookout leases, 114; lookout maintenance and safety, 147–48, 182, 189–90; Texas National Forest Study, 175, 179
Forest Fire Fighters Service (FFFS), 139
forest fires. *See* wildland fires; piney woods, fires in
forest improvements. *See* structures, lookout yards
forest infrastructure. *See* lookout towers; structures, lookout yards
Forest Patrolmen (later lookout observers, lookoutmen), 34, 61, 63, 110–11, 153, 162, 165; living at guard compounds, 127–30; nonproductive time, 59, 206; qualifications and responsibilities, 15, 151–52. *See also* crewleader and crewman
Forse Mountain Lookout, 101, 108, 118, 266
Foster, John, 140–41
Foster, John H., 14, 16, 28, 31, 61, 123, 177
Four Notch Lookout, 36, 108, 118, 129, 130, 135, 158, 205, 232, 326n70, 334n288
Fox, Joe, 104, 209
Foxe, David, vii
Fred Lookout, 108, 173, 233
Frost Lumber Industries, 33
Frost, Robert, 2
Frost, Sherman (Jack) L., vii, 145, 195; acting director, TFS, 153–54; TNF

Index **353**

land acquisitions, 84, 327n110; "Texas National Forest Study," 175–78, 180
Fulton, John B., 327n110

Gay, Roy, 19, 191–94
Geneva Lookout, 101, 108, 113, 118, 266
Gerland, Jonathan, vii
Gilchrist, Gibb, 154
Gilmer Company, 32
Gilmont Lookout, 47, 173, 213, 217, 292, 321n71, 341n57
Grapeland Lookout, 174–75, 188, 213, 217, 220, 267
Graves, Henry S., 110
Grayburg Lookout, 47, 188, 213, 217, 233, 321n71
Great Depression, 68–69
Greeley, William B., 54
Green, Arthur, 170, 190, 342n1
Green, Faye, 37
Griesenbeck, Jack, 205
Grimshaw, George, 57, 58
Groben, W. Ellis, 2, 131–33, 135–36
guard compound. *See* structures, lookout yards
Gulf Coast Building and Construction Trades Council, 199
Gum Springs Lookout, 101, 108, 118, 267
Hagen, Jeff, 50
Hagood, Johnson, 79, 106
Hall, Herbert, 165
Hallsville Lookout, 108, 166, 205, 213, 217, 293
Hammon, Gordon A., 327n110
Haney, Robert, 198
Hanna, C. W., 74, 76–77, 91
Hardegree, William P., 78
Hare, Herbert, 165
Harrison, Hubert, 175
Hartman, William (Bill) T., 19, 123, 144, 164
Havard, Obie W., 186
Have You Heard, vii, 50
Hawkins, Benjamin D., 58, 113, 140, 171

Hi Point Lookout, 47, 108, 118, 199–200, 202, 268
Hightower, George, 165
hogs, feral: impact on forests, 7–8, 64
Holder, Albert, 190–193
Hooks Lookout, 108, 124, 141–44, 149, 173, 209–10, 305
Horn, Cleo, 151
Horn, William, 151
Hortense Lookout, 108, 138, 149, 156, 234, 319n2
Horton Hill Lookout, 108, 213, 217, 268
Houston County Timber Company, 10, 32, 85; 88–89, 97, 328n133
Huffman, Betty, 146–49
Huffman, John W., 29, 127, 129, 144, 148–49
Hughes, Lettie, 147–48
Hull, James, 104
Humble Lookout, 108, 138, 149, 234
Hunter, Jeff, 168–69, 210
Huntington Lookout, 108, 158, 213, 216, 269
Huntington Tree Cab, 33
Hyman, Carolyn Francis, 1, 84

Indian Mounds Nursery (Alto), 122, 138
Industrial Lumber Company, 63
International Creosoting Company, 42
International Derrick & Equipment Company (IDECO), 32, 41, 53, 311, 319n9
Irwin, Robert F., 180
Ivie, Knox, 38–39, 40, 57, 58, 96, 113, 132–33, 141, 150, 184, *185*

Jackson Hill Lookout, 36, 108, 118, 124–25, 158, 207, 269
Jacobs, Jimmie, 33
James, William P., 157
Jared, Alonzo, 28
Jeep, fire fighting. *See* fire fighting tools
Jefferson Lookout, 108, 145, 155, 169, 210, 213, 217, 293–94

Jester, Beaufort, 153
Job Corps Center, New Waverly, 107, 199
Johnson, Bobby, 96
Johnson, Chester, 143
Johnson, George W., 14
Johnston, Lawrence, 164
Jones State Forest (W. Goodrich Jones State Forest). *See* Texas State Forests
Jones, Albert, *181*
Jones, E. E., 66
Jones, Ivan H., 33, 75, 113
Jones, John Maxwell, 10–11
Jones, William Goodrich, conservationist, 10–14, 65, 121; family and upbringing. 10–11. *See also* Texas State Forests
Jungmichel, Charles, 199

Karnack Lookout, 47, 108–109, 138, 141, 144, 173, 188–89, 213, 294–95
Keenan Lookout, 100, 108, 213, 217, 235
Kelley, Fount, 96
Kelley, Ider, 165
Kenefick Lookout, 108, 188, 213, 235
Kerouac, Jack, 59, 177
Kildare Lookout, 108, 213, 295
Kimbro, James, 147–48
Kincel, Neal W., 151
Kirby Forest Lookout, 108, *128*, 140, 186, 213, 217, 220–21, 236
Kirby Lumber Company, 9, 42, 82, 89, 114, 193, 328n133
Kirby State Forest (John Henry Kirby Memorial Forest). *See* Texas State Forests
Kirby, John Henry, 65, 129. *See also* Texas State Forests
Kirbyville Lookout. *See* Siecke State Forest (Kirbyville) Lookout
Kircher, Joseph E., 85–87, 127, 130
Kirkland, Winston, 113, 115–16, 123
Kountze Lookout, 108, 145, 146, 165, 213, 237
Kramer, Paul, 198, 201, 203, 206–207, 211

Lacey, Columbus, 37, 124
Laird, John W., 83
Lake Livingston, 201
Langston Pole Cab. *See* Walker Mountain
Langston, Alton and Mary, 39, 114
Latch Lookout, 47, 188, 213, 296, 321n71
law enforcement, fire law, 17, 61–62
Lawrence, Millard, 113, 150, 156
Lay, Daniel, 28
Leatherwood, Bill, 115–16, 140
Lee C. Moore Corporation, 41, 48–49, 188
Leesburg Lookout, 47, 49–50, 188, 204, 213, 305, 321n71
Leopold, Aldo, 110, 187
Lewis, R. T., 201
Liberty Hill Lookout (State), 47, 64, 186, 192–93, 213, 237, 319n2
Liberty Hill Lookout (USFS), 58, 108–109, 113, 173, 180–81, 220, 238
Lindale Lookout, 47, 201–203, 213, 296
live-in lookout. *See* lookout towers
Livingston Lookout, 46–47, 108–109, 149, 172–73, 188–90, 213, 238–39
loblolly pine (*Pinus taeda*), 8–9, 121; in Lost Pines, 198
Local Experienced Men (LEM), 70, 166
logging trams. *See* commercial logging
logging. *See* commercial logging
Lone Star Trail, 1, 129
Long Bell Lumber Company, 89, 328n133
Longhorn Ordnance Plant, 141
longleaf pine (*Pinus palustris*), 6–8; 23–24, 28, 64, 121
Lookout towers: abandonment and sale of, 211–12, 214–17; appropriations during Emergency Conservation Work (ECW) era, 76; as aircraft warning posts. *See* Aircraft Warning Service (AWS); cabins (cabs), 42, 46–48, 169,

172; categorization; construction and design attributes, 40, 44–45, 51, 172–74, 186; construction by Civilian Conservation Corps (CCC), core network, 76, 94, 101, 107, 112–14; construction by CCC, network expansion, 79, 81, 114–18, 210, 320n40; construction, post-CCC expansion, 151, 170, 172, 188, 198–99, 201–202; construction, pre-ECW era, 25, 31–32, 36, 63, 107, 117; creosoted wood towers, 43, 48, 54, 56–57, 138, 152; decline, 170, 198–201, 205–208, 211, 212, 215; educational value, 164, 190, 193–94, 210, 218; emergency lookout towers, 40, 41; highway signs, 190; isolation, safety, and depredations, 104, 147, 189–90, 192, 197–98, 203–205, 208–209, 211, 215; leases, 112–13, 115; limitations, 171; live-in towers, 40; maintenance, 104, 111, 159, 174, 182, 184–86, 189, 194, 199–200; manufacturers, 41–42; observation only towers, 40; piers. *See* piers, lookout; preservation, 104, 215, 218–21; primary lookout, 40; procurement, 41; project lookout, 40; reorganization, 149, 154; replacement, 182, 184, 186; secondary lookout, 40; yard. *See also* structures, lookout yard

Louisiana Forestry Commission, 30, 32

Love's Lookout, 101–105, 108, 111, 116, 213, 217, 297

Lowery, E. O., 182–84, 188–89

Lowery, Mervis, 59, 182

Lufkin, 24; as CCC headquarters, 74–75, 77–78, 101; as base for USFS field surveys, 84

Lufkin Lookout, 43, 108, 138, 270

Lyell, Charles, 28

Magnolia Lookout, 47, 172–73, 213, 239, 321n71, 341n57

Magnusson, George, 100

Mandeville, Albert, 171

map quadrangles: Austin, 317n5; Beaumont, 23, 46, 107, 112, 115, 177, 223–53; Palestine 23–24, 94, 107, 112, 115, 177, 254–86; Seguin, 317–18n5; Texarkana, 25, 115, 302–307; Tyler, 25, 115, 287–301

Mark Twain National Forest (Missouri), 127

Martin, Henry, 172–73

mast, observation. *See* pole cab

Masterson State Forest. *See* Texas State Forests

Mauriceville Lookout, 47, 173, 186, 213, 240, 321n71, 341n57

Maydelle Lookout, 101–102, 109, 188–89, 213, 217, 270–71

Mayflower Lookout, 55, 109, 213, 217, 220, 271

McCann, Connie F., 90–91, 119–20

McClintic-Marshall Corporation, 181

McCurdy, Charles, 98

McFarlane, W. D., 143

McGowan, Thomas, 146

McKinney, Carl S., 91

Meadows Lookout, 47, 188, 213, 297, 321n71

mechanization, fire control. *See* fire fighting tools

Mercy Lookout, 57, 108, 138, 199, 240

Mewshaw sawmill, 102

Milam Tree Cab, 32, 37, 208

Miles, Bruce, 104, 205, 211, 215

Mims Chapel Lookout, 108, 116–17, 205, 213, 298

Montalba Lookout, 201–202, 203, 213, 215, 217, 272

Moore Plantation. *See* Sabine National Forest

Moores Grove Lookout, 26, 108, 199, 241

Morgan, Ralph, 327n110

Moscow Lookout, 47, 109, 188–89, 213, 241–42

Moss Hill Lookout, 47, 58, 109, 113, 118, 158, 199, 202, 272

Moursund, Carl, 169

Moye, Perry, 38

Mt. Enterprise, 23
Mt. Enterprise Lookout, 108, 116, 190, 213, 273
Mt. Magazine, Arkansas, 131
Mt. Vernon Lookout, 108, 273, 347n1
Mt. Zion Lookout, 108, 118, 152, 180, 274, 332n230
Muir, Bluford, 130
Munson, H. F., 66, 323n37
Mutscher, Gus, 199

Nashville Bridge Company, 41–42, 174–75, 311
National Creosoting Company, 42
National Forest Purchases, Texas. *See* US Department of Agriculture (USDA) Forest Service
National Forest Reservation Commission, 87–88, 176
National Forests and Grasslands in Texas (NFGT). *See* US Department of Agriculture (USDA) Forest Service
National Lumber and Creosoting Company, 43
Neblett Lookout, 26, 108, 118, 242
Neches Lookout, 101–102, 108, 167, 213, 217, 220, 274
Negley Lookout, 108, 114–15, 140, 168, 213, 217, 220, 306
New Boston water tower, 40, 41, 210
New Deal. *See* Great Depression
New Waverly Job Corps Center. *See* Job Corps Center, New Waverly
Newton Lookout, 109, 138, 141, 243
Nogalus Lookout, 43, 58, 82, 96, 108, 152, 158, 173, 180–81, 207, 216, 275
nonproductive time. *See* Forest Patrolmen
Norman, W. C., 79

Oates, William, vii
O'Banion, Joshua, vii
Offenbach, Henrietta, 10–11
Office of Economic Opportunity, 199
Olson, John, 207
Onalaska Lookout, 47, 57, 58, 109, 201–202, 213, 243–44

Operation Sage Brush, 111
O'Quinn, Jeff, 61–62, 65, 150, 154, 165
Ott, Francis, 145
Outstanding Crewleader Awards. *See* crewleaders and crewmen
Ozark Nursery (Arkansas), 122

Page, Johnnie, 179
Page, O. D., 58
Palmetto State Park, 125
Parker, Tim, 103
Parker, William H., 149–50
Parrish, Barney, 151, 156
Paxton Lookout, 108, 172–73, 213, 275
Peach Tree Village Lookout, 40, 109, 138, 149, 244
Peck, Harold, 165
Perkins, Cleo, 167
Perricone quadruplets (Anthony, Bernard, Carl and Donald), 38
Peters, J. Girvin, 14
Phelps, Walter, 165
Pickering Lumber Company, 89, 328n133
piers, lookout, 51; creosoted wood, 54–57; cross category, 52–54; design and dimensions, 44, 311; diamond category, 52–54, 55; graphing (cross-plotting), 55; water tower, 57, 95
Pinchot, Gifford, 11–12, 110
Pinckney Lookout, 101, 108–109, 118, 192, 193, 213, 245
Pine Mills Lookout, 47–49, 188, 213, 217, 220, 298, 321n71
Pine Springs Tree Cab, 173
Piney Woods (Piney) Lookout, 43, 56, 108, 152, 158, 173, 180, 207, 216, 276
Piney Woods (Weldon) Lookout, 109, 213, 276, 326n70
Piney Woods: biological crossroads, 21; climate, 21; community, 163, 166; composition (forest communities), 8–9; definition of, 5–6; early reforestation, 65, 121–22; erosion of, 19, 85; federal land

purchase in, 62–63; 83–85; fire in, 17–19, 26–29, 152–54, 162, 196; fires, public reporting, 202, 205–206, 211; forest protection 14–15, 31–33, 61, 63, 196; geology, 23; guardians of, 161–65, 190–93; ice storms, 148, 152, 157; mapping, 25, 122–25; railroad loggers in, 8; sabotage, timber, World War II, 139

Pineywoods Pickups, 18

Pitts, Enoch, 145–46

Pitts, Mabel, 145–46

Platt, Noah, 164

pole cab (observation mast), 39–40

Pollok Tree Cab, 40, 154–55

Pool Lookout, 26, 47, 108, 113, 118, 132–35, 158, 188, 245

Post Oak Forest, 21

Powell, William, 152

Poynor Lookout, 47, 201–203, 213, 299

prescribed fire, 27–28

primary lookout tower. *See* lookout towers

project lookout tower. *See* lookout towers

protection units, 31–33, 36, 60–61, 64

Purvis, Claud, 165–66

radio communication, 136, 153–55, 157–58, 184, 190, 205, 207, 209; operator, 168

Randall, Hunters, 21

Rankin Ridge Lookout (South Dakota), 127

Ratcliff Lake Recreation Area. *See* Davy Crockett National Forest

Ratcliff Lookout, 10, 57, 64, 95–98, 113, 158, 207, 220–21, 277, 319n2, 319n9

Red Hills Lake Recreation Area. *See* Sabine National Forest

Red River Arsenal and Lone Star Ordnance Plant, 41, 141–43

Redding, Robert, 49

Redditt, John S., 83

Redwater Lookout, 47, 108, 144, 149, 172–73, 210, 213, 217, 306

Reed, George C., 155

Rich, Joe, 165

Riley, Allen, 110–11

Robinson, L. B., 327n110

Rocky Hill Lookout, 40, 108, 173, 277

Rocky Hill Tree Cab, 33, 40

Rod, A. J., 200

Roosevelt, Franklin Delano, 69, 76, 88, 121, 175–76

Roosevelt, Theodore, 12

Rose, G. W., 145

Rusk Lookout, 101, 108–109, 213, 217, 278

Sabine National Forest, 24, 37, 158, 181, 199, 201; Moore Plantation, 37; Red Hills Lake Recreation Area, 208. *See also* US Department of Agriculture (USDA) Forest Service

sabotage, timber, World War II. *See* Piney Woods

Saf-T-Climb, 174

Salem Lookout, 47, 173, 213, 279, 321n71, 341n57

Sam Houston National Forest, 23, 26, 92, 112, 158, 178, 180; Big Thicket Work Center, 181; Double Lake Recreation Area, 126, 181; fire in, 27; protection units in, 33; Raven District, 26, 135; unitizing, 86. *See also* US Department of Agriculture (USDA) Forest Service

Sam Houston State Teachers College, 106

Sam Rayburn Reservoir, 200

San Augustine Lookout, 108, 138, 213, 279

San Jacinto Monument, 149

Saratoga Lookout, 43, 46–47, 109, 171–73, 188–89, 213, 246

Schenk, Carl, 11

Schoen, Paul W., 89

Scottsville Lookout, 108, 145, 188–89, 299

secondary lookout tower. *See* lookout towers

seen area, 60, 200, 211

358 Index

Senate Concurrent Resolution (S.C.R.) No. 73, 83–84, 87, 177
Shady Grove Lookout, 109, 158, 213, 216–17, 280
Sheffield, James V., 63
Sheffield's Ferry Lookout, 67, 108, 149, 246, 317n114
Shelton, Larry, vii
Shepherd Lookout, 58, 108–109, 113, 118, 188, 213, 217, 247
shortleaf pine (*Pinus echinata*), 8–9, 25
Siecke State Forest (E. O. Siecke State Forest). *See* Texas State Forests
Siecke State Forest (Kirbyville) Lookout, 29, 31, 58, 63–64, 66–67, 104, 108, 111, 138, 144, 148–49, 174, 186, 213, 217, 220–21, 236, 317n114
Siecke, Eric O.: as state forester, 16–18, 31, 57, 61, 64, 142–44, 149; during CCC program, 18, 71, 74–76, 93, 97; erecting lookouts, 42, 79; experimental use of fire, 28; influence on USFS during TNF land acquisitions, 86–87; law enforcement, 62
Sloan, Carter S., 327n110
Smart School Lookout, 45, 108, 145, 156, 213, 248
Smith Ferry Lookout, 40, 109, 164, 213, 248
Smith Jr., Albert, 327n110
Smith Sr., Dwight T., 106–107
Smith, DeFord, 131
Smith, R. V., 142–43, 210
Smith, W. Hickman (Uncle Hick), 63–64
Smith, William M., 103
Smithland Lookout, 47, 141, 173, 213, 300
Smithville Lookout, 47, 200–203, 213, 309
Smokey Bear, 28, 162
Snap Lookout, 108, 213, 300
Snyder, Gary, 59, 177
Soda Lookout, 40, 46–47, 108, 111, 249

Southern Forest Experimental Station, 28
Southern Pine Association, 41, 43
Southern Pine Lumber Company, 9, 32, 39, 97
Southland Paper Mills, 94
spiked trees, 37–38, 40, 63
Splendora Lookout, 100, 109, 213, 249
Springfield, C. L., 168–69
Spurger Lookout, 109, 213, 250
Stanford, Atma, 202
Stark, Peter, 86, 176–77
Starrville Lookout, 47, 188, 213, 301, 321n71
State Relief Board, 70
Stone, Allan W., 28
Strickland, Earl, 166–67
Stripling, Zachary T., 33
structures, lookout yards, 57–58, 125–36, 180–81, garages, 35; toolboxes, 35; warehouses, 35–36
Stuart Nursery (Louisiana), 121–22
Stuart, Robert Y., 176
Suiter, John, 2

Tabernacle Lookout, 109, 147–48, 158, 213, 217, 220, 280
Tadmor Lookout, 54, 56–57, 108, 281, 347n2
Talco Oil Field, 48, 173
Taylor, William, 166
Teco split ring connectors, 43
telephone network and communications, 32, 36, 63–64, 76, 79, 96, 101, 116, 135, 138, 142–43, 148, 151–52, 156–57, 207
Temple Lumber Company, 32, 89, 328n133
Tenaha Lookout, 43, 108, 133, 171, 188, 281
Texas Creosoting Company, 42
Texas Department of Forestry (1915 to 1926). *See* Texas Forest Service
Texas Forest News (News), 17, 75, 324n42
Texas Forest Service (TFS), 3, 316n104; during Conroe fires

(1947), 152–54; initially Texas Department of Forestry (1915 to 1926), 13, 15–16, 62; innovations, 156; nurseries, 82, 121–22; reorganization (1949), 159; technical agency, administrator of CCC projects, 70–71, 74–78, 81, 93, 106, 137–38. *See also* Texas State Forests

Texas Forestry Association (TFA), 13, 17, 207

Texas Forestry Museum, 50, 136, 154, 202, 220, 282

Texas National Forests (TNF). *See also* US Department of Agriculture (USDA) Forest Service; Angelina National Forest; Davy Crockett National Forest; Sabine National Forest; Sam Houston National Forest

Texas Parks and Wildlife Department, 201

Texas State Forests: Fairchild (formerly State Forest No. 3), 18, 65, 78, 101–102, 189; fire on, 28; Jones (formerly State Forest No. 2), 14, 18, 50–51, 53, 64–65; 99, 121, 184–86, 323n29; Kirby (formerly State Forest No. 4), 18, 128, 129, 140, 145, 146, 221; Masterson 18; nurseries, *see* Texas Forest Service; purchase of, 62, 64; research on, 64–66; Siecke (formerly State Forest No. 1), 18, 31, 64, 66–67, 113, 121, 127, 144, 148, 186, 221, 323n29

Texas State Parks Board, 70, 83, 141, 178

Thigpen, John, 165

Thornton, Mark, 211

Tidwell, Clyde, 103

Timpson Lookout, 35, 108, 111, 162, 195, 213–14, 282

Todd, Daniel O., 181

toolbox. *See* structures, lookout yard

Travis, Allen, 165

tree cab, 32, 37–38, 39, 40, 107

triangulate, 32, 36

Trinity County Lumber Company, 89, 328n133

Truett, Joe, 28

Turner, Emmett, 143, 209–10

Turner, John M., 66, 113, 323n137

Tyler State Park, 77, 126

Tyler: as CCC headquarters, 77–78, 101

Tyrrell Park (Beaumont), 66

Undritz, F. R., 92

Union Hill Lookout, 108, 166–67, 213, 307

US Coast and Geodetic Survey (USC&GS), 43, 123–25, 144; benchmarks, 25, 171, 173, 182, 192, 206

US Department of Agriculture (USDA) Forest Service, *also* US Forest Service (USFS): acquisition of lands in Texas, 62–63, 83–89; administrative changes, Region 8, 87–88; Angelina Purchase Unit, 85, 88; as CCC technical agency, 70, 76, 81, 88, 95; Davy Crockett Purchase Unit, 85–86, 88; federal land purchases in the south, 83; lookout contracts, Texas, 41, 76, 164; lookout network, 24, 112, 180; lookouts transferred from TFS, 97, 112–13, 117–18, 125, 136, 199; lookouts, decommissioning, 98, 199, 207; proclamation boundaries and forest inauguration, 84, 86, 88, 180; purpose of, 178–79; revenue from, 86–87, 178; Sabine Purchase Unit, 85, 88; Sam Houston Purchase Unit, 85–86, 88; San Jacinto Purchase Unit, 86; Tejas Purchase Unit, 86, 328n116; Texas National Forest Study, 175–81; Texas National Forests, definition and administration, 1, 24, 86

US Department of Labor, 70

US Department of the Interior, 70

US Department of War (War

360 Index

Department), 70, 143; CCC administration, 74; CCC wages distributed, 71
US Forest Service (USFS). *See* US Department of Agriculture (USDA) Forest Service
US Geological Survey (USGS), 22
Utley, Dan, vii, 110

Vidor Lookout, 47–48, 136, 171, 173, 213, 250
Votaw Lookout, 38, 109, 204, 213, 251

Wagoner, Ed, 50, 207
Wakefield Lookout, 45, 108, 194, 213, 283
Walker Mountain (Langston Pole Cab), 39, 40, 101, 113
Webster, C. B., 66
Weches Lookout, 36, 40, 43, 47–48, 96–97, 108–109, 113, 138, 158, 173, 213, 283
Weeks Act, 13–14, 31, 83, 87, 176–77; expansion of 17
Weeks, John, 13. *See* Weeks Act
Werner Sawmill Company, 116
West, Jeff, 66
Westbrook, Jerry, 168
Westbrook, Juanita, 168
Westbrook, Lawrence, 75
Whalen, Philip, 59
Whipple, B. B., 113
White City Lookout, 101, 108, 113, 118, 284
White House Conference of Governors on Conservation (1908), 12–13
White Rock, potential lookout site, 115
White, William E., 33–34; as chief of fire control, 57, 75, 97, 110–11, 113, 210; as director, 116–17, 149–50, 152–53; during CCC program, 74, 76, 93, 138; during World War II, 139, 141–43, 149; erecting lookouts, 42, 45, 82, 113, 115, 171, 184-85, 192
Wicker, Al, 166
wildland fires, definition of, 20; fire types classified by TFS, 20; policy decisions, 211; presuppression activities, 63; TFS response to, 206
Wilhite Lookout, 47, 201–203, 213, 307
Williams Tree Cab, 40
Williams, Arthur, 165
Williams, B. F., 75
Williams, Bob, 29, 152
Williams, Simeon, 194
Willis Formation, 23, 100
Willis Lookout, 26, 29, 109, 158, 213, 217, 251
Willow Springs Collection, 112–13, 118
Willow Springs Lookout, 108, 112–13, 213, 217, 252, 332n231
Wilson, Charles, 203
Wilson, R. A., 66
Winchester Lookout, 47, 201–203, 213, 217, 310
Winston Ranch, 186–87, 284
Winston, Simon, vii
Wolf Hill Lookout, 43, 138, 172–73, 213, 285
women: 4-H forestry projects, 165; CCC roles, 81; emergency work, 145–48; regular employment, 168
Wood, William, 37
Woodville, 23
Woodville Lookout, 58, 104, 109–110, 146, 186, 213, 217, 220–21, 252

Yellowpine Lookout, 32, 64, 97, 113, 133, 138, 173, 180–81, 286, 319n2, 319n9
Young, Don, 117, 125; during CCC program, 96; during World War II, 141–43

Zion Hill Lookout, 109, 165, 213, 253